Springer INdAM Series

Volume 37

Springer INdAM Series

This series will publish textbooks, multi-authors books, thesis and monographs in English language resulting from workshops, conferences, courses, schools, seminars, doctoral thesis, and research activities carried out at INDAM - Istituto Nazionale di Alta Matematica, http://www.altamatematica.it/en. The books in the series will discuss recent results and analyze new trends in mathematics and its applications.

THE SERIES IS INDEXED IN SCOPUS

More information about this series at http://www.springer.com/series/10283

Dražen Adamović · Paolo Papi
Editors

Affine, Vertex and W-algebras

Springer

Editors
Dražen Adamović
Faculty of Science and Department of Mathematics
University of Zagreb
Zagreb, Croatia

Paolo Papi
Department of Mathematics
Sapienza University of Rome
Rome, Italy

ISSN 2281-518X ISSN 2281-5198 (electronic)
Springer INdAM Series
ISBN 978-3-030-32908-2 ISBN 978-3-030-32906-8 (eBook)
https://doi.org/10.1007/978-3-030-32906-8

This Springer imprint is published by the registered company Springer Nature Switzerland AG
The registered company address is: Gewerbestrasse 11, 6330 Cham, Switzerland

Preface

This volume is based on the INdAM Workshop "*Affine, Vertex and W-algebras*" held in Rome from 11 to 15 December 2017. This meeting was devoted to recent developments in the theory of vertex algebras, with particular emphasis on affine vertex algebras, affine W-algebras, and W-algebras appearing in physical theories such as logarithmic conformal field theory.

It is widely accepted in the mathematical community that the best way to study the representation theory of affine Kac–Moody algebras is by investigating the representation theory of the associated affine vertex and W-algebras. In this volume, this general idea can be seen at work from several points of view.

Most of the state-of-the-art topics dealt with at the INdAM Conference are represented in this volume, including

- The idea of *fusion*, which is developed through its connections with mathematical physics and with classical problems in vertex algebra theory (conformal embeddings);
- Relationships with finite-dimensional Lie theory (Kostant's theory of Lie pairs, Vogan's theory of the algebraic Dirac operator) and with Lie pseudoalgebras;
- Permutation orbifolds (in the case of fermionic vertex superalgebras);
- Higher Zhu algebras;
- Connections with combinatorics, which naturally arise in the construction of bases for standard representations of symplectic affine Lie algebras and for principal subspaces of generalized Verma modules;
- De Sole-Kac work on integrable Hamiltonian equations and Poisson vertex algebras, appearing through the notes of a postgraduate course held by De Sole.

Zagreb, Croatia — Dražen Adamović
Rome, Italy — Paolo Papi

Contents

About the Editors

Dražen Adamović is Full Professor of Mathematics at the University of Zagreb, Croatia. He received his Ph.D. in Mathematics from the University of Zagreb. He is the author of more than 50 peer-reviewed research publications on the representation theory of vertex algebras, W-algebras, and infinite-dimensional Lie algebras with special emphasis on vertex algebras appearing in conformal field theory.

Paolo Papi is Full Professor of Geometry at Sapienza University of Rome, Italy. He received his Ph.D. in Mathematics from the University of Pisa. He is the author of more than 40 peer-reviewed research publications on Lie theory, algebraic combinatorics, representation theory of Lie algebras and superalgebras, with special emphasis on combinatorics of root systems and infinite-dimensional structures (affine Lie algebras, vertex algebras).

Kostant Pairs of Lie Type and Conformal Embeddings

Dražen Adamović, Victor G. Kac, Pierluigi Möseneder Frajria, Paolo Papi and Ozren Perše

Abstract We deal with some aspects of the theory of conformal embeddings of affine vertex algebras, providing a new proof of the Symmetric Space Theorem and a criterion for conformal embeddings of equal rank subalgebras. We study some examples of embeddings at the critical level. We prove a criterion for embeddings at the critical level which enables us to prove equality of certain central elements.

Keywords Symmetric space theorem · Conformal embedding · Pair of Lie type · Vertex operator algebras

1 Introduction

In this paper we deal with three aspects related to the notion of conformal embedding of affine vertex algebras. We provide the following three contributions:

D. Adamović · O. Perše
Department of Mathematics, Faculty of Science, University of Zagreb, Bijenička 30, 10 000 Zagreb, Croatia
e-mail: adamovic@math.hr

O. Perše
e-mail: perse@math.hr

V. G. Kac
Department of Mathematics, MIT, 77 Mass. Ave, Cambridge, MA 02139, USA
e-mail: kac@math.mit.edu

P. Möseneder Frajria
Polo regionale di Como, Politecnico di Milano, Via Valleggio 11, 22100 Como, Italy
e-mail: pierluigi.moseneder@polimi.it

P. Papi (✉)
Dipartimento di Matematica, Sapienza Università di Roma, P.le A. Moro 2, 00185 Roma, Italy
e-mail: papi@mat.uniroma1.it

D. Adamović and P. Papi (eds.), *Affine, Vertex and W-algebras*, Springer INdAM Series 37, https://doi.org/10.1007/978-3-030-32906-8_1

1. a proof of the Symmetric Space Theorem via Kostant's theory of *pairs of Lie type*;
2. a central charge criterion for conformal embeddings of equal rank subalgebras;
3. a detailed study of some relevant examples of embeddings at the critical level.

To illustrate these topics, let us recall that if $\mathfrak{g}$ is a semisimple finite-dimensional complex Lie algebra and $\mathfrak{k} \subset \mathfrak{g}$ a reductive subalgebra, the embedding $\mathfrak{k} \hookrightarrow \mathfrak{g}$ is called conformal if the central charge of the Sugawara construction for the affinization $\widehat{\mathfrak{g}}$, acting on a integrable $\widehat{\mathfrak{g}}$-module of non negative integer level k, equals that for $\widehat{\mathfrak{k}}$. Then necessarily $k = 1$ [8]. Maximal conformal embeddings were classified in [8, 21], and related decompositions can be found in [9, 16, 17]. In the vertex algebra framework the definition can be rephrased as follows: the simple affine vertex algebras $V_1(\mathfrak{g})$ and its vertex subalgebra $\widetilde{V}(k, \mathfrak{k})$ generated by $\{x_{(-1)}\mathbf{1} \mid x \in \mathfrak{k}\}$ have the same Sugawara conformal vector.

In [4] we generalized the previous situation to study when the simple affine vertex algebra $V_k(\mathfrak{g})$ and its subalgebra of $\widetilde{V}(k, \mathfrak{k})$ have the same Sugawara conformal vector for some non-critical level k, not necessarily 1, assuming $\mathfrak{k}$ to be a maximal equal rank reductive subalgebra. The non equal rank case has been studied in [6].

The conformal embeddings in $V_k(so(V))$ with $k \in \mathbb{Z}_+$ and V a finite dimensional representation are the subject of the Symmetric Space Theorem (see Theorem 2), which is at the basis of the early history of the theory. In our subsequent work on conformal embeddings (see in particular [6, Sect. 3]) we used a special case of it, but we later realized that it is indeed possible to use Kostant's theory quoted above (together with other tools from the theory of affine and vertex algebras) to provide a new proof of the theorem in full generality. This is done in Sect. 3.

In Sect. 4 we improve our results from [4], where we mainly studied conformal embeddings of maximal equal rank subalgebras, by providing full details on a criterion for conformal embedding in which we drop the maximality assumption.

The final section is devoted to certain instances of the following general problem: provide the explicit decomposition of $V_k(\mathfrak{g})$ regarded as a $\widetilde{V}(k, \mathfrak{k})$-module. This problem was studied in [3, 4, 6]. When k is not a non-negative integer, it was noted that the following situations can occur:

- $V_k(\mathfrak{g})$ is a semisimple $\widetilde{V}(k, \mathfrak{k})$–module with finite or infinite decompositions.
- $\widetilde{V}(k, \mathfrak{k})$ is not simple and $V_k(\mathfrak{g})$ contains subsingular vectors.
- $\widetilde{V}(k, \mathfrak{k})$ is a vertex algebra at the critical level.

There are many examples for which explicit decompositions are still unknown. Such cases are also connected with some recent physics papers and conjectures. In Sect. 5 we present some results in the case when $\widetilde{V}(k, \mathfrak{k})$ is a vertex algebra at the critical level. We show that then $\widetilde{V}(k, \mathfrak{k})$ is never simple by constructing certain non-trivial central elements in $\widetilde{V}(k, \mathfrak{k})$. In most cases, $\mathfrak{k} = \mathfrak{g}_1 \oplus \mathfrak{g}_2$ ($\mathfrak{g}_1, \mathfrak{g}_2$ being simple ideals) and the embedded vertex subalgebra is a certain quotient of $(V^{-h^\vee}(\mathfrak{g}_1) \otimes V^{-h^\vee}(\mathfrak{g}_2))/\langle s \rangle$ where s is a linear combination of quadratic Sugawara central elements.

2 Conformal Embeddings

Let $\mathfrak{g}$ be a simple Lie algebra. Let $\mathfrak{h}$ be a Cartan subalgebra, Δ the $(\mathfrak{g}, \mathfrak{h})$-root system, Δ^+ a set of positive roots and ρ the corresponding Weyl vector. Let $(\cdot, \cdot)$ denote the normalized bilinear invariant form (i.e., $(\alpha, \alpha) = 2$ for any long root).

We denote by $V^k(\mathfrak{g})$ the universal affine vertex algebra of level k. If $k + h^\vee \neq 0$ then $V^k(\mathfrak{g})$ admits a unique simple quotient that we denote by $V_k(\mathfrak{g})$. More generally, if $\mathfrak{a}$ is a reductive Lie algebra that decomposes as $\mathfrak{a} = \mathfrak{a}_0 \oplus \cdots \oplus \mathfrak{a}_s$ with $\mathfrak{a}_0$ abelian and $\mathfrak{a}_i$ simple ideals for $i > 0$ and $\mathbf{k} = (k_1, \ldots, k_s)$ is a multi-index of levels, we let

$$V^{\mathbf{k}}(\mathfrak{a}) = V^{k_0}(\mathfrak{a}_0) \otimes \cdots \otimes V^{k_s}(\mathfrak{a}_s), \quad V_{\mathbf{k}}(\mathfrak{a}) = V_{k_0}(\mathfrak{a}_0) \otimes \cdots \otimes V_{k_s}(\mathfrak{a}_s).$$

We let $\mathbf{1}$ denote the vacuum vector of both $V^{\mathbf{k}}(\mathfrak{a})$ and $V_{\mathbf{k}}(\mathfrak{a})$.

If $j > 0$, let $\{x_i^j\}$, $\{y_i^j\}$ be dual bases of $\mathfrak{a}_j$ with respect to the normalized invariant form of $\mathfrak{a}_j$ and $h_j^\vee$ its dual Coxeter number. For $\mathfrak{a}_0$, let $\{x_i^j\}$, $\{y_i^j\}$ be dual bases with respect to any chosen nondegenerate form and set $h_0^\vee = 0$.

Assuming that $k_j + h_j^\vee \neq 0$ for all j, we consider $V^{\mathbf{k}}(\mathfrak{a})$ and all its quotients, including $V_{\mathbf{k}}(\mathfrak{a})$, as conformal vertex algebras with conformal vector $\omega_{\mathfrak{a}}$ given by the Sugawara construction:

$$\omega_{\mathfrak{a}} = \sum_{j=0}^{s} \frac{1}{2(k_j + h_j^\vee)} \sum_{i=1}^{\dim \mathfrak{a}_j} :x_i^j y_i^j: .$$

Recall that the central charge of $\omega_{\mathfrak{g}}$ for a simple or abelian Lie algebra $\mathfrak{g}$ is, assuming $k + h^\vee \neq 0$,

$$c(\mathfrak{g}, k) = \frac{k \dim \mathfrak{g}}{k + h^\vee}. \tag{1}$$

If $\mathfrak{a}$ is a reductive Lie algebra, which decomposes as $\mathfrak{a} = \mathfrak{a}_0 \oplus \cdots \oplus \mathfrak{a}_s$ then the central charge of $\omega_{\mathfrak{a}}$ is, for a multiindex $\mathbf{k} = (k_0, \ldots, k_s)$,

$$c(\mathfrak{a}, \mathbf{k}) = \sum_{j=0}^{s} c(\mathfrak{a}_j, k_j). \tag{2}$$

2.1 AP-Criterion

Assume that $\mathfrak{k}$ is a Lie subalgebra in a simple Lie algebra $\mathfrak{g}$ which is reductive in $\mathfrak{g}$. Then $\mathfrak{k}$ decomposes as

$$\mathfrak{k} = \mathfrak{k}_0 \oplus \cdots \oplus \mathfrak{k}_t.$$

where $\mathfrak{k}_1, \dots \mathfrak{k}_t$ are the simple ideals of $\mathfrak{k}$ and $\mathfrak{k}_0$ is the center of $\mathfrak{k}$. Let $\mathfrak{p}$ be the orthocomplement of $\mathfrak{k}$ in $\mathfrak{g}$ w.r.t to $(\cdot, \cdot)$ and let

$$\mathfrak{p} = \bigoplus_{i=1}^{N} \left(\bigotimes_{j=1}^{t} V(\mu_i^j) \right)$$

be its decomposition as a $\mathfrak{k}$-module. Let $(\cdot, \cdot)_j$ denote the normalized invariant bilinear form on $\mathfrak{k}_j$.

We denote by $\widetilde{V}(k, \mathfrak{k})$ the vertex subalgebra of $V_k(\mathfrak{g})$ generated by $\{x_{(-1)}\mathbf{1} \mid x \in \mathfrak{k}\}$. Note that $\widetilde{V}(k, \mathfrak{k})$ is an affine vertex algebra, more precisely it is a quotient of $\otimes V^{k_j}(\mathfrak{k}_j)$, with the levels k_j determined by k and the ratio between $(\cdot, \cdot)$ and $(\cdot, \cdot)_j$.

Theorem 1 (AP-criterion) [2] *$\widetilde{V}(k, \mathfrak{k})$ is conformally embedded in $V_k(\mathfrak{g})$ if and only if*

$$\sum_{j=0}^{t} \frac{(\mu_i^j, \mu_i^j + 2\rho_0^j)_j}{2(k_j + h_j^\vee)} = 1 \tag{3}$$

for any $i = 1, \dots, s$.

We reformulate the criterion highlighting the dependence from the choice of the form on $\mathfrak{k}$. Fix an invariant nondegenerate symmetric form $B_\mathfrak{k}$ on $\mathfrak{k}$. Then $(\cdot, \cdot)_{|\mathfrak{k}_j \times \mathfrak{k}_j} = n_j (B_\mathfrak{k})_{|\mathfrak{k}_j \times \mathfrak{k}_j}$. Let $\{X_i\}$ be an orthonormal basis of $\mathfrak{k}_j$ with respect to $B_\mathfrak{k}$ and let $C_{\mathfrak{k}_j} = \sum_i X_i^2$ be the corresponding Casimir operator. Let $2g_j^{B_\mathfrak{k}}$ be the eigenvalue for the action of $C_{\mathfrak{k}_j}$ on $\mathfrak{k}_j$ and $\gamma_{ji}^{B_\mathfrak{k}}$ the eigenvalue of the action of $C_{\mathfrak{k}_j}$ on $V(\mu_i^j)$. Then $\widetilde{V}_k(\mathfrak{k})$ is conformally embedded in $V_k(\mathfrak{g})$ if and only if

$$\sum_{j=0}^{t} \frac{\gamma_{ji}^{B_\mathfrak{k}}}{2(n_j k + g_j^{B_\mathfrak{k}})} = 1 \tag{4}$$

for any $i = 1, \dots, s$.

Corollary 1 *Assume $\mathfrak{k}$ is simple, so that $\mathfrak{k} = \mathfrak{k}_1$. Then there is $k \in \mathbb{C}$ such that $\widetilde{V}(k, \mathfrak{k})$ is conformally embedded in $V_k(\mathfrak{g})$ if and only if $C_\mathfrak{k}$ acts scalarly on $\mathfrak{p}$.*

Proof If $\widetilde{V}(k, \mathfrak{k})$ is conformally embedded in $V_k(\mathfrak{g})$ then, by (4), $\gamma_i^1 = 2(k + g_1)$ is independent of i. If $C_\mathfrak{k}$ acts scalarly on $\mathfrak{p}$, then, solving (4) for k, one finds a level where, by AP-criterion, conformal embedding occurs. □

3 Symmetric Space Theorem

The Symmetric Space Theorem has originally been proved in [13]; an algebraic approach can be found in [10]. We recall this theorem in a vertex algebra formulation.

Theorem 2 (Symmetric Space Theorem) *Let K be a compact, connected, nontrivial Lie group with complexified Lie algebra $\mathfrak{k}$ and let $\nu : K \to End(V)$ be a finite dimensional faithful representation admitting a K-invariant symmetric nondegenerate form.*

Then there is a conformal embedding of $\widetilde{V}(k, \mathfrak{k})$ in $V_k(so(V))$ with $k \in \mathbb{Z}_+$ if and only if $k = 1$ and there is a Lie algebra structure on $\mathfrak{r} = \mathfrak{k} \oplus V$ making $\mathfrak{r}$ semisimple, $(\mathfrak{r}, \mathfrak{k})$ is a symmetric pair, and a nondegenerate invariant form on $\mathfrak{r}$ is given by the direct sum of an invariant form on $\mathfrak{k}$ with the chosen K-invariant form on V.

In this section we give a proof of the above theorem using Kostant's theory of pairs of Lie type and Kac's *very strange formula* as main tools.

We start by showing that if $\mathfrak{r} = \mathfrak{k} \oplus V$ gives rise to a symmetric pair $(\mathfrak{r}, \mathfrak{k})$, then $\widetilde{V}(1, \mathfrak{k})$ embeds conformally in $V_1(so(V))$.

It suffices to show that

$$\sum_j c(\mathfrak{k}_j, k_j) = c(so(V), k). \tag{5}$$

In fact, the action of $(\omega_{so(V)} - \omega_{\mathfrak{k}})_{(n)}$ on $V_1(so(V))$ defines a unitary representation of the Virasoro algebra. The Virasoro algebra has no nontrivial unitary representations with zero central charge, hence, if (5) holds, then $(\omega_{so(V)} - \omega_{\mathfrak{k}})_{(n)}$ acts trivially on $V_1(so(V))$ for all n. Since $V_1(so(V))$ is simple, this implies that $\omega_{so(V)} = \omega_{\mathfrak{k}}$.

The normalized invariant form on $so(V)$ is given by $(X, Y) = \frac{1}{2}tr(XY)$. Let $\mathfrak{k}_{re}$ be the Lie algebra of K. Since V admits a nondegenerate symmetric invariant form, there is a real subspace U of V which is K-stable and $V = U \oplus \sqrt{-1}U$. Choose an orthonormal basis $\{u_i\}$ of U with respect to a K-invariant inner product on U. Then the matrix $\pi(X)$ of the action of $X \in \mathfrak{k}_{re}$ on V in this basis is real and antisymmetric. It follows that $tr(\pi(X)^2) \neq 0$ unless $X = 0$, thus $(\cdot, \cdot)_{|\mathfrak{k}\times\mathfrak{k}}$ is nondegenerate.

Using $(\cdot, \cdot)_{|\mathfrak{k}\times\mathfrak{k}}$ as invariant form on $\mathfrak{k}$ we have

$$\sum_j c(\mathfrak{k}_j, k_j) = \sum_j \frac{\dim \mathfrak{k}_j}{1 + g_j},$$

while it is easy to check that $c(so(V), 1) = \frac{1}{2} \dim V$. Thus we need to check that

$$\sum_j \frac{\dim \mathfrak{k}_j}{1 + g_j} - \frac{1}{2} \dim V = 0. \tag{6}$$

Let σ denote the involution defining the pair. We can assume that σ is indecomposable, for, otherwise, one can restrict to each indecomposable ideal.

Let $\kappa(\cdot,\cdot)$ be the Killing form on $\mathfrak{r}$. Since σ is assumed indecomposable, the center $\mathfrak{k}_0$ of $\mathfrak{k}$ has dimension at most one. It follows that $\kappa(\cdot,\cdot)_{|\mathfrak{k}_j\times\mathfrak{k}_j} = r_j(\cdot,\cdot)_{|\mathfrak{k}_j\times\mathfrak{k}_j}$ for some $r_j \in \mathbb{C}$, $j \ge 0$. Recall that $(x, y) = \frac{1}{2}tr_V(xy)$. Note that, if a_i, b_i is a pair of dual bases w.r.t $(\cdot,\cdot)_{|\mathfrak{k}_j\times\mathfrak{k}_j}$

$$
\begin{aligned}
r_j \dim \mathfrak{k}_j &= \sum_{i=1}^{\dim \mathfrak{k}_j} \kappa(a_i, b_i) = 2g_j \dim \mathfrak{k}_j + \sum_{i=1}^{\dim \mathfrak{k}_j} tr_V(a_i b_i) \\
&= 2g_j \dim \mathfrak{k}_j + \sum_{i=1}^{\dim \mathfrak{k}_j} 2(a_i, b_i) = 2(g_j+1) \dim \mathfrak{k}_j,
\end{aligned}
$$

hence $r_j = 2(g_j + 1)$.

If we choose $B_{\mathfrak{k}} = \kappa(\cdot,\cdot)_{|\mathfrak{k}\times\mathfrak{k}}$ as invariant form on $\mathfrak{k}$ then $C_{\mathfrak{k}_j}^{B_{\mathfrak{k}}} = \frac{1}{r_j} C_{\mathfrak{k}_j}$ so the eigenvalue of the action of $C_{\mathfrak{k}_j}^{B_{\mathfrak{k}}}$ on $\mathfrak{k}_j$ is $\frac{g_j}{g_j+1}$.

We write now the *very strange* formula [15, (12.3.6)] in the form given in [18, (55)] for the involution σ using the Killing form as invariant form on $\mathfrak{r}$. Since σ is an involution we have that $\rho_\sigma = \rho_{\mathfrak{k}}$ and $z(\mathfrak{r}, \sigma) = \frac{\dim V}{16}$, hence the formula reads:

$$
\kappa(\rho_{\mathfrak{k}}, \rho_{\mathfrak{k}}) = \frac{\dim \mathfrak{r}}{24} - \frac{\dim V}{16},
$$

On the other hand, using the strange formula for $\mathfrak{k}$ (in the form of (54) in [18]), we have

$$
\kappa(\rho_{\mathfrak{k}}, \rho_{\mathfrak{k}}) = \sum_{j=1}^{t} \frac{g_j \dim \mathfrak{k}_j}{24(1+g_j)}.
$$

Equating these last two expressions one derives (6).

We now prove the converse implication: assume that there is a conformal embedding of $\widetilde{V}(k, \mathfrak{k})$ in $V_k(so(V))$ with $k \in \mathbb{Z}_+$. Let $\nu : \mathfrak{k} \to so(V)$ be the embedding. Let also $\tau : \bigwedge^2 V \to so(V)$ be the $\mathfrak{k}$-equivariant isomorphism such that $\tau(u)(v) = -2i(v)(u)$, where i is the contraction map, extended to $\bigwedge V$ as an odd derivation. More explicitly $\tau^{-1}(X) = \frac{1}{4}\sum_i X(v_i) \wedge v_i$, where $\{v_i\}$ is an orthonormal basis of V.

Choose an invariant nondegenerate symmetric form $B_{\mathfrak{k}}$ on $\mathfrak{k}$. Recall that $\mathfrak{p}$ is the orthocomplement of $\mathfrak{k}$ in $so(V)$ with respect to $(\cdot,\cdot)$. Let $p_{\mathfrak{k}}$, $p_{\mathfrak{p}}$ be the orthogonal projections of $\bigwedge^2 V$ onto $\mathfrak{k}$ and $\mathfrak{p}$ respectively.

We claim that we can choose $B_{\mathfrak{k}}$ so that

$$
2(n_j k + g_j^{B_{\mathfrak{k}}}) = 1 \tag{7}
$$

(see Sect. 2 for notation). Indeed, it is clear that $n_j g_j = g_j^{B_{\mathfrak{k}}}$, hence it is enough to choose

$$n_j = \frac{1}{2(k+g_j)}.$$

With this choice for $B_{\mathfrak{k}}$ we have that, by (4),

$$\sum_j \gamma_{ji}^{B_{\mathfrak{k}}} = 1.$$

In particular, $C_{\mathfrak{k}}^{B_{\mathfrak{k}}}$ acts on $\otimes_{j=1}^{t} V(\mu_i^j)$ as the identity for all i. Thus $C_{\mathfrak{k}}^{B_{\mathfrak{k}}}$ acts as the identity on $\mathfrak{p}$.

Set $\nu_* = \tau^{-1} \circ \nu$ and let $Cl(V)$ denote the Clifford algebra of $(V, \langle \cdot, \cdot \rangle)$. Extend ν_* to a Lie algebra homomorphism $\nu_* : \mathfrak{k} \to Cl(V)$, hence to a homomorphism of associative algebras $\nu_* : U(\mathfrak{k}) \to Cl(V)$. Consider $\nu_*(C_{\mathfrak{k}}^{B_{\mathfrak{k}}}) = \sum_i \nu_*(X_i)^2$. Also recall that a pair $(\mathfrak{k}, \nu)$ consisting of a Lie algebra $\mathfrak{k}$ with a bilinear symmetric invariant form $B_{\mathfrak{k}}$ and of a representation $\nu : \mathfrak{k} \to so(V)$ is said to be *of Lie type* if there is a Lie algebra structure on $\mathfrak{r} = \mathfrak{k} \oplus V$, extending that of $\mathfrak{k}$, such that (1) $[x, y] = \nu(x)y$, $x \in \mathfrak{k}$, $y \in V$ and (2) the bilinear form $B_{\mathfrak{r}} = B_{\mathfrak{k}} \oplus \langle \cdot, \cdot \rangle$ is $ad_{\mathfrak{r}}$-invariant. In [19, Theorem 1.50] Kostant proved that $(\mathfrak{k}, \nu)$ is a pair of Lie type if and only if there exists $v \in (\bigwedge^3 \mathfrak{p})^{\mathfrak{k}}$ such that

$$\nu_*(C_{\mathfrak{k}}^{B_{\mathfrak{k}}}) + v^2 \in \mathbb{C}. \tag{8}$$

Moreover he proved in [19, Theorem 1.59] that v can be taken to be 0 if and only if $(\mathfrak{k} \oplus V, \mathfrak{k})$ is a symmetric pair.

Lemma 1 *Assume that $\widetilde{V}(k, \mathfrak{k})$ embeds conformally in $V_k(so(V))$ with $k \in \mathbb{Z}_+$. Then*

$$\sum_i \tau^{-1}(X_i) \wedge \tau^{-1}(X_i) = 0$$

where $\{X_i\}$ is an orthonormal basis of $\mathfrak{k}$ with respect to $B_{\mathfrak{k}}$.

Proof Let $\{v_i\}$ be an orthonormal basis of V. Then

$$\begin{aligned}
\sum_i \tau^{-1}(X_i) \wedge \tau^{-1}(X_i) &= \frac{1}{16} \sum_{i,j,r} X_i(v_j) \wedge v_j \wedge X_i(v_r) \wedge v_r \\
&= -\frac{1}{16} \sum_{i,j,r} X_i(v_j) \wedge X_i(v_r) \wedge v_j \wedge v_r = -\frac{1}{32} \sum_{i,j,r} X_i^2(v_j \wedge v_r) \wedge v_j \wedge v_r \\
&+ \frac{1}{32} \sum_{i,j,r} X_i^2(v_j) \wedge v_r \wedge v_j \wedge v_r + \frac{1}{32} \sum_{i,j,r} v_j \wedge X_i^2(v_r) \wedge v_j \wedge v_r \\
&= -\frac{1}{32} \sum_{i,j,r} X_i^2(v_j \wedge v_r) \wedge v_j \wedge v_r
\end{aligned}$$

Recall that we chose the form $B_{\mathfrak{k}}$ so that $C_{\mathfrak{k}}^{B_{\mathfrak{k}}}$ acts as the identity on $\mathfrak{p}$, thus we can write

$$
\begin{aligned}
&\sum_{i,j,r} X_i^2(v_j \wedge v_r) \wedge v_j \wedge v_r \\
&= \sum_{i,j,r} X_i^2(p_{\mathfrak{k}}(v_j \wedge v_r)) \wedge v_j \wedge v_r + \sum_{i,j,r} X_i^2(p_{\mathfrak{p}}(v_j \wedge v_r)) \wedge v_j \wedge v_r \\
&= \sum_{j,r,s} 2g_s^{B_{\mathfrak{k}}} p_{\mathfrak{k}_s}(v_j \wedge v_r) \wedge v_j \wedge v_r + \sum_{j,r} p_{\mathfrak{p}}(v_j \wedge v_r) \wedge v_j \wedge v_r \\
&= \sum_{j,r,s} (2g_s^{B_{\mathfrak{k}}} - 1) p_{\mathfrak{k}_s}(v_j \wedge v_r) \wedge v_j \wedge v_r + \sum_{j,r} v_j \wedge v_r \wedge v_j \wedge v_r \\
&= \sum_{j,r,s} (2g_s^{B_{\mathfrak{k}}} - 1) p_{\mathfrak{k}_s}(v_j \wedge v_r) \wedge v_j \wedge v_r.
\end{aligned}
$$

We now compute $p_{\mathfrak{k}_s}(v_j \wedge v_r)$ explicitly. We extend $\langle \cdot, \cdot \rangle$ to $\bigwedge^2 V$ by determinants. Note that

$$
\begin{aligned}
\langle \tau^{-1}(X_i), \tau^{-1}(X_j) \rangle &= \frac{1}{16} \sum_{r,s} det \begin{pmatrix} \langle X_i(v_r), X_j(v_s) \rangle & \langle X_i(v_r), v_s \rangle \\ \langle v_r, X_j(v_s) \rangle & \langle v_r, v_s \rangle \end{pmatrix} \\
&= \frac{1}{16} \sum_r (\langle X_i(v_r), X_j(v_r) \rangle - \langle v_r, X_j(X_i(v_r)) \rangle) \\
&= -\frac{1}{8} \sum_r \langle X_j(X_i(v_r)), v_r \rangle = -\frac{1}{8} tr(X_j X_i).
\end{aligned}
$$

Recall that $(X, Y) = \frac{1}{2} tr_V(XY)$, hence, if $X_i, X_j \in \mathfrak{k}_s$, then $tr_V(X_j X_i) = 2(X_j, X_i) = 2n_s B_{\mathfrak{k}}(X_j, X_i)$, so that

$$
\langle \tau^{-1}(X_i), \tau^{-1}(X_j) \rangle = -\frac{1}{4} n_s \delta_{ij};
$$

therefore

$$
\begin{aligned}
p_{\mathfrak{k}_s}(v_j \wedge v_r) &= -\frac{4}{n_s} \sum_{X_t \in \mathfrak{k}_s} \langle v_j \wedge v_r, \tau^{-1}(X_t) \rangle \tau^{-1}(X_t) \\
&= -\frac{1}{n_s} \sum_{t,k} \langle v_j \wedge v_r, X_t(v_k) \wedge v_k \rangle \tau^{-1}(X_t) \\
&= -\frac{1}{n_s} \sum_{t,k} det \begin{pmatrix} \langle v_j, X_t(v_k) \rangle & \langle v_j, v_k \rangle \\ \langle v_r, X_t(v_k) \rangle & \langle v_r, v_k \rangle \end{pmatrix} \tau^{-1}(X_t) \\
&= -\frac{2}{n_s} \sum_t \langle v_j, X_t(v_r) \rangle \tau^{-1}(X_t).
\end{aligned}
$$

Substituting we find that

$$\sum_i \tau^{-1}(X_i) \wedge \tau^{-1}(X_i) = \sum_s \frac{2g_s^{B_\mathfrak{k}} - 1}{16n_s} \sum_{j,r=1}^{\dim V} \sum_{X_t \in \mathfrak{k}_s} \langle v_j, X_t(v_r)\rangle \tau^{-1}(X_t) \wedge v_j \wedge v_r.$$

By (7), we have $\frac{2g_s^{B_\mathfrak{k}}-1}{n_s} = -2k$, hence

$$\begin{aligned}
\sum_i \tau^{-1}(X_i) \wedge \tau^{-1}(X_i) &= -\frac{k}{32} \sum_s \sum_{j,r=1}^{\dim V} \sum_{X_t \in \mathfrak{k}_s} \langle v_j, X_t(v_r)\rangle \tau^{-1}(X_t) \wedge v_j \wedge v_r \\
&= -\frac{k}{32} \sum_{j,r=1}^{\dim V} \sum_{t=1}^{\dim \mathfrak{k}} \langle v_j, X_t(v_r)\rangle \tau^{-1}(X_t) \wedge v_j \wedge v_r \\
&= -\frac{k}{32} \sum_{j,r,t,k} \langle v_j, X_t(v_r)\rangle X_t(v_k) \wedge v_k \wedge v_j \wedge v_r \\
&= -\frac{k}{32} \sum_{r,t,k} X_t(v_k) \wedge v_k \wedge X_t(v_r) \wedge v_r \\
&= -\frac{k}{2} \sum_t \tau^{-1}(X_t) \wedge \tau^{-1}(X_t).
\end{aligned}$$

Since $k \in \mathbb{Z}_+$, we have $\sum_t \tau^{-1}(X_t) \wedge \tau^{-1}(X_t) = 0$ as desired. □

Lemma 2 *If $\widetilde{V}(k, \mathfrak{k})$ embeds conformally in $V_k(so(V))$ with $k \in \mathbb{Z}_+$, then $V^\mathfrak{k} = \{0\}$.*

Proof It is well known that V is the irreducible module for $so(V)$ with highest weight the fundamental weight ω_1 (ordering the simple roots of $so(V)$ in the standard way). Next we notice that if $k \in \mathbb{Z}_+$, then $L(k\Lambda_0 + \omega_1)$ is an integrable $\widehat{so(V)}$–module of level k. Since the category of $V_k(so(V))$–modules coincides with the integrable, restricted $\widehat{so(V)}$–module of level k (cf. [11]), we conclude that $L(k\Lambda_0 + \omega_1)$ is a $V_k(so(V))$–module. The top component of $L(k\Lambda_0 + \omega_1)$ is isomorphic to V as a $so(V)$-module. Since the vectors in $V^\mathfrak{k}$ are $\widetilde{V}(k, \mathfrak{k})$-singular vectors in $L(k\Lambda_0 + \omega_1)$, $(\omega_\mathfrak{k})_0$ acts trivially on them. Since $(\omega_{so(V)})_0$ acts as

$$\frac{\langle \omega_1, \omega_1 + 2\rho\rangle}{2(k + h^\vee)} I_V = \frac{\dim V - 1}{2(k + \dim V - 2)} I_V$$

on V, we have that either $\dim V = 1$ or $V^\mathfrak{k} = \{0\}$. The former possibility is excluded since it implies $so(V) = \{0\}$, thus $\mathfrak{k} = \{0\}$. We are assuming that K is connected and non-trivial, hence this is not possible. □

We are now ready to prove the converse implication of the Symmetric Space Theorem:

Proposition 1 *Assume that $\widetilde{V}(k, \mathfrak{k})$ embeds conformally in $V_k(so(V))$ with $k \in \mathbb{Z}_+$. Then $k = 1$, there is a Lie algebra structure on $\mathfrak{r} = \mathfrak{k} \oplus V$ making $\mathfrak{r}$ semisimple, $(\mathfrak{r}, \mathfrak{k})$*

is a symmetric pair, and a nondegenerate invariant form on $\mathfrak{r}$ is given by the direct sum of $B_{\mathfrak{k}}$ and $\langle\cdot,\cdot\rangle$.

Proof In [19, Proposition 1.37] it is shown that $\nu_*(C_{\mathfrak{k}})$ might have nonzero components only in degrees 0, 4 w.r.t. the standard grading of $\bigwedge V \cong Cl(V)$. Recall that, if $y \in V$ and $w \in Cl(V)$, then $y \cdot w = y \wedge w + i(y)w$, hence $\nu_*(C_{\mathfrak{k}}) = \sum_i \tau^{-1}(X_i) \wedge \tau^{-1}(X_i) + a$ with $a \in \wedge^0 V = \mathbb{C}$. By Lemma 1, we have that $\sum_t \tau^{-1}(X_t) \wedge \tau^{-1}(X_t) = 0$, hence $\nu_*(C_{\mathfrak{k}}) \in \mathbb{C}$. By [19, Theorem 1.50], $\mathfrak{r} = \mathfrak{k} \oplus V$ has the structure of a Lie algebra and $B_{\mathfrak{k}} \oplus \langle\cdot,\cdot\rangle$ defines a nondegenerate invariant form on $\mathfrak{r}$. Moreover, by [19, Theorem 1.59], the pair $(\mathfrak{r}, \mathfrak{k})$ is a symmetric pair. By Lemma 2 we have $V^{\mathfrak{k}} = \{0\}$. We can therefore apply [19, Theorem 1.60] and deduce that $\mathfrak{r}$ is semisimple. Finally, we have to show that $k = 1$. If the embedding is conformal then $\omega_{so(V)} - \omega_{\mathfrak{k}}$ is in the maximal ideal of $V^k(so(V))$, hence there must be a singular vector in $V^k(so(V))$ of conformal weight 2. This implies that there is λ, with λ either zero or a sum of at most two roots of $so(V)$, such that

$$\frac{(2\rho_{so(V)} + \lambda, \lambda)}{2(k + \dim V - 2)} = 2.$$

Here $\rho_{so(V)}$ is a Weyl vector for $so(V)$. The previous equation can be rewritten as

$$\frac{(2\rho_{so(V)} + \lambda, \lambda) - 2(k + \dim V - 2)}{k + \dim V - 2} = 2.$$

Since $(\rho_{so(V)}, \alpha) \leq \dim V - 3$ for any root α and $\|\lambda\|^2 \leq 8$, we see that the above equality implies that $\frac{2(\dim V-2)+4-2k}{k+\dim V-2} \geq 2$, so $k \leq 1$.

4 Classification of Equal Rank Conformal Embeddings

It is natural to investigate conformal embeddings beyond the integrable case. Recall [4, Theorem 3.1] that if $\mathfrak{k}$ is maximal equal rank in $\mathfrak{g}$, then

$$\widetilde{V}(k, \mathfrak{k}) \text{ is conformal in } V_k(\mathfrak{g}) \iff c(\mathfrak{k}) = c(\mathfrak{g}) \tag{9}$$

Here we provide a complete classification of conformal embeddings in the case when $rk(\mathfrak{k}) = rk(\mathfrak{g})$, making more precise both the statement and the proof of [4, Proposition 3.3].

That result gives a numerical criterion to reduce the detection of conformal embeddings from any subalgebra $\mathfrak{k}$ reductive in $\mathfrak{g}$ to a maximal one: let

$$\mathfrak{k} = \mathfrak{k}_0 \subset \mathfrak{k}_1 \subset \mathfrak{k}_2 \subset \cdots \subset \mathfrak{k}_t = \mathfrak{g} \tag{10}$$

be a sequence of equal rank subalgebras with $\mathfrak{k}_i$ maximal in $\mathfrak{k}_{i+1}$. Let $\mathfrak{k}_i = \oplus_{j=0}^{n_i}\mathfrak{k}_{i,j}$ be the decomposition of $\mathfrak{k}_i$ into simple ideals $\mathfrak{k}_{i,j}$, $j \geq 1$ and a center $\mathfrak{k}_{i,0}$. Since $\mathfrak{k}_{i-1}$ is maximal and equal rank in $\mathfrak{k}_i$, there is an index $j_0 \geq 1$ such that $\mathfrak{k}_{i-1} = \oplus_{j\neq j_0}\mathfrak{k}_{i,j} \oplus \tilde{\mathfrak{k}}_{i-1}$ with $\tilde{\mathfrak{k}}_{i-1}$ maximal in $\mathfrak{k}_{i,j_0}$ (note that $\tilde{\mathfrak{k}}_{i-1}$ is not simple, in general).

Theorem 3 *With notation as above, $\widetilde{V}(k,\mathfrak{k})$ is conformally embedded in $V_k(\mathfrak{g})$ if and only if for any $i = 1,\dots,t$ we have*

$$c(\tilde{\mathfrak{k}}_{i-1}) = c(\mathfrak{k}_{i,j_0}). \tag{11}$$

Proof Assume first that $\widetilde{V}(k,\mathfrak{k})$ is conformally embedded in $V_k(\mathfrak{g})$. We prove (11) by induction on t. The base $t = 1$ corresponds to $\mathfrak{k}$ maximal in $\mathfrak{g}$. Since $\omega_{\mathfrak{k}} = \omega_{\mathfrak{g}}$ it is clear that $c(\mathfrak{k}) = c(\mathfrak{g})$ which is condition (11) in this case.

Assume now $t > 1$. Since the form $(\cdot,\cdot)$ is nondegenerate when restricted to $\mathfrak{k}_{t-1}$, the orthocomplement $\mathfrak{p}$ of $\mathfrak{k}$ in $\mathfrak{g}$ can be written as $\mathfrak{p} = \mathfrak{p}\cap\mathfrak{k}_{t-1} \oplus V$ with V the orthocomplement of $\mathfrak{k}_{t-1}$ in $\mathfrak{g}$. Let $\overline{V}(k,\mathfrak{k})$ be the vertex subalgebra of $V_k(\mathfrak{k}_{t-1})$ generated by $a_{(-1)}\mathbf{1}$, $a \in \mathfrak{k}$. Since $\omega_{\mathfrak{k}} = \omega_{\mathfrak{g}}$, we have that $\omega_{\mathfrak{k}} x_{(-1)}\mathbf{1} = x_{(-1)}\mathbf{1}$ for all $x \in \mathfrak{p}\cap\mathfrak{k}_{t-1}$, hence, by Theorem 1, $\overline{V}(k,\mathfrak{k})$ is conformally embedded in $V_k(\mathfrak{k}_{t-1})$, thus, by the induction hypothesis, (11) holds for $i = 1,\dots,t-1$. In particular we have $c(\mathfrak{k}) = c(\mathfrak{k}_{t-1})$. Since $c(\mathfrak{k}) = c(\mathfrak{g})$, we get (11) also for $i = t$.

We show on case by case basis that condition (11) is sufficient.

Type A. Recall the following facts.

1. A maximal equal rank reductive subalgebra of A_n is of type $A_{h-1} \times A_{n-h} \times Z$ with Z a one-dimensional center.
2. Possible non-integrable conformal embeddings occur at level -1 except when $n = 1$ or $h = n-1$ and at level $-\frac{n+1}{2}$ except when $n = 1$ or $h = n-h-1$ [4, Theorem 5.1 (1)].
3. $\widetilde{V}_{-1}(A_{h-1} \times A_{n-h} \times Z)$ is simple if $n > 5, h > 2, n-h > 1$ [4, Theorem 5.1 (1)].
4. $\widetilde{V}_{-1}(A_2 \times A_2 \times Z)$ is simple [4, Theorem 5.2 (1)].
5. $\widetilde{V}_{-1}(A_{h-1} \times A_{n-h} \times Z)$ is simple if $h = 2$ or $n-h = 1$ (see proof of Theorem 5.3 (2) in [4]).

Let $\mathfrak{k} = \mathfrak{k}_0 \subset \mathfrak{k}_1 \subset \mathfrak{k}_2 \subset \cdots \subset \mathfrak{k}_t = \mathfrak{g}$ be a chain as in (10). It is clear that (11) implies $c(\mathfrak{k}_i) = c(\mathfrak{k}_{i-1})$. In particular $c(\mathfrak{k}_{t-1}) = c(\mathfrak{g})$, hence by (2) either $k = -\frac{n+1}{2}$ or $k = -1$. In the former case we claim that $t = 1$. If by contradiction $t > 1$, then $c(\tilde{\mathfrak{k}}_{t-2}) = c(\mathfrak{k}_{t-1,j_0})$. By (1), we know that $\mathfrak{k}_{t-1}$ is of type A_{r-1}, hence $k = -1$ or $k = -r/2$. In turn $\frac{n+1}{2} = -1$ or $\frac{n+1}{2} = -r/2$. Both cases are impossible, since $n > 1$. Therefore $t = 1$ and by (9) $\widetilde{V}_k(\mathfrak{k}_{t-1})$ embeds conformally in $V_k(\mathfrak{g})$.

Assume now that $k = -1$. Again by (9) we have that $\widetilde{V}_k(\mathfrak{k}_{t-1})$ embeds conformally in $V_k(\mathfrak{g})$. But (3)–(5) imply that $\widetilde{V}_k(\mathfrak{k}_{t-1}) = V_k(\mathfrak{k}_{t-1})$ and we can proceed inductively.

Types B, D. Recall that the maximal equal rank reductive subalgebras of $so(n)$ are $so(s) \times so(n-s)$ and $sl(n) \times Z \hookrightarrow so(2n)$. The only non-integrable conformal level is $2 - n/2$ in the first case and -2 in the second.

Assume $n \geq 7, n \neq 8$. Consider a chain as in (10). The condition $c(\mathfrak{k}_{t-1}) = c(\mathfrak{g})$ implies that $k = 2 - n/2$ or $\mathfrak{k}_{t-1}$ is of type $A_{n/2-1} \times Z$, n even and $k = -2$. We claim that in both cases $t = 1$. If by contradiction $t > 1$, then $c(\tilde{\mathfrak{k}}_{t-2}) = c(\mathfrak{k}_{t-1,j_0})$. Assume $\mathfrak{k}_{t-1} = so(s) \times so(n-s)$, we may assume that $\tilde{\mathfrak{k}}_{t-1,j_0}$ is a simple component of $so(s)$; in particular $s \geq 3$. If $s = 3$ then we have equality of central charges only at level 1, which implies $n = 0$. If $s = 4$ then $so(4) = sl(2) \times sl(2)$ and again $1 = 2 - n/2$. If $s = 5$ we have $-1/2 = 2 - n/2$ hence $n = 5$. If $s = 6$, the level is either -1 or -2; correspondingly $2 - n/2 = -1$, i.e. $n = 6$ of $2 - n/2 = -2$, i.e. $n = 8$. If $s \geq 7$, then the level is $2 - s/2 = 2 - n/2$, which gives $s = n$, or $\tilde{\mathfrak{k}}_{t-2} = sl(s/2) \times Z$. In this last case $-2 = 2 - n/2$, or $n = 8$. So $t = 1$.

In the other case $\tilde{\mathfrak{k}}_{t-2} \subset A_{n-1}$; since the central charges are equal, we should have either $-2 = -1$ or $-n/2 = -2$, which is not possible. So again $t = 1$ and we can finish the proof using (9).

Finally, in case $n = 8$, we observe that the only possible chain is

$$A_2 \times Z_1 \times Z \subset A_3 \times Z \subset D_4 \tag{12}$$

with $k = -2$ and Z_1, Z one-dimensional subalgebras. We need only to check that the hypothesis of Theorem 1 hold in this case. We use the following setup: from the explicit realization

$$so(8, \mathbb{C}) = \left\{ \begin{pmatrix} A & B \\ C & -A^t \end{pmatrix} \mid A, B, C \in gl(4, \mathbb{C}),\ B = -B^t,\ C = -C^t \right\}$$

we have

$$\begin{aligned} sl(4) \times Z &= \left\{ \begin{pmatrix} A & 0 \\ 0 & -A^t \end{pmatrix} \mid A \in gl(4, \mathbb{C}) \right\} \\ &= \left\{ \begin{pmatrix} A & 0 \\ 0 & -A^t \end{pmatrix} \mid A \in sl(4, \mathbb{C}) \right\} \oplus \mathbb{C} \begin{pmatrix} I & 0 \\ 0 & -I \end{pmatrix}, \end{aligned}$$

$$\begin{aligned} sl(3) \times Z_1 &= \left\{ \begin{pmatrix} A & 0 & 0 & 0 \\ 0 & -tr(A) & 0 & 0 \\ 0 & 0 & -A^t & 0 \\ 0 & 0 & 0 & tr(A) \end{pmatrix} \mid A \in gl(3, \mathbb{C}) \right\} \\ &= \left\{ \begin{pmatrix} A & 0 & 0 & 0 \\ 0 & -tr(A) & 0 & 0 \\ 0 & 0 & -A^t & 0 \\ 0 & 0 & 0 & tr(A) \end{pmatrix} \mid A \in sl(3, \mathbb{C}) \right\} \oplus \mathbb{C} \begin{pmatrix} I & 0 & 0 & 0 \\ 0 & -3 & 0 & 0 \\ 0 & 0 & -I & 0 \\ 0 & 0 & 0 & 3 \end{pmatrix}. \end{aligned}$$

Let $(\cdot, \cdot)$ be the normalized invariant form of $so(8)$. With notation as in (3), in this case, we have $\mathfrak{k}_1 = sl(3)$, $\mathfrak{k}_0 = Z \times Z_1$, $(\cdot, \cdot)_1 = (\cdot, \cdot)_{|sl(3) \times sl(3)}$ and, by definition, $(\cdot, \cdot)_0 = (\cdot, \cdot)_{|\mathfrak{k}_0 \times \mathfrak{k}_0}$.

Let $\varepsilon_i \in \mathfrak{h}^*$ be defined by setting $\varepsilon_i(E_{j,j} - E_{4+j,4+j}) = \delta_{i,j}$, $1 \le i, j \le 4$. Set $\eta \in Z^*$ be defined by setting $\eta(\begin{pmatrix} I & 0 \\ 0 & -I \end{pmatrix}) = 1$ and $\eta_1 \in Z_1^*$ be defined by η_1 $(\begin{pmatrix} I & 0 & 0 & 0 \\ 0 & -3 & 0 & 0 \\ 0 & 0 & -I & 0 \\ 0 & 0 & 0 & 3 \end{pmatrix}) = 1$. It is clear that, as $sl(4) \times Z$-module, $so(8, \mathbb{C}) = (sl(4, \mathbb{C}) \times Z) \oplus (\bigwedge^2 \mathbb{C}^4) \oplus (\bigwedge^2 \mathbb{C}^4)^*$. As $gl(3) \times Z$-module we have

$$\bigwedge^2 \mathbb{C}^4 = \bigwedge^2 \mathbb{C}^3 \otimes V(2\eta) \oplus \mathbb{C}^3 \otimes V(2\eta)$$

and

$$(\bigwedge^2 \mathbb{C}^4)^* = (\bigwedge^2 \mathbb{C}^3)^* \otimes V(-2\eta) \oplus (\mathbb{C}^3)^* \otimes V(-2\eta).$$

As $sl(3) \times Z_1 \times Z$-module, we have, setting $\omega_1 = \frac{2}{3}\varepsilon_1 - \frac{1}{3}\varepsilon_2 - \frac{1}{3}\varepsilon_3$ and $\omega_2 = \frac{1}{3}\varepsilon_1 + \frac{1}{3}\varepsilon_2 - \frac{2}{3}\varepsilon_3$

$$\bigwedge^2 \mathbb{C}^4 = V_{sl(3)}(\omega_2) \otimes V(2\eta_1) \otimes V(2\eta) \oplus V_{sl(3)}(\omega_1) \otimes V(-2\eta_1) \otimes V(2\eta),$$

$$(\bigwedge^2 \mathbb{C}^4)^* = V_{sl(3)}(\omega_1) \otimes V(-2\eta_1) \otimes V(-2\eta) \oplus V_{sl(3)}(\omega_2) \otimes V(2\eta_1) \otimes V(-2\eta).$$

We extend η and η_1 to $\mathfrak{h}$ by setting

$$\eta(\begin{pmatrix} I & 0 & 0 & 0 \\ 0 & -3 & 0 & 0 \\ 0 & 0 & -I & 0 \\ 0 & 0 & 0 & 3 \end{pmatrix}) = \eta_1(\begin{pmatrix} I & 0 \\ 0 & -I \end{pmatrix}) = 0, \quad \eta_{|\mathfrak{h}\cap sl(3)} = (\eta_1)_{|\mathfrak{h}\cap sl(3)} = 0.$$

Explicitly

$$\eta = \frac{1}{4}\sum_{i=1}^{4} \varepsilon_i, \quad \eta_1 = \frac{1}{12}\sum_{i=1}^{3} \varepsilon_i - \frac{1}{4}\varepsilon_4.$$

Since $\begin{pmatrix} I & 0 \\ 0 & -I \end{pmatrix}$ and $\begin{pmatrix} I & 0 & 0 & 0 \\ 0 & -3 & 0 & 0 \\ 0 & 0 & -I & 0 \\ 0 & 0 & 0 & 3 \end{pmatrix}$ are orthogonal and $(Z \times Z_1)^\perp = \mathfrak{h} \cap sl(3)$, then, for $\lambda = x\eta + y\eta_1$, $\mu = x'\eta + y'\eta_1 \in (Z \times Z_1)^*$, we have

$$(\lambda, \mu)_0 = (x\eta + y\eta_1, x'\eta + y'\eta_1). \tag{13}$$

We are now ready to apply Theorem 1 and check (3). Consider the component $V_{sl(3)}(\omega_2) \otimes V(2\eta_1) \otimes V(2\eta)$ of $\mathfrak{p}$. Then, in this case, $\rho_0^1 = 2\epsilon_1 - 2\epsilon_3$ and $\rho_0^0 = 0$ so that

$$\sum_{j=0}^{1} \frac{(\mu_i^j, \mu_i^j + 2\rho_0^j)_j}{2(k_j + h_j^\vee)} = \frac{(\omega_2, \omega_2 + 2\rho_0^1)}{2(-2+3)} + \frac{(2\eta_1 + 2\eta, 2\eta_1 + 2\eta)}{-4} = 4/3 - 1/3 = 1.$$

The other components of $\mathfrak{p}$ are handled by similar computations. It follows from Theorem 1 that $sl(3) \times Z_1 \times Z_2$ embeds conformally in D_4.

Type C. Recall that the maximal equal rank reductive subalgebras are of type $C_h \times C_{n-h}$, $h \geq 1, n-h \geq 1$ or $A_{n-1} \times Z$. In both cases then level $-1/2$ occurs, whereas in the first case we have also level $-1-n/2$, $h \neq n-h$. Reasoning as in the previous cases, one deals with level $-1-n/2$. For the other level we can use Theorems 2.3, 5.1, 5.2 (2) from [4] to obtain the simplicity of $\tilde{V}(\mathfrak{k}_i)$.

Exceptional types. In these cases, starting with a chain of subalgebras (10), we directly check that either $t = 1$, so that we can conclude by (9), or $t = 2$ and we are in one of the following cases:

$$\begin{array}{ll}
A_1 \times A_1 \times A_1 \times D_4 \hookrightarrow A_1 \times D_6 \hookrightarrow E_7 & k = -4, \\
A_1 \times D_5 \times Z \hookrightarrow A_1 \times D_6 \hookrightarrow E_7 & k = -4, \\
A_1 \times A_4 \times Z \hookrightarrow A_1 \times A_5 \hookrightarrow E_6 & k = -3, \\
A_1 \times A_1 \times A_3 \times Z \hookrightarrow A_1 \times A_5 \hookrightarrow E_6 & k = -3, \\
D_4 \times Z \times Z \hookrightarrow D_5 \times Z \hookrightarrow E_6 & k = -3, \\
A_1 \times A_1 \times A_3 \times Z \hookrightarrow D_5 \times Z \hookrightarrow E_6 & k = -3, \\
A_1 \times A_1 \times B_2 \hookrightarrow A_1 \times C_3 \hookrightarrow F_4 & k = -5/2.
\end{array}$$

We can then check the conformality of each composite embedding using the AP-criterion as done for (12). □

Definition 1 We say that a chain of vertex algebras $U_1 \subset U_2 \subset \cdots \subset U_n$ is conformal if U_i is conformally embedded into U_{i+1} for all $i = 1, \ldots, n-1$.

Remark 1 1. The conformality of the embedding given in (12) can be derived in a different way using Lemma 10.4 of [6]. More precisely, result from [6] implies that the vertex subalgebra $\widetilde{V}(-2, sl(4) \times Z)$ of $V_{-2}(D_4)$ must be simple. Since $\widetilde{V}(-2, sl(3) \times Z)$ is a simple vertex subalgebra of $\widetilde{V}(-2, A_3)$ we actually have conformal chain of *simple* affine vertex algebras

$$V_{-2}(A_2) \otimes M(1) \otimes M(1) \subset V_{-2}(A_3) \otimes M(1) \subset V_{-2}(D_4), \tag{14}$$

where $M(1)$ is the Heisenberg vertex algebra of rank one.

The determination of branching rules for conformal embeddings in (14) is an important open problem which is also recently discussed in the physics literature [12, Sect. 5].

2. Similar arguments can be applied in the cases $A_1 \times A_1 \times A_3 \times Z \hookrightarrow D_5 \times Z \hookrightarrow E_6$; $D_4 \times Z \times Z \hookrightarrow D_5 \times Z \hookrightarrow E_6$ at $k = -3$. It was proved in [3] that $\widetilde{V}(-3, D_5 \times Z)$ is a simple vertex algebra and therefore, by equality of central charges, we have conformal chains of vertex algebras

$$\widetilde{V}(-3, A_1 \times A_1 \times A_3) \otimes M(1) \subset V_{-3}(D_5) \otimes M(1) \subset V_{-3}(E_6),$$

$$\widetilde{V}(-3, D_4 \times Z) \otimes M(1) \subset V_{-3}(D_5) \otimes M(1) \subset V_{-3}(E_6).$$

3. There are however cases where conformal embeddings in non-simple vertex algebras occur. This is quite possible (see [6, Remark 2.1]). One instance of this phenomenon occurs in E_7. By using similar arguments as in [6, Sect. 8] one can show that the affine vertex subalgebra $\widetilde{V}(-4, A_1 \times D_6)$ of $V_{-4}(E_7)$ is not simple. We however checked the conformality of the chains

$$\widetilde{V}(-4, A_1 \times D_5 \times Z) \hookrightarrow \widetilde{V}(-4, A_1 \times D_6) \hookrightarrow V_{-4}(E_7),$$

$$\widetilde{V}(-4, A_1 \times D_5 \times Z) \hookrightarrow \widetilde{V}(-4, A_1 \times D_6) \hookrightarrow V_{-4}(E_7)$$

using the AP criterion.

5 Embeddings at the Critical Level

The classification of maximal conformal embeddings $\mathfrak{k} = \oplus_i \mathfrak{g}_i \hookrightarrow \mathfrak{g}$ was studied in detail in [6]. On the other hand we detected in [6] some border cases where we can not speak of conformal embeddings since the embedded affine vertex subalgebras have critical levels. In this section we provide some results on this case. Instead of considering equality of conformal vectors, we shall here consider equality of central elements. We prove that the embedded vertex subalgebra is the quotient of $\left(\bigotimes_i V^{-h_i^\vee}(\mathfrak{g}_i)\right)/\langle s\rangle$ where s is a linear combination of quadratic Sugawara central elements. In particular the embedded vertex subalgebra is not the tensor product of two affine vertex algebras.

It seems that the embeddings at the critical levels were also investigated in physical literature. In particular, Y. Tachikawa presented in [22] conjectures on existence of certain embeddings of affine vertex algebras at the critical level inside of larger vertex algebras: see Remark 5.

5.1 A Criterion

We shall describe a criterion for embeddings at the critical level.

Proposition 2 *Assume that there is a central element $s \in \bigotimes_j V^{k_j}(\mathfrak{k}_j)$ that is not a scalar multiple of* $\mathbf{1}$ *and, in* $V^k(\mathfrak{g})$,

$$s_{(n)}v = 0 \quad \textit{for all } v \in \mathfrak{p} \textit{ and } n \geq 1.$$

Then $s = 0$ in $V_k(\mathfrak{g})$. Moreover there is a non-trivial homomorphism of vertex algebras

$$\Phi^{(s)} : \left(\bigotimes_j V^{k_j}(\mathfrak{k}_j)\right) \Big/ \langle s \rangle \to V_k(\mathfrak{g}).$$

Proof We need to prove that s belongs to the maximal graded ideal in $V^k(\mathfrak{g})$. It suffices to prove that

$$x_{(n)}s = 0, \; n > 0, \; x \in \mathfrak{g}. \tag{15}$$

Since s is in the center of $\bigotimes_j V^{k_j}(\mathfrak{k}_j)$, we get that

$$x_{(n)}s = 0 \quad \text{for all } x \in \mathfrak{k}, \; n \geq 0.$$

Assume now that $x \in \mathfrak{p}$ and $n \geq 1$. We have,

$$x_{(n)}s = -[s_{(-1)}, x_{(n)}]\mathbf{1} = -\sum_{j \geq 0} \binom{-1}{j} (s_{(j)}x)_{(n-1-j)}\mathbf{1} = -(s_{(0)}x)_{(n-1)}\mathbf{1} = 0.$$

The claim follows. □

5.2 Embeddings for Lie Algebras of Type A

Let M_ℓ be the Weyl vertex algebra. Recall that the symplectic affine vertex algebra $V_{-1/2}(C_\ell)$ is a subalgebra of M_ℓ. This realization provides a nice framework for studying embeddings of affine vertex algebras in $V_{-1/2}(C_\ell)$. Some interesting cases were studied in [6]. In this section we study the embeddings in M_ℓ of critical level affine vertex algebras.

Consider the case $\ell = n^2$. Then the embeddings $sl(n) \times sl(n) \hookrightarrow sl(\ell)$, $sl(n) \times sl(n) \times \mathbb{C} \hookrightarrow gl(\ell)$ induce vertex algebra homomorphisms

$$\Phi : V^{-n}(sl(n)) \otimes V^{-n}(sl(n)) \to V_{-1}(sl(\ell)) \subset M_\ell,$$
$$\overline{\Phi} : V^{-n}(sl(n)) \otimes V^{-n}(sl(n)) \otimes M(1) \to V_{-1}(gl(\ell)) \subset M_\ell.$$

We first want to describe $\mathrm{Im}(\Phi)$. We shall see that, quite surprisingly, $\mathrm{Im}(\Phi)$ does not have the form $\widetilde{V}^{-n}(sl(n)) \otimes \widetilde{V}^{-n}(sl(n))$, where $\widetilde{V}^{-n}(sl(n))$ is a certain quotient of $V^{-n}(sl(n))$.

Let $\{x_i\}$, $\{y_i\}$ be dual bases of $sl(n)$ with respect to the trace form, and let

$$S = \sum_{i=1}^{n^2-1} : x_i y_i : .$$

Then S is a central element of the vertex algebra $V^{-n}(sl(n))$. Define $S_1 = S \otimes 1$, $S_2 = 1 \otimes S$, so that

$$s := S_1 - S_2 \in V^{-n}(sl(n)) \otimes V^{-n}(sl(n)).$$

Lemma 3 *We have:*

(1) $\Phi(S_1) = \Phi(S_2)$ *in* $V_{-1}(sl(\ell)) \subset M_\ell$.
(2) $\Phi(S_1) \neq 0$ *in* $V_{-1}(sl(\ell))$.

Proof First we notice that $sl(\ell) = sl(n) \times sl(n) \bigoplus \mathfrak{p}$ where $\mathfrak{p} \simeq L_{sl(n)}(\theta) \otimes L_{sl(n)}(\theta)$ as a $sl(n) \times sl(n)$-module

Since the quadratic Casimir operator acts on $L_{sl(n)}(\theta)$ as

$$(\theta, \theta + 2\rho)_{sl(n)} = 2n \neq 0,$$

we get, for $x \in \mathfrak{p}$,

$$(S_i)_{(1)} x_{(-1)} \mathbf{1} = 2n x_{(-1)} \mathbf{1} \qquad (i = 1, 2).$$

This shows that $\Phi(S_i) \neq 0$ in $V_{-1}(sl(\ell))$. Then $s = S_1 - S_2$ is a central element of $V^{-n}(sl(n)) \otimes V^{-n}(sl(n))$ which satisfies the conditions of Proposition 2. So $s = 0$ in $V_{-1}(sl(\ell))$. The claim follows. □

Remark 2 Note that the proof of Lemma 3 is similar to the proof of the AP criterion. But instead of proving the equality of conformal vectors, we prove the equality of central elements.

Let

$$\widetilde{V}(sl(n) \times sl(n)) = \left(V^{-n}(sl(n)) \otimes V^{-n}(sl(n))\right) / \langle s \rangle,$$

where $\langle s \rangle$ is ideal in $V^{-n}(sl(n)) \otimes V^{-n}(sl(n))$ generated by s. Using Lemma 3 we get:

Proposition 3 *We have:*

(1) *$Im(\Phi)$ is a quotient of $\widetilde{V}(sl(n) \times sl(n))$.*
(2) *$Im(\Phi)$ is not simple.*

5.3 A Conjecture

We will now present a conjecture on $\mathrm{Im}(\Phi)$.

Recall that the center of $V^{-n}(sl(n))$ (the Feigin–Frenkel center) is generated by the central element $S^{(2)} = S$ and by the higher rank Sugawara elements $S^{(3)}, \ldots, S^{(n)}$. Define the following quotient of $V^{-n}(sl(n)) \otimes V^{-n}(sl(n))$:

$$\mathcal{V}(sl(n) \times sl(n)) = \left(V^{-n}(sl(n)) \otimes V^{-n}(sl(n))\right) / \langle s_2, s_3, \ldots, s_n \rangle,$$

where $s_k = S^{(k)} \otimes \mathbf{1} - \mathbf{1} \otimes S^{(k)}$.
Conjecture $\mathrm{Im}(\Phi) \cong \mathcal{V}(sl(n) \times sl(n))$.

5.4 Maximal Embeddings in Other Simple Lie Algebras

We now investigate whether similar phenomena as the one discussed in Sect. 5.2 can occur for other types.

We list the maximal embeddings $\mathfrak{k} \hookrightarrow \mathfrak{g}$ with $\mathfrak{k}$ semisimple but not simple such that there is $k \in \mathbb{C}$ with $k_j + h_j^\vee = 0$ for all j:

1. $sp(2n) \times sp(2n) \hookrightarrow so(4n^2)$ at level $k = -1 - 1/n$;
2. $so(n) \times so(n) \hookrightarrow so(n^2)$ at level $k = -1 + 2/n$;
3. $sp(2n) \times so(2n+2) \hookrightarrow sp(4n(n+1))$ at level $k = -1/2$;
4. $so(n) \times so(n) \hookrightarrow so(2n)$ at level $k = 2 - n$;
5. $sp(2n) \times sp(2n) \hookrightarrow sp(4n)$ at level $k = -1 - n$;
6. $sl(5) \times sl(5) \hookrightarrow E_8$ at level $k = -5$;
7. $sl(3) \times sl(3) \times sl(3) \hookrightarrow E_6$ at level $k = -3$;

Some of them were detected in [6, Theorem 4.1]. In this situation we have a homomorphism of vertex algebras

$$\Phi : \bigotimes_j V^{-h_j^\vee}(\mathfrak{k}_j) \to V_k(\mathfrak{g}).$$

Write $\mathfrak{k} = \oplus_{j=1}^t \mathfrak{k}_i$ with $\mathfrak{k}_i$ simple ideals. Let $\{x_i^j\}$, $\{y_i^j\}$ be dual bases of $\mathfrak{k}_j$ with respect to $(\cdot,\cdot)_j$ and let Let $S_j = \sum_i : x_i^j y_i^j :$ be the quadratic Sugawara central element in $V^{-h_j^\vee}(\mathfrak{k}_j)$. Set $s_j = \mathbf{1} \otimes \cdots \otimes S_j \otimes \cdots \otimes \mathbf{1}$. Write $\mathfrak{p} \oplus_i V_i$ for the complete decomposition of $\mathfrak{p}$ as a $\mathfrak{k}$-module. If $V_i = \otimes_j L_{\mathfrak{k}_j}(\mu_{ij})$, set $\lambda_{ij} = (\mu_{ij}, \mu_{ij} + \rho_0^j)_j$.

Proposition 4 *If z is a non zero vector $(z_1, \dots, z_t) \in \mathbb{C}^t$ such that $\sum_j \lambda_{ij} z_j = 0$ for all i, set $s_z = \sum_j z_j s_j$ and*

$$\widetilde{V}(\mathfrak{k}) = \left(\otimes_j V^{-h_j^\vee}(\mathfrak{k}_j)\right) \Big/ \langle s_z \rangle.$$

Then $\mathrm{Im}(\Phi)$ *is a non-simple quotient of* $\widetilde{V}(\mathfrak{k})$.

Proof Note that for all j there is i such that $\lambda_{ij} \neq 0$, for, otherwise, $\mathfrak{k}_j$ would be an ideal in $\mathfrak{g}$. It follows that $\Phi(s_j) \neq 0$, so $\mathrm{Im}(\Phi)$ is not simple. Since s satisfies the hypothesis of Lemma 3, the proof follows. □

We now apply Proposition 4 to the cases listed above.

In case (1), as shown in the proof of Theorem 4.1 of [6] we have, as $sp(2n) \times sp(2n)$-module,

$$so(4n^2) = S^2(\mathbb{C}^{2n}) \otimes \bigwedge^2 \mathbb{C}^{2n} \oplus \bigwedge^2 \mathbb{C}^{2n} \otimes S^2(\mathbb{C}^{2n}),$$

so

$$\mathfrak{p} = \begin{cases} (L_{sp(2n)}(\theta) \otimes L_{sp(2n)}(\omega_2)) \oplus (L_{sp(2n)}(\omega_2) \otimes L_{sp(2n)}(\theta)) & \text{if } n > 1 \\ \{0\} & \text{if } n = 0 \end{cases}.$$

Let Λ be the matrix (λ_{ij}). Then, if $n > 1$,

$$\Lambda = \begin{pmatrix} 2(n+1) & 2n \\ 2n & 2(n+1) \end{pmatrix}$$

so the hypothesis of Proposition 4 are never satisfied.

Similarly, in case (2), we have, as $so(n) \times so(n)$-module,

$$so(n^2) = \bigwedge^2 \mathbb{C}^n \otimes S^2(\mathbb{C}^n) \oplus S^2(\mathbb{C}^n) \otimes \bigwedge^2 \mathbb{C}^n,$$

so

$$\mathfrak{p} = \begin{cases} L_{so(n)}(\theta) \otimes L_{so(n)}(2\omega_1) \oplus L_{so(n)}(2\omega_1) \otimes L_{so(n)}(\theta) & \text{if } n > 4 \\ L_{A_1}(\theta)^{\otimes 3} \otimes \mathbb{C} \oplus L_{A_1}(\theta)^{\otimes 2} \otimes \mathbb{C} \otimes L_{A_1}(\theta) & \\ \qquad \oplus L_{A_1}(\theta) \otimes \mathbb{C} \otimes L_{A_1}(\theta)^{\otimes 2} \oplus \mathbb{C} \otimes L_{A_1}(\theta)^{\otimes 3} & \text{if } n = 4 \\ L_{A_1}(\theta) \otimes L_{A_1}(2\theta) \oplus L_{A_1}(2\theta) \otimes L_{A_1}(\theta) & \text{if } n = 3 \end{cases}.$$

Then, if $n > 4$,

$$\Lambda = \begin{pmatrix} 2(n-1) & 2n \\ 2n & 2(n-1) \end{pmatrix}$$

while, for $n = 4$,

$$\Lambda = \begin{pmatrix} 4 & 4 & 4 & 0 \\ 4 & 4 & 0 & 4 \\ 4 & 0 & 4 & 4 \\ 0 & 4 & 4 & 4 \end{pmatrix}$$

and

$$\Lambda = \begin{pmatrix} 4 & 12 \\ 12 & 4 \end{pmatrix}$$

when $n = 3$. In all cases the hypothesis of Proposition 4 are not satisfied.

In case (3) we have

$$sp(4n(n+1)) = S^2(\mathbb{C}^{2n}) \otimes S^2(\mathbb{C}^{2n+2}) \oplus \bigwedge^2 \mathbb{C}^{2n} \otimes \bigwedge^2 \mathbb{C}^{2n+2},$$

hence

$$\mathfrak{p} = \begin{cases} L_{sp(2n)}(\theta) \otimes L_{so(2n+2)}(2\omega_1) \oplus L_{sp(2n)}(\omega_2) \otimes L_{so(2n+2)}(\theta) & \text{if } n > 2 \\ L_{sp(4)}(\theta) \otimes L_{A_3}(2\omega_2) \oplus L_{sp(4)}(\omega_2) \otimes L_{A_3}(\theta) & \text{if } n = 2 \\ L_{A_1}(\theta)^{\otimes 3} & \text{if } n = 1 \end{cases}.$$

In the case $n = 1$, we have that, if $(z_1, z_2, z_3)) \in \mathbb{C}^3 \setminus \{(0, 0, 0)\}$ with $z_1 + z_2 + z_3 = 0$, then Proposition 4 holds.

If $n \geq 2$ then

$$Ł = \begin{pmatrix} 2(n+1) & 4(n+1) \\ 2n & 4n \end{pmatrix},$$

thus the hypothesis of Proposition 4 are satisfied with $s = s_1 - \frac{1}{2}s_2$, which acts trivially on both components of $\mathfrak{p}$.

In cases (4) and (5) $\mathfrak{p}$ is irreducible as $\mathfrak{k}$-module, so, since $\mathfrak{k}$ is not simple, the hypothesis of Proposition 4 are clearly satisfied.

In case (6) the decomposition of $\mathfrak{p}$ as $\mathfrak{k}$-module is $\mathfrak{p} = (L_{sl(5)}(\omega_1) \otimes L_{sl(5)}(\omega_3)) \oplus (L_{sl(5)}(\omega_2) \otimes L_{sl(5)}(\omega_4)) \oplus (L_{sl(5)}(\omega_3) \otimes L_{sl(5)}(\omega_1)) \oplus (L_{sl(5)}(\omega_4) \otimes L_{sl(5)}(\omega_2))$ hence

$$\Lambda = \begin{pmatrix} 4 & 36/5 \\ 36/5 & 4 \\ 36/5 & 4 \\ 4 & 36/5 \end{pmatrix},$$

so the hypothesis of Proposition 4 are not satisfied.

Finally in case (7) above, $\mathfrak{p} = L_{sl(3)}(\omega_1)^{\otimes 3} \bigoplus L_{sl(3)}(\omega_2)^{\otimes 3}$ hence Proposition 4 holds: we can take $s_z = z_1 s_1 + z_2 s_2 + z_3 s_3$ with $(z_1, z_2, z_3)) \in \mathbb{C}^3 \setminus \{(0, 0, 0)\}$ such that $z_1 + z_2 + z_3 = 0$ and we obtain a family of non-trivial homomorphisms

$$\Phi_z : V^{-2}(A_1)^{\otimes 3}/\langle s_z\rangle \to V_{-3}(E_6).$$

Remark 3 For the embedding $so(n) \times so(n) \hookrightarrow so(2n)$ at level $k = 2 - n$, the case $n = 4$ is of particular interest. We have an embedding

$$sl(2) \times sl(2) \times sl(2) \times sl(2) \hookrightarrow so(8)$$

at $k = -2$. Then $\mathfrak{p} = L_{sl(2)}(\omega_1)^{\otimes 4}$ and we can take

$$s_z = z_1(S_1 \otimes \mathbf{1} \otimes \mathbf{1} \otimes \mathbf{1}) + z_2(\mathbf{1} \otimes S_2 \otimes \mathbf{1} \otimes \mathbf{1}) + z_3(\mathbf{1} \otimes \mathbf{1} \otimes S_3 \otimes \mathbf{1}) + z_4(\mathbf{1} \otimes \mathbf{1} \otimes \mathbf{1} \otimes S_4)$$

for every element $z = (z_1, \ldots, z_4) \in \mathbb{C}^4 \setminus \{(0, 0, 0, 0)\}$, so that $z_1 + z_2 + z_3 + z_4 = 0$. We have a family of non-trivial homomorphisms

$$\Phi_z : V^{-2}(A_1)^{\otimes 4}/\langle s_z\rangle \to V_{-2}(D_4).$$

Similarly, for the embedding $sl(2) \times sl(2) \times sl(2) \hookrightarrow sp(8)$ at $k = -1/2$, we get a family of homomorphism

$$V^{-2}(A_1)^{\otimes 3}/\langle s_z\rangle \to V_{-1/2}(C_4),$$

where $z = (z_1, z_2, z_3) \in \mathbb{C}^3 \setminus \{(0, 0, 0)\}$, and $z_1 + z_2 + z_3 = 0$.

Remark 4 Consider the embedding

$$A_3 \times A_3 \times A_1 \hookrightarrow D_6 \times A_1 \hookrightarrow E_7 \quad \text{at } k = -4.$$

Then the embedding $A_3 \times A_3 \times A_1 \hookrightarrow E_7$ at $k = -4$ also gives a subalgebra of $V_{-4}(E_7)$ at the critical level for $A_3 \times A_3$.

Remark 5 In [20, 22] some very interesting conjectures were presented. In particular, it was stated that vertex algebras for $V_{-2}(D_4)$, $V_{-3}(E_6)$, $V_{-4}(E_7)$, $V_{-6}(E_8)$ are related to certain affine vertex algebras at the critical level. In the Remarks above we have detected subalgebras at the critical level inside $V_{-2}(D_4)$, $V_{-3}(E_6)$, $V_{-4}(E_7)$ which we believe are the vertex algebras appearing in [22, Conjecture 5].

A proof of the conjecture from [20] was announced in [7]. We hope that this new result can be used in describing decomposition of some embeddings at the critical level.

Acknowledgements Dražen Adamović and Ozren Perše are partially supported by the Croatian Science Foundation under the project 2634 and by the QuantiXLie Centre of Excellence, a project cofinanced by the Croatian Government and European Union through the European Regional Development Fund–the Competitiveness and Cohesion Operational Programme (KK.01.1.1.01.0004).

References

1. Adamović, D., Perše, O.: The vertex algebra $M(1)^+$ and certain affine vertex algebras of level -1. SIGMA **8**, 040 (2012), 16 pp
2. Adamović, D., Perše, O.: Some general results on conformal embeddings of affine vertex operator algebras. Algebr. Represent. Theory **16**(1), 51–64 (2013)
3. Adamović, D., Perše, O.: Fusion rules and complete reducibility of certain modules for affine Lie algebras. J. Algebra Appl. **13**, 1350062 (2014)
4. Adamović, D., Kac, V.G., Moseneder Frajria, P., Papi, P., Perše, O.: Finite vs infinite decompositions in conformal embeddings. Commun. Math. Phys. **348**, 445–473 (2016)
5. Adamović, D., Kac, V.G., Moseneder Frajria, P., Papi, P., Perše, O.: Conformal embeddings of affine vertex algebras in minimal W-algebras II: decompositions. Japanese J. Math. **12**(2), 261–315 (2017)
6. Adamović, D., Kac, V.G., Moseneder Frajria, P., Papi, P., Perše, O.: On the classification of non-equal rank affine conformal embeddings and applications. Sel. Math. New Ser. **24**, 2455–2498 (2018)
7. Arakawa, T.: Representation theory of W–algebras and Higgs branch conjecture. arXiv:1712.07331, to appear in Proceedings of the ICM (2018)
8. Arcuri, R.C., Gomez, J.F., Olive, D.I.: Conformal subalgebras and symmetric spaces. Nucl. Phys. B **285**(2), 327–339 (1987)
9. Cellini, P., Kac, V.G., Möseneder Frajria, P., Papi, P.: Decomposition rules for conformal pairs associated to symmetric spaces and abelian subalgebras of $\mathbb{Z}_2$-graded Lie algebras. Adv. Math. **207**, 156–204 (2006)
10. Daboul C.: Algebraic proof of the symmetric space theorem. J. Math. Phys. **37**(7), 3576–3586 (1996)
11. Frenkel, I.B., Zhu, Y.: Vertex operator algebras associated to representations of affine and Virasoro algebras. Duke Math. J. **66**(1), 123–168 (1992)
12. Gaiotto, D.: Twisted compactifications of 3d N=4 theories and conformal blocks. arXiv:1611.01528
13. Goddard, P., Nahm, W., Olive, D.: Symmetric spaces, Sugawara energy momentum tensor in two dimensions and free fermions. Phys. Lett. B **160**, 111–116
14. Kac, V.G.: Lie superalgebras. Adv. Math. **26**(1), 8–96 (1977)
15. Kac, V.G.: Infinite Dimensional Lie Algebras, 3rd edn. Cambridge University Press, Cambridge (1990)
16. Kac, V.G., Sanielevici, M.: Decomposition of representations of exceptional affine algebras with respect to conformal subalgebras. Phys. Rev. D **37**(8), 2231–2237 (1988)
17. Kac, V.G., Wakimoto, M.: Modular and conformal invariance constraints in representation theory of affine algebras. Adv. Math. **70**, 156–236 (1988)
18. Kac, V.G., Möseneder Frajria, P., Papi, P.: Dirac operators and the very strange formula for Lie superalgebras. In: Papi, P., Gorelik, M. (eds.) Advances in Lie Superalgebras. Springer INdAM Series, vol. 7. Springer, Berlin (2014)
19. Kostant, B.: A cubic Dirac operator and the emergence of Euler number multiplets for equal rank subgroupps. Duke Math. J. **100**(3), 447–501 (1999)
20. Moore, G.W., Tachikawa, Y.: On 2d TQFTs whose values are holomorphic symplectic varieties. In: Proceeding of Symposia in Pure Mathematics, vol. 85 (2012). arXiv:1106.5698
21. Schellekens, A.N., Warner, N.P.: Conformal subalgebras of Kac-Moody algebras. Phys. Rev. D (3) **34**(10), 3092–3096 (1986)
22. Tachikawa, Y.: On some conjectures on VOAs, preprint

Twisted Dirac Index and Applications to Characters

Dan Barbasch and Pavle Pandžić

Abstract We present recent joint work with Peter Trapa on the notion of twisted Dirac index and its applications to (twisted) characters and to extensions of modules in a short and informal way. We also announce some further generalizations with applications to Lefschetz numbers and automorphic forms.

Keywords $(\mathfrak{g}, K)$-module · Dirac cohomology · Dirac index · Characters · Twisted characters · Euler-Poincaré pairing · Lefschetz numbers

1 Introduction

The main purpose of this paper is to present recent work in [8] in a short and informal way, as well as to announce some further applications [9]. We start by explaining the background.

If a Lie group G acts on a manifold X, then it also acts on functions on X, via

$$(g \cdot f)(x) = f(g^{-1} \cdot x).$$

In this way one gets typical examples of representations of G, like $C^\infty(X)$ (a smooth representation of G), or, in case X has a G-invariant measure, $L^2(X)$ (a unitary representation of G).

D. Barbasch was supported by NSA grant H98230-16-1-0006. P. Pandžić was supported by the QuantiXLie Center of Excellence, a project cofinanced by the Croatian Government and European Union through the European Regional Development Fund—the Competitiveness and Cohesion Operational Programme (KK.01.1.1.01.0004).

D. Barbasch
Department of Mathematics, Cornell University, Ithaca, NY 14850, USA
e-mail: barbasch@math.cornell.edu

P. Pandžić (✉)
Department of Mathematics, Faculty of Science, University of Zagreb,
Bijenička 30, 10000 Zagreb, Croatia
e-mail: pandzic@math.hr

D. Adamović and P. Papi (eds.), *Affine, Vertex and W-algebras*,
Springer INdAM Series 37, https://doi.org/10.1007/978-3-030-32906-8_2

(A representation (π, V) of G is a complex topological vector space V, with a continuous G-action by linear operators $\pi(g)$, $g \in G$. (π, V) is smooth if the map

$$g \mapsto \pi(g)v$$

is smooth for every $v \in V$. (π, V) is unitary if V is a Hilbert space and $\pi(g)$ is a unitary operator for every $g \in G$.)

A basic problem in harmonic analysis is to decompose representations as above into irreducible representations. ((π, V) is irreducible if V has no closed G-invariant subspaces other than 0 or V.)

To work on this problem, one first needs to describe the irreducible representations. This is for example well known when $G = \mathbb{T}$ is the circle group

$$\mathbb{T} = \{z \in \mathbb{C} \bigm| |z| = 1\}.$$

The irreducible modules of $\mathbb{T}$ are 1-dimensional, spanned by the functions $e^{it} \mapsto e^{int}$ on $\mathbb{T}$, $n \in \mathbb{Z}$, and

$$L^2(\mathbb{T}) = \widehat{\bigoplus}_{n\in\mathbb{Z}} \mathbb{C} f_n$$

(Fourier series).

Similarly, for $G = \mathbb{R}$, the irreducible unitary representations are 1-dimensional, spanned by the functions $t \mapsto e^{ixt}$ on $\mathbb{R}$, $x \in \mathbb{R}$, and

$$L^2(\mathbb{R}) = \int_{x\in\mathbb{R}}^{\oplus} \mathbb{C} f_x$$

(Fourier transformation).

One important application of decompositions as above is the connection with differential equations. Namely, if Δ is a G-invariant differential operator on X, then any eigenspace of Δ (when it exists) is G-invariant. Conversely, Δ acts by scalars on irreducible G-subspaces. So decomposing the representation is related to finding Δ-eigenspaces.

In the following we restrict our attention to real reductive Lie groups G. Instead of recalling the (somewhat technical) definition, we mention that the main examples are closed (Lie) subgroups of $GL(n, \mathbb{C})$, stable under the Cartan involution

$$\Theta(g) =^t \bar{g}^{-1}.$$

For example, G can be $SL(n, \mathbb{R})$, $U(p, q)$, $Sp(2n, \mathbb{R})$, $O(p, q)$.

We assume that $K = G^\Theta$ is a maximal compact subgroup of G. In the above examples, K is respectively $SO(n) \subset SL(n, \mathbb{R})$; $U(p) \times U(q) \subset U(p, q)$; $U(n) \subset Sp(2n, \mathbb{R})$, $O(p) \times O(q) \subset O(p, q)$.

2 $(\mathfrak{g}, K)$-Modules

2.1 *Passing from Representations of G to* $(\mathfrak{g}, K)$*-Modules*

To study algebraic properties of representations, it is convenient to introduce their algebraic analogs, $(\mathfrak{g}, K)$-modules.

Let V be an admissible representation of G, i.e.,

$$\dim \operatorname{Hom}_K(V_\delta, V) < \infty$$

for all irreducible finite-dimensional K-representations V_δ.

Let V_K be the space of K-finite vectors in V, i.e., the vectors v such that the span of the orbit $\pi(K)v$ is finite-dimensional.

Then V_K has an action of the Lie algebra $\mathfrak{g}_0$ (K-finite implies smooth, because each $v \in V_K$ satisfies an elliptic differential equation). Since V is complex, $\mathfrak{g} = (\mathfrak{g}_0)_{\mathbb{C}}$ also acts on V_K.

A $(\mathfrak{g}, K)$-module is a vector space M, with a Lie algebra action of $\mathfrak{g}$ and a locally finite action of K, which are compatible, i.e., induce the same action of the Lie algebra of K, $\mathfrak{k}_0$. (If K is disconnected, we also require that the action map $\mathfrak{g} \otimes M \to M$ is K-equivariant. In case K is connected, this property is automatically satisfied.)

Every M as above can be decomposed under K as

$$M = \bigoplus_{\delta \in \hat{K}} m_\delta V_\delta.$$

M is a Harish-Chandra module if it is finitely generated over the universal enveloping algebra $U(\mathfrak{g})$ of $\mathfrak{g}$ and all $m_\delta < \infty$.

2.2 *Example:* $G = SU(1, 1) \cong SL(2, \mathbb{R})$

The Lie algebra $\mathfrak{g} = \mathfrak{sl}(2, \mathbb{C})$ of two by two matrices of trace 0 has a basis

$$h = \begin{pmatrix} 1 & 0 \\ 0 & -1 \end{pmatrix}, \quad e = \begin{pmatrix} 0 & 1 \\ 0 & 0 \end{pmatrix}, \quad f = \begin{pmatrix} 0 & 0 \\ 1 & 0 \end{pmatrix},$$

with commutation relations

$$[h, e] = 2e, \quad [h, f] = -2f, \quad [e, f] = h.$$

Since ih spans the Lie algebra of K, h diagonalizes on $(\mathfrak{g}, K)$-modules and has integer eigenvalues. The possible irreducible $(\mathfrak{g}, K)$-modules are described by the following pictures:

$$\begin{array}{cccc} \bullet & \bullet & \bullet & \cdots \\ k & k+2 & k+4 & \cdots \end{array} \tag{1}$$

$$\begin{array}{cccc} \cdots & \bullet & \bullet & \bullet \\ \cdots & -k-4 & -k-2 & -k \end{array} \tag{2}$$

$$\begin{array}{cccc} \bullet & \bullet & \cdots & \bullet \\ -n & -n+2 & \cdots & n \end{array} \tag{3}$$

$$\begin{array}{ccccc} \cdots & \bullet & \bullet & \bullet & \cdots \\ \cdots & i-2 & i & i+2 & \cdots \end{array} \tag{4}$$

where $k > 0$, $n \geq 0$ and i are integers.

Each dot represents a 1-dim eigenspace for h. Numbers under the dots are the corresponding eigenvalues.

In each picture, e raises the eigenvalue by 2, and f lowers the eigenvalue by 2.

The pictures (1), (2) and (3) define unique modules. Namely, we know that $ef - fe = h$, and we know ef or fe at one point. From this, it is easy to determine ef and fe at all points, and this tells us how to define the action on a basis, in a unique way.

For the picture (4), there are however many corresponding modules. Here we do not know ef or fe at any point. But we would know them if we knew $ef + fe$. We use

$$\mathrm{Cas}_{\mathfrak{g}} = \frac{1}{2}h^2 + ef + fe,$$

which commutes with $\mathfrak{g}$ and so acts by a scalar on any irreducible module. Fixing this scalar determines the module. (Not all values are allowed; in some cases the module may not be irreducible, but break up into submodules corresponding to the pictures (1)–(3).)

2.3 *Infinitesimal Character*

In general, one can define the so called Casimir element $\mathrm{Cas}_{\mathfrak{g}}$ in the center of the enveloping algebra $U(\mathfrak{g})$. The construction is as follows:

Fix a nondegenerate invariant symmetric bilinear form B on $\mathfrak{g}$ (e.g. $\operatorname{tr} XY$);

Take dual bases b_i, d_i of $\mathfrak{g}$ with respect to B;

Write

$$\mathrm{Cas}_{\mathfrak{g}} = \sum b_i d_i.$$

It is easy to see that $\mathrm{Cas}_{\mathfrak{g}}$ is indeed central in $U(\mathfrak{g})$, and that it does not depend on the choice of basis b_i.

The center $Z(\mathfrak{g})$ of $U(\mathfrak{g})$ is a polynomial algebra; one of the generators is $\mathrm{Cas}_{\mathfrak{g}}$.

All elements of $Z(\mathfrak{g})$ act as scalars on irreducible modules (a version of Schur's lemma). This defines the infinitesimal character of M; it is an algebra homomorphism $\chi_M : Z(\mathfrak{g}) \to \mathbb{C}$.

Harish-Chandra proved that $Z(\mathfrak{g}) \cong P(\mathfrak{h}^*)^W$, so infinitesimal characters correspond to $\mathfrak{h}^*/W$. (Here $\mathfrak{h}$ is a Cartan subalgebra of $\mathfrak{g}$; in examples, it typically consists of the diagonal matrices in $\mathfrak{g}$. W is the Weyl group of $(\mathfrak{g}, \mathfrak{h})$; it is a finite reflection group.)

3 Dirac Operators

3.1 Dirac Operator on $\mathbb{R}^n$

We start by recalling the classical Dirac operator on $\mathbb{R}^n$. The idea behind the construction is to look for D such that $D^2 = -\sum \partial_i^2$ (or $D^2 = \sum \pm\partial_i^2$).

If $D = \sum e_i \partial_i$, we get

$$e_i^2 = -1; \qquad e_i e_j + e_j e_i = 0, \quad i \neq j.$$

So the coefficients should belong to the Clifford algebra $C(\mathbb{R}^n)$, which is defined exactly by these relations between the generators e_i.

Identifying $\partial_i \leftrightarrow e_i$, we get

$$D = \sum e_i \otimes e_i \in D_{cc}(\mathbb{R}^n) \otimes C(\mathbb{R}^n).$$

($D_{cc}(\mathbb{R}^n)$ is the algebra of constant coefficient differential operators on $\mathbb{R}^n$.)

3.2 Dirac Operator for G

To obtain a similar operator attached to our group G, we first describe the relevant Clifford algebra. Let

$$\mathfrak{g} = \mathfrak{k} \oplus \mathfrak{p}$$

be the Cartan decomposition of $\mathfrak{g}$; here $\mathfrak{k}$ and $\mathfrak{p}$ are the ± 1 eigenspaces of the Cartan involution. In particular, $\mathfrak{k}$ is the complexified Lie algebra of the maximal compact subgroup K of G.

Let $C(\mathfrak{p})$ be the Clifford algebra of $\mathfrak{p}$ with respect to B: it is the associative algebra with 1, generated by $\mathfrak{p}$, with relations

$$xy + yx + 2B(x, y) = 0, \qquad x, y \in \mathfrak{p}.$$

The Dirac operator attached to G is defined as follows. Let b_i be any basis of $\mathfrak{p}$ and let d_i be the dual basis with respect to B. The Dirac operator is

$$D = \sum_i b_i \otimes d_i \qquad \in U(\mathfrak{g}) \otimes C(\mathfrak{p}).$$

It is easy to see that D is independent of the choice of basis b_i and K-invariant.

Moreover, D^2 is the spin Laplacean (Parthasarathy [36]):

$$D^2 = -\operatorname{Cas}_{\mathfrak{g}} \otimes 1 + \operatorname{Cas}_{\mathfrak{k}_\Delta} + \text{constant}.$$

Here $\operatorname{Cas}_{\mathfrak{g}}$, $\operatorname{Cas}_{\mathfrak{k}_\Delta}$ are the Casimir elements of $U(\mathfrak{g})$, $U(\mathfrak{k}_\Delta)$, where $\mathfrak{k}_\Delta$ is the diagonal copy of $\mathfrak{k}$ in $U(\mathfrak{g}) \otimes C(\mathfrak{p})$ defined by

$$\mathfrak{k} \hookrightarrow \mathfrak{g} \hookrightarrow U(\mathfrak{g}) \quad \text{and} \quad \mathfrak{k} \to \mathfrak{so}(\mathfrak{p}) \hookrightarrow C(\mathfrak{p}).$$

3.3 Dirac Cohomology

We use D to define the Dirac cohomology of a $(\mathfrak{g}, K)$-module M. Let S be a spin module for $C(\mathfrak{p})$. (S is constructed as $S = \bigwedge \mathfrak{p}^+$ for $\mathfrak{p}^+ \subset \mathfrak{p}$ a maximal isotropic subspace, with $\mathfrak{p}^+ \subset C(\mathfrak{p})$ acting by wedging, and an opposite isotropic subspace $\mathfrak{p}^-$ acting by contractions.)

The Dirac operator D acts on $M \otimes S$, and we define the Dirac cohomology of M as

$$H_D(M) = \ker D / \operatorname{Im} D \cap \ker D.$$

$H_D(M)$ is a module for the spin double cover $\widetilde{K}$ of K. It is finite-dimensional if M is of finite length.

If M is unitary, then D is self adjoint with respect to an inner product. It follows that

$$H_D(M) = \ker D = \ker D^2,$$

and $D^2 \geq 0$ (Parthasarathy's Dirac inequality).

3.4 Example: $G = SU(1,1) \cong SL(2,\mathbb{R})$

Let $G = SU(1,1) \cong SL(2,\mathbb{R})$ with $(\mathfrak{g}, K)$-modules described as in Example 2.2.

The modules corresponding to the pictures (1)–(3) have $H_D \neq 0$. For each of these modules, the weights of H_D are equal to the highest weight of the module plus 1 and/or the lowest weight of the module minus 1.

The modules corresponding to the picture (4) all have $H_D = 0$.

3.5 *Vogan's Conjecture*

Let $\mathfrak{h} = \mathfrak{t} \oplus \mathfrak{a}$ be a fundamental Cartan subalgebra of $\mathfrak{g}$, i.e., $\mathfrak{t}$ is a Cartan subalgebra of $\mathfrak{k}$ and $\mathfrak{a}$ is the centralizer of $\mathfrak{t}$ in $\mathfrak{p}$. View $\mathfrak{t}^* \subset \mathfrak{h}^*$ via extension by 0 over $\mathfrak{a}$.

The following result was conjectured by Vogan [40], and proved by Huang-Pandžić [18].

Theorem 1 *Assume M has infinitesimal character and $H_D(M)$ contains a $\widetilde{K}$-type E_γ of highest weight $\gamma \in \mathfrak{t}^*$.*

Then the infinitesimal character of M is, up to conjugation by the Weyl group $W_\mathfrak{g}$, equal to $\gamma + \rho_\mathfrak{k}$.

The above Vogan's conjecture also has a structural version. Let $\zeta : Z(\mathfrak{g}) \to Z(\mathfrak{k}) \cong Z(\mathfrak{k}_\Delta)$ be the homomorphism corresponding under Harish-Chandra isomorphisms to the restriction map $P(\mathfrak{h}^*)^{W_\mathfrak{g}} \to P(\mathfrak{t}^*)^{W_\mathfrak{k}}$.

Then one shows that for any $z \in Z(\mathfrak{g})$, there is $a \in (U(\mathfrak{g}) \otimes C(\mathfrak{p}))^K$ such that

$$z \otimes 1 = \zeta(z) + Da + aD.$$

This implies the module version from Theorem 1, since $Da + aD$ acts as zero on $H_D(M)$.

4 Motivation

The following are some of the applications of Dirac operators and Dirac cohomology to $(\mathfrak{g}, K)$-modules:

- The notion of Dirac cohomology is related to unitarity of $(\mathfrak{g}, K)$-modules, which remains to be an important unsolved problem. For example, Dirac inequality has proved very useful in some known partial classifications of unitary modules, see for example [13, 41]. Vogan's conjecture can be interpreted as an improvement of the Dirac inequality, and one can hope to find further improvements.
- Irreducible unitary $(\mathfrak{g}, K)$-modules M with $H_D \neq 0$ are interesting; for example, they include discrete series representations, $A_\mathfrak{q}(\lambda)$ modules, unitary highest weight modules, and some unipotent representations. These representations should form a nice part of the unitary dual. Their Dirac cohomology has been computed in many cases [6, 7, 17, 21].
- Dirac cohomology is related to classical topics like the generalized Weyl character formula, the generalized Bott-Borel-Weil Theorem, construction of the discrete series representations and multiplicities of automorphic forms [19].
- Dirac cohomology is related to other notions, like $\mathfrak{n}$-cohomology [22], $(\mathfrak{g}, K)$-cohomology [17], characters and branching [23].

- There are nice constructions of representations with $H_D \neq 0$. For example, Parthasarathy [36] and Atiyah-Schmid [4] constructed the discrete series representations using spin bundles on G/K. A similar but more algebraic construction is the "algebraic Dirac induction" as defined by Pandžić–Renard [34] and further developed by Prlić [37, 38].
- There is a translation principle for the Euler characteristic of H_D, i.e., the Dirac index [31]. We will study the Dirac index in detail in the next section.
- There are applications to characteristic cycles [32, 33].

There has also been a number of generalizations to other settings, such as

- quadratic subalgebras and "cubic" Dirac operators [14, 26, 27];
- affine Lie algebras and twisted affine cubic Dirac operators [24];
- Lie superalgebras [20, 25, 44];
- graded affine Hecke algebras and p-adic groups [5];
- noncommutative equivariant cohomology [1, 28];
- symplectic reflection algebras, rational Cherednik algebras [12, 16];
- the quantum group $U_q(\mathfrak{sl}(2, \mathbb{C}))$ [35].

5 Dirac Index in the Equal Rank Case

5.1 *Dirac Index*

For many purposes, like the study of characters or the translation principle, it is good to replace the Dirac cohomology by its Euler characteristic, the Dirac index.

To define the Dirac index, assume $\operatorname{rank}\mathfrak{g} = \operatorname{rank}\mathfrak{k}$. Then $\dim\mathfrak{p}$ is even, so the $C(\mathfrak{p})$-module S is graded:

$$S = S^+ \oplus S^- = \textstyle\bigwedge^{\text{even}}\mathfrak{p}^+ \oplus \bigwedge^{\text{odd}}\mathfrak{p}^+.$$

Since D interchanges $M \otimes S^+$ and $M \otimes S^-$, the Dirac cohomology splits as

$$H_D(M) = H_D(M)^+ \oplus H_D(M)^-.$$

The Dirac index of M is defined as the virtual $\widetilde{K}$-module

$$I(M) = H_D(M)^+ - H_D(M)^-$$

If M is admissible and has infinitesimal character, then

$$M \otimes S^+ - M \otimes S^- = I(M).$$

It follows that

$$\operatorname{ch} M(\operatorname{ch} S^+ - \operatorname{ch} S^-) = \operatorname{ch} I(M),$$

where ch denotes the $\widetilde{K}$-character, and

$$\operatorname{ch} M = \frac{\operatorname{ch} I(M)}{\operatorname{ch} S^+ - \operatorname{ch} S^-}$$

as functions on the regular part of the Cartan subgroup $T \subseteq K$.

5.2 *Example: $G = SU(1, 1)$ $(\cong SL(2, \mathbb{R}))$*

Let $G = SU(1, 1)$ $(\cong SL(2, \mathbb{R}))$ and consider the modules described in Example 2.2.

For M corresponding to the picture (1), the K-character is

$$e^{ikt} + e^{i(k+2)t} + e^{i(k+4)t} + \cdots = \frac{e^{ikt}}{1 - e^{i2t}} = \frac{e^{i(k-1)t}}{e^{-it} - e^{it}}.$$

Since $I(M) = H_D(M)$ has a single weight $k - 1$, and S has weights -1 and 1, this agrees with the general result.

5.3 *Characters*

Let us review some facts about characters starting with the basic case of a finite-dimensional representation.

If (π, M) is a finite-dimensional representation of G (or K, or $\widetilde{K}$), define $\operatorname{ch} M = \Theta_\pi : G \to \mathbb{C}$ by

$$\operatorname{ch} M(g) = \operatorname{tr} \pi(g).$$

Then $\operatorname{ch} M(gg_1g^{-1}) = \operatorname{ch} M(g_1)$, so the function ch M is determined on maximal tori. Moreover, the following properties hold:

1. $\operatorname{ch}(M_1 \oplus M_2) = \operatorname{ch} M_1 + \operatorname{ch} M_2$;
2. $\operatorname{ch}(M_1 \otimes M_2) = \operatorname{ch} M_1 \operatorname{ch} M_2$;
3. inequivalent M have linearly independent characters;
4. the character is given on a maximal torus in the group by the Weyl character formula.

On the other hand, if (π, V) is an admissible infinite-dimensional representation of G, then the operators $\pi(g)$, $g \in G$, are not trace class. However,

$$\pi(f) = \int_G f(g)\pi(g)dg$$

is trace class for any $f \in C_c^\infty(G)$, and $f \mapsto \operatorname{tr} \pi(f)$ is a distribution on G. Harish-Chandra proved that this distribution is given by a locally L^1 function on G, which is analytic on the regular set.

One can also define the K-character of the G-representation (π, V). It is a distribution on K, sending $f \in C_c^\infty(K)$ to $\operatorname{tr} \pi(f)$. It can be identified with the formal K-character of the underlying $(\mathfrak{g}, K)$-module; the formal K-character of an admissible representation $M = \oplus_\delta m_\delta V_\delta$ of K or $\widetilde{K}$ is defined as

$$\operatorname{ch} M = \sum_\delta m_\delta \operatorname{ch} V_\delta.$$

The K-character is not a function on K, but it is a function equal to the G-character on the set $G^{\text{reg}} \cap K$. This is a mild extension of the results of [15].

6 Dirac Index in the Unequal Rank Case

6.1 *The Extended Group*

If the rank of K is smaller than the rank of G, then the Dirac index as defined above does not make sense.

Namely, if $\dim \mathfrak{p}$ is even, we can still write $S = S^+ \oplus S^-$, but S^+ and S^- are now isomorphic $\widetilde{K}$-modules, and $I(M) = 0$ for any M. On the other hand, if $\dim \mathfrak{p}$ is odd, S^+ and S^- are not defined. There are now two spin modules, different as $C(\mathfrak{p})$-modules, but isomorphic as $\widetilde{K}$-modules, so one can not use them in place of $S^\pm$.

In [8] we go around this difficulty by considering the extended group

$$G^+ = G \rtimes \{1, \theta\},$$

where θ is the Cartan involution of G. The maximal compact subgroup of G^+ is

$$K^+ = K \times \{1, \theta\}.$$

Representations of G^+ correspond to $(\mathfrak{g}, K^+)$-modules, which can be thought of as $(\mathfrak{g}, K)$-modules with a compatible action of θ.

If (π, M) is an irreducible $(\mathfrak{g}, K)$-module, we can consider M^θ, the same space as M, but with action $\pi \circ \theta$. If $M^\theta \cong M$, then M can be made into a $(\mathfrak{g}, K^+)$-module, in two ways. If M^θ is not isomorphic to M, then $M \oplus M^\theta$ can be made into an irreducible $(\mathfrak{g}, K^+)$-module, in a unique way; these modules are however not so interesting for us, since their twisted Dirac indices defined below are always equal to 0.

6.2 Example: $G = SL(2, \mathbb{C})$

Let $G = SL(2, \mathbb{C})$. Then $K = SU(2)$, and the K-type structure of an irreducible $(\mathfrak{g}, K)$-module M is

$$F_k \oplus F_{k+2} \oplus F_{k+4} \oplus \cdots$$

for some $k \geq 0$, with F_n denoting the irreducible K-module with highest weight n and dimension $n + 1$.

The neighbouring K-types are connected by the action of $\mathfrak{p}$, so if M admits a compatible action of θ, θ can only act by $1, -1, 1, -1, \ldots$, or by $-1, 1, -1, 1, \ldots$.

(For a full description of M, one needs an extra complex parameter corresponding to the infinitesimal character, and M admits an action of θ precisely when this parameter transforms correctly under θ.)

6.3 Twisted Characters

For an irreducible $(\mathfrak{g}, K^+)$-module M, we denote by $M^{\pm}$ the ± 1 eigenspaces of θ on M, and consider the twisted K-character

$$\mathrm{ch}_\theta(M) = M^+ - M^-.$$

This encodes the twisted character of the corresponding G^+-representation π, i.e., the character on $\theta G \subset G^+$, as function on the regular part of the twisted Cartan subgroup θT. Such twisted characters were studied in [11].

To define the twisted Dirac index, we consider the involution $\gamma = \theta \otimes 1$ of $U(\mathfrak{g}) \otimes C(\mathfrak{p})$. Since $\gamma(D) = -D$, γ acts on $H_D(M)$ for any $(\mathfrak{g}, K^+)$-module M, and $H_D(M)$ splits into eigenspaces for γ:

$$H_D(M) = H_D(M)^+ \oplus H_D(M)^-.$$

The twisted Dirac index of M is the virtual $\widetilde{K}$-module

$$I_\theta(M) = H_D(M)^+ - H_D(M)^-.$$

For any admissible M with infinitesimal character, we show

$$I_\theta(M) = M^+ \otimes S - M^- \otimes S.$$

This implies that the twisted character on θT^{reg} is equal to

$$\frac{\mathrm{ch}\, I_\theta(M)}{\mathrm{ch}\, S}.$$

In Example 6.2, the spin module is $S = F_1$, and one easily checks that

$$(F_k - F_{k+2} + F_{k+4} - F_{k+6} + \cdots) \otimes F_1$$

equals F_{k-1} if $k > 0$, and zero if $k = 0$. This agrees with $I_\theta(M)$.

7 Applications to Extensions of Modules

In the equal rank case, the Euler-Poincaré pairing of $(\mathfrak{g}, K)$-modules M and N is defined as the virtual vector space

$$EP(M, N) = \sum_{i=1}^{s} (-1)^i \operatorname{Ext}^i_{(\mathfrak{g},K)}(M, N).$$

In [8] we prove

$$EP(M, N) = \operatorname{Hom}_{\widetilde{K}}(I(M), I(N)).$$

Moreover, the two sides of this equality are also equal to the so called elliptic pairing of M and N.

In the unequal rank case, both sides of the above equality are 0, for any M and N, but there is a nontrivial twisted analogue which we now describe.

If M and N are $(\mathfrak{g}, K^+)$-modules, we replace the virtual vector space $EP(M, N)$ (which is always zero), by the virtual $\{1, \theta\}$-module

$$EP_\theta(M, N) = \sum_{i=1}^{s} (-1)^i \operatorname{Ext}^i_{(\mathfrak{g},K)}(M, N).$$

We show that the trace of θ on $EP_\theta(M, N)$ equals

$$c \dim \operatorname{Hom}_{\widetilde{K}}(I_\theta(M), I_\theta(N)),$$

where $c = 1$ if $\dim \mathfrak{p}$ is even, and $c = 2$ if $\dim \mathfrak{p}$ is odd.

This is also equal to the "twisted elliptic pairing" studied by Arthur, Labesse, Waldspurger and others [2, 3, 29, 30, 42, 43].

8 Twisted Dirac Index and Lefschetz Numbers

In [9] we extend the above results to the setting when the twisting is not only by θ but by a finite order automorphism τ of G commuting with θ. Any such τ naturally corresponds to an automorphism γ of the pair $(U(\mathfrak{g}) \otimes C(\mathfrak{p}), \widetilde{K})$, which anticommutes with D and hence acts on the Dirac cohomology of any Harish-Chandra module M for the extended group $G \rtimes \langle\tau\rangle$. The τ-twisted Dirac index of M is defined as the function

$$k \mapsto \operatorname{tr}(k\gamma; H_D(M))$$

on the spin double cover $\widetilde{K}$ of K. We show that the τ-twisted Dirac index has similar properties as the ordinary or θ-twisted Dirac indices, and we show that it does not vanish for certain interesting representations in case when the group G^τ of fixed points of τ is an equal rank group.

We next show how Lefschetz numbers, which are defined as traces of τ on the Euler characteristic of $(\mathfrak{g}, K)$-cohomology, can be obtained from the τ-twisted Dirac index. The latter is however more general because it can be nonzero for modules with zero $(\mathfrak{g}, K)$-cohomology.

We then revisit the theory of [29, 30], which attaches to Lefschetz numbers certain functions called pseudocoefficients. This enables relations with the trace formula and consequently with automorphic forms; see e.g. [39]. Our work in progress aims at obtaining analogous results for the τ-twisted Dirac index.

References

1. Alekseev, A., Meinrenken, E.: Lie theory and the Chern-Weil homomorphism. Ann. Sci. Ecole. Norm. Sup. **38**, 303–338 (2005)
2. Arthur, J.: A Paley-Wiener theorem for real reductive groups. Acta Math. **150**(1–2), 1–89 (1983)
3. Arthur, J.: The invariant trace formula II. Global theory. J. Am. Math. Soc. **1**(3), 501–554 (1988)
4. Atiyah, M., Schmid, W.: A geometric construction of the discrete series for semisimple Lie groups. Invent. Math. **42**, 1–62 (1977)
5. Barbasch, D., Ciubotaru, D., Trapa, P.: Dirac cohomology for graded affine Hecke algebras. Acta Math. **209**(2), 197–227 (2012)
6. Barbasch, D., Pandžić, P.: Dirac cohomology and unipotent representations of complex groups. In: Connes, A. Gorokhovsky, A. Lesch, M. Pflaum, M. Rangipour, B. (eds.) Noncommutative Geometry and Global Analysis. Contemporary Mathematics, vol. 546, pp. 1–22. American Mathematical Society, Providence (2011)
7. Barbasch, D., Pandžić, P.: Dirac cohomology of unipotent representations of $Sp(2n, \mathbb{R})$ and $U(p, q)$. J. Lie Theory **25**(1), 185–213 (2015)
8. Barbasch, D., Pandžić, P., Trapa, P.: Dirac index and twisted characters. Trans. Am. Math. Soc. **371**(3), 1701–1733 (2019)
9. Barbasch, D., Pandžić, P.: Dirac index, Lefschetz numbers and the trace formula, in preparation
10. Barbasch, D., Speh, B.: Cuspidal representations of reductive groups. arXiv:0810.0787
11. Bouaziz, A.: Sur les charactères des groupes de Lie réductifs non connexes. J. Funct. Anal. **70**(1), 1–79 (1987)
12. Ciubotaru, D.: Dirac cohomology for symplectic reflection algebras. Sel. Math. **22**(1), 111–144 (2016)
13. Enright, T., Howe, R., Wallach, N.: A classification of unitary highest weight modules. Representation theory of reductive groups (Park City, Utah). Progress in Mathematics, vol. 40, pp. 97–143. Birkhäuser, Boston (1982)
14. Goette, S.: Equivariant η-invariants on homogeneous spaces. Math. Z. **232**, 1–42 (1998)

15. Harish-Chandra, *The Characters of Semisimple Groups*, Trans. Am. Math. Soc. **83**(1), 98–163 (1956)
16. Huang, J.-S., Wong, K.D.: A Casselman-Osborne theorem for rational Cherednik algebras. Transform. Groups **23**(1), 75–99 (2018)
17. Huang, J.-S., Kang, Y.-F., Pandžić, P.: Dirac cohomology of some Harish-Chandra modules. Transform. Groups **14**(1), 163–173 (2009)
18. Huang, J.-S., Pandžić, P.: Dirac cohomology, unitary representations and a proof of a conjecture of Vogan. J. Am. Math. Soc. **15**, 185–202 (2002)
19. Huang, J.-S., Pandžić, P.: Dirac Operators in Representation Theory. Mathematics: Theory and Applications. Birkhäuser, Boston (2006)
20. Huang, J.-S., Pandžić, P.: Dirac cohomology for Lie superalgebras. Transform. Groups **10**, 201–209 (2005)
21. Huang, J.-S., Pandžić, P., Protsak, V.: Dirac cohomology of Wallach representations. Pac. J. Math. **250**(1), 163–190 (2011)
22. Huang, J.-S., Pandžić, P., Renard, D.: Dirac operators and Lie algebra cohomology. Represent. Theory **10**, 299–313 (2006)
23. Huang, J.-S., Pandžić, P., Zhu, F.: Dirac cohomology, K-characters and branching laws. Am. J. Math. **135**(5), 1253–1269 (2013)
24. Kac, V.P., Möseneder Frajria, P., Papi, P.: Multiplets of representations, twisted Dirac operators and Vogan's conjecture in affine setting. Adv. Math. **217**, 2485–2562 (2008)
25. Kac, V.P., Möseneder Frajria, P., Papi, P.: Dirac operators and the very strange formula for Lie superalgebras. In: Advances in Lie Superalgebras, Springer INdAM Series, vol. 7, pp. 121–148
26. Kostant, B.: A cubic Dirac operator and the emergence of Euler number multiplets of representations for equal rank subgroups. Duke Math. J. **100**, 447–501 (1999)
27. Kostant, B.: Dirac cohomology for the cubic Dirac operator. Studies in Memory of Issai Schur. Progress in Mathematics, vol. 210, pp. 69–93 (2003)
28. Kumar, S.: Induction functor in non-commutative equivariant cohomology and Dirac cohomology. J. Algebra **291**, 187–207 (2005)
29. Labesse, J.P.: Pseudo-coefficients très cuspidaux et K-theorie. Math. Ann. **291**, 607–616 (1991)
30. Labesse, J.-P.: Cohomologie, stabilization et changement de base. Astérisque (257), (1999)
31. Mehdi, S., Pandžić, P., Vogan, D.: Translation principle for Dirac index. Am. J. Math. **139**(6), 1465–1491 (2017)
32. Mehdi, S., Pandžić, P., Vogan, D., Zierau, R.: Dirac index and associated cycles of Harish-Chandra modules. arXiv:1712.04169
33. Mehdi, S., Pandžić, P., Vogan, D., Zierau, R.: Computing the associated cycles of certain Harish-Chandra modules. Glas. Mat. **53**(73), 2, 275–330 (2018)
34. Pandžić, P., Renard, D.: Dirac induction for Harish-Chandra modules. J. Lie Theory **20**(4), 617–641 (2010)
35. Pandžić, P., Somberg, P.: Dirac operator and its cohomology for the quantum group $U_q(\mathfrak{sl}(2))$. J. Math. Phys. **58**(4), 041702 (2017), 13 pp
36. Parthasarathy, R.: Dirac operator and the discrete series. Ann. Math. **96**, 1–30 (1972)
37. Prlić, A.: Algebraic Dirac induction for nonholomorphic discrete series of $SU(2, 1)$. J. Lie Theory **26**(3), 889–910 (2016)
38. Prlić, A.: Construction of discrete series representations of $SO_e(4, 1)$ via algebraic Dirac induction. arXiv:1811.01610
39. Rohlfs, J., Speh, B.: Automorphic representations and Lefschetz numbers. Ann. Sci. Ec. Norm. Super. 4e sér. 22, 473–499(1989)
40. Vogan, D.: Dirac operators and unitary representations, 3 talks at MIT Lie groups seminar, (Fall 1997)
41. Vogan, D., Zuckerman, G.: Unitary representations with nonzero cohomology. Compos. Math. **53**, 51–90 (1984)
42. Waldspurger, J.-L.: Les facteurs de transfert pour les groupes classiques: une formulaire. Manuscr. Math. **133**(1–2), 41–82 (2010)
43. Waldspurger, J.-L.: La formule des traces locale tordue. arXiv:1205.1100
44. Xiao, W.: Dirac operators and cohomology for Lie superalgebra of type I. J. Lie Theory **27**(1), 111–121 (2017)

The Level One Zhu Algebra for the Heisenberg Vertex Operator Algebra

Katrina Barron, Nathan Vander Werf and Jinwei Yang

Abstract The level one Zhu algebra for the Heisenberg vertex operator algebra is calculated, and implications for the use of Zhu algebras of higher level for vertex operator algebras are discussed. In particular, we show the Heisenberg vertex operator algebra gives an example of when the level one Zhu algebra, and in fact all its higher level Zhu algebras, do not provide new indecomposable non simple modules for the vertex operator algebra beyond those detected by the level zero Zhu algebra.

Keywords Vertex operator algebra · Heisenberg algebra · Conformal field theory

1 Introduction

In this paper, we give a detailed calculation of the level one Zhu algebra for the vertex operator algebra associated to the Heisenberg algebra. Theoretically, higher level Zhu algebras for a vertex operator algebra V are potentially important tools in studying the indecomposable non simple V-modules. However, until recently there seemed to be no specific examples of higher level Zhu algebras, and there were some conceptual gaps in our understanding of the relationship between higher level Zhu algebras and the indecomposable non simple V-modules. In [3], the authors prove several results about the relationship between higher level Zhu algebras and various V-modules, and we use the examples of the level one Zhu algebras for both the vertex operator algebra associated to the Heisenberg algebra (calculated here in this paper) and the vertex operator algebra associated to the Virasoro algebra (calculated in [4])

K. Barron (✉)
Department of Mathematics, University of Notre Dame, Notre Dame, IN 46556, USA
e-mail: kbarron@nd.edu

N. Vander Werf
Department of Mathematics and Statistics, University of Nebraska, Kearney, NE 68849, USA
e-mail: vanderwerfnp@unk.edu

J. Yang
Department of Mathematics, University of Alberta, Edmonton, AB T6G 2G1, Canada
e-mail: yookinwi@gmail.com

D. Adamović and P. Papi (eds.), *Affine, Vertex and W-algebras*,
Springer INdAM Series 37, https://doi.org/10.1007/978-3-030-32906-8_3

to illustrate these relationships. In particular, in [3] we show how the nature of the level n Zhu algebra in relation to the level $n-1$ Zhu algebra determines whether new information can be gleaned from the level n Zhu algebra not already detected by the level $n-1$ Zhu algebra, and the nature of that information.

In this sense, the Heisenberg vertex operator algebra $M_a(1)$ is used as a foil to show that for such a vertex operator algebra, there is no new information about the $M_a(1)$-modules coming from the higher level Zhu algebras (even if we did not already have a complete classification of the indecomposable $\mathbb{N}$-gradable modules for $M_a(1)$, which we do). However, for other vertex operator algebras, such as that associated to the Virasoro or $\mathcal{W}$-algebras, the higher level Zhu algebras do give a wealth of added information about the category of modules, and in fact how to construct new modules. The problem comes then in actually constructing the higher level Zhu algebras for these interesting cases. Since $M_a(1)$ is arguably the most tractable vertex operator algebra to calculate with, we give here the details of our calculation of the level one Zhu algebra of $M_a(1)$ just from the vertex operator algebra structure of $M_a(1)$ alone so as to develop techniques we can use to calculate the level one Zhu algebra for other vertex operator algebras. And in fact in [4], we use and build on the techniques we developed in this paper to calculate the level one Zhu algebra for the Virasoro vertex operator algebra.

In [12], Zhu introduced an associative algebra, which we denote by $A_0(V)$, for V a vertex operator algebra. This Zhu algebra has proven to be very useful in understanding the module structure of V for certain classes of modules and certain types of vertex operator algebras. In particular, in the case that the vertex operator algebra is rational, Frenkel and Zhu, in [6], showed that there is a bijection between the module category for a vertex operator algebra and the module category for the Zhu algebra associated with this vertex operator algebra. Subsequently, in [5], Dong, Li, and Mason introduced higher level Zhu algebras, $A_n(V)$ for $n \in \mathbb{Z}_+$, and proved many important fundamental results about these algebras. Dong, Li, and Mason presented several statements that generalize the results of Frenkel and Zhu from the level zero algebras to these higher level algebras, results that mainly focused on the semi-simple setting, e.g., the case of rational vertex operator algebras; see [3] for more details.

For an irrational vertex operator algebra, instead of the irreducible modules, indecomposable modules are the fundamental objects in the module category. To date, it has proven difficult to find examples of vertex operator algebras that have certain nice finiteness properties but non semi-simple representation theory, i.e., so called C_2-cofinite irrational vertex operator algebras, which is an important setting for logarithmic conformal field theory. And in general, in the indecomposable non simple setting, the correspondence between the category of such V-modules and the category of $A_n(V)$-modules is not well understood. In [3], we are able to obtain correspondences for certain module subcategories and begin a more systematic study of the settings for which the higher level Zhu algebras become effective tools for understanding and constructing V-modules.

As mentioned earlier, in theory, the higher level Zhu algebras should prove to be important tools for detecting indecomposable $\mathbb{N}$-gradable modules, in particular those that have an increase in the Jordan block size at a higher level than degree zero with respect to the $\mathbb{N}$-grading. Whereas the zero level Zhu algebra has been used, for instance in [1, 2], to determine that important examples, such as the $\mathcal{W}_p$ triplet vertex operator algebras, do have indecomposable modules, more information about the indecomposables for such vertex operator algebras is given by higher level Zhu algebras, including information necessary to compute the fusion rules for the module category in these non semi-simple settings, e.g. [11] (see also [10]). For C_2-cofinite rational vertex operator algebras, Zhu showed in [12] that the characters (reflective of the $L(0)$ eigenspaces structure) of the irreducible modules are closed under the action of modular transformations. However, as Miyamoto showed in [9], for C_2-cofinite irrational vertex operator algebras, pseudo-characters for indecomposable non-simple modules must be introduced and considered along with the characters in order to preserve modular invariance, where these so called pseudo-characters detect the generalized eigenspaces of the module. It is precisely the higher level Zhu algebras which give information about these pseudo-characters. However despite these important applications of higher level Zhu algebras, so far in practice, higher level Zhu algebras have not been well understood, and techniques for calculating them have not been developed. This paper is a step in filling in the gaps in the examples of higher level Zhu algebras.

In this paper, in Sect. 2, we follow [3, 5] in defining the higher level Zhu algebras $A_n(V)$ for a vertex operator algebra V, the functor Ω_n from the category of $\mathbb{N}$-gradable V-modules to the category of $A_n(V)$-modules, and the functor L_n from the category of $A_n(V)$-modules to the category of $\mathbb{N}$-gradable V-modules, for $n \in \mathbb{N}$. We then recall some properties about these algebras and their relationship to the modules for V from [3].

In Sect. 3, we recall the definition of the Heisenberg vertex operator algebra $M_a(1)$ and a few facts about its irreducible modules, as well as the construction of $A_0(M_a(1))$ from [6].

In Sect. 4, we construct $A_1(M_a(1))$ using only the internal vertex operator algebra structure of $M_a(1)$, and some minimal facts about the irreducible modules, namely that it has ones with nontrivial degree one subspaces.

In Sect. 5, we recall the classification of indecomposable $M_a(1)$-modules from [8], and remark on how all indecomposables can be induced by the functor L_0 using $A_0(M_a(1))$. We note that it is the structure of $A_1(M_a(1))$ in regards to $A_0(M_a(1))$, i.e. that $A_0(M_a(1))$ is isomorphic to a direct summand of $A_1(M_a(1))$, that is at play here, illustrating Corollary 1 below, recalled from [3]. And thus $M_a(1)$ gives an example of a vertex operator algebra for which $A_1(V)$ (and in fact $A_n(V)$ for $n > 0$) gives no new information about the V-modules not already contained in $A_0(V)$.

2 The Algebras $A_n(V)$, and the Functors Ω_n and L_n

In this section, we recall the definition and some properties of the algebras $A_n(V)$ for $n \in \mathbb{N}$, first introduced in [12] for $n = 0$, and then generalized to $n > 0$ in [5].

We then recall the functors Ω_n and L_n defined in [5], and we recall some results from [3].

For $n \in \mathbb{N}$, let $O_n(V)$ be the subspace of V spanned by elements of the form

$$u \circ_n v = \mathrm{Res}_x \frac{(1+x)^{\mathrm{wt}\, u+n} Y(u,x)v}{x^{2n+2}} \tag{1}$$

for all homogeneous $u \in V$ and for all $v \in V$, and by elements of the form $(L(-1) + L(0))v$ for all $v \in V$. The vector space $A_n(V)$ is defined to be the quotient space $V/O_n(V)$.

Remark 1 For $n = 0$, since $v \circ_0 \mathbf{1} = v_{-2}\mathbf{1} + (\mathrm{wt}\, v)v = L(-1)v + L(0)v$, it follows that $O_0(V)$ is spanned by elements of the form (1). But this is not necessarily true of $O_n(V)$ for $n > 0$.

We define the following multiplication on V

$$u *_n v = \sum_{m=0}^{n} (-1)^m \binom{m+n}{n} \mathrm{Res}_x \frac{(1+x)^{\mathrm{wt}\, u+n} Y(u,x)v}{x^{n+m+1}},$$

for $v \in V$ and homogeneous $u \in V$, and for general $u \in V$, $*_n$ is defined by linearity. It is shown in [5] that with this multiplication, the subspace $O_n(V)$ of V is a two-sided ideal of V, and $A_n(V)$ is an associative algebra, called the *level n Zhu algebra*.

For every homogeneous element $u \in V$ and $m \geq k \geq 0$, elements of the form

$$\mathrm{Res}_x \frac{(1+x)^{\mathrm{wt}\, u+n+k} Y(u,x)v}{x^{m+2n+2}} \tag{2}$$

lie in $O_n(V)$. This fact follows from the $L(-1)$-derivative property for V. This implies that $O_n(V) \subset O_{n-1}(V)$. In fact, from Proposition 2.4 in [5], we have that the map

$$\begin{aligned} A_n(V) &\longrightarrow A_{n-1}(V) \\ v + O_n(V) &\mapsto v + O_{n-1}(V) \end{aligned} \tag{3}$$

is a surjective algebra homomorphism.

From Lemma 2.1 in [5], we have that

$$u *_n v - v *_n u - \mathrm{Res}_x (1+x)^{\mathrm{wt}\, u-1} Y(u,x)v \in O_n(V), \tag{4}$$

and from Theorem 2.3 in [5], we have that $\omega + O_n(V)$ is a central element of $A_n(V)$.

Next we recall the definitions of various V-module structures. We assume the reader is familiar with the notion of weak V-module for a vertex operator algebra V (cf. [7]).

Definition 1 An $\mathbb{N}$-*gradable weak* V-*module* (also often called an *admissible* V-*module* as in [5]) W for a vertex operator algebra V is a weak V-module that is $\mathbb{N}$-gradable, $W = \coprod_{k\in\mathbb{N}} W(k)$, with $v_m W(k) \subset W(k + \mathrm{wt} v - m - 1)$ for homogeneous $v \in V$, $m \in \mathbb{Z}$ and $k \in \mathbb{N}$, and without loss of generality, we can and do assume $W(0) \neq 0$, unless otherwise specified. We say elements of $W(k)$ have *degree* $k \in \mathbb{N}$.

An $\mathbb{N}$-*gradable generalized weak* V-*module* W is an $\mathbb{N}$-gradable weak V-module that admits a decomposition into generalized eigenspaces via the spectrum of $L(0) = \omega_1$ as follows: $W = \coprod_{\lambda\in\mathbb{C}} W_\lambda$ where $W_\lambda = \{w \in W \mid (L(0) - \lambda\, id_W)^j w = 0 \text{ for some } j \in \mathbb{Z}_+\}$, and in addition, $W_{n+\lambda} = 0$ for fixed λ and for all sufficiently small integers n. We say elements of W_λ have *weight* $\lambda \in \mathbb{C}$.

A *generalized* V-*module* W is an $\mathbb{N}$-gradable generalized weak V-module where $\dim W_\lambda$ is finite for each $\lambda \in \mathbb{C}$.

An *(ordinary)* V-*module* is a generalized V-module such that the generalized eigenspaces W_λ are in fact eigenspaces, i.e., $W_\lambda = \{w \in W \mid L(0)w = \lambda w\}$.

We will often omit the term "weak" when referring to $\mathbb{N}$-gradable weak and $\mathbb{N}$-gradable generalized weak V-modules.

The term *logarithmic* is also often used in the literature to refer to $\mathbb{N}$-gradable generalized weak modules or generalized modules.

Remark 2 An $\mathbb{N}$-gradable V-module with $W(k)$ of finite dimension for each $k \in \mathbb{N}$ is not necessarily a generalized V-module since the generalized eigenspaces might not be finite dimensional.

We define the *generalized graded dimension* of a generalized V-module $W = \coprod_{\lambda\in\mathbb{C}} W_\lambda$ to be

$$\mathrm{gdim}_q\, W = q^{-c/24} \sum_{\lambda\in\mathbb{C}} (\dim W_\lambda)\, q^\lambda. \tag{5}$$

Next, we recall the functors Ω_n and L_n, for $n \in \mathbb{N}$, defined and studied in [5]. Let W be an $\mathbb{N}$-gradable V-module, and let

$$\Omega_n(W) = \{w \in W \mid v_i w = 0 \text{ if } \mathrm{wt}\, v_i < -n \text{ for } v \in V \text{ of homogeneous weight}\}. \tag{6}$$

It was shown in [5] that $\Omega_n(W)$ is an $A_n(V)$-module via the action $o(v + O_n(V)) = v_{\mathrm{wt}\, v-1}$ for $v \in V$. In particular, this action satisfies $o(u *_n v) = o(u)o(v)$ for $u, v \in A_n(V)$.

Furthermore, it was shown in [3, 5] that there is a bijection between the isomorphism classes of irreducible $A_n(V)$-modules which cannot factor through $A_{n-1}(V)$ and the isomorphism classes of irreducible $\mathbb{N}$-gradable V-modules with nonzero degree n component.

In order to define the functor L_n from the category of $A_n(V)$-modules to the category of $\mathbb{N}$-gradable V-modules, we need several notions, including the notion of the universal enveloping algebra of V, which we now define.

Let

$$\hat{V} = \mathbb{C}[t, t^{-1}] \otimes V / D\mathbb{C}[t, t^{-1}] \otimes V, \tag{7}$$

where $D = \frac{d}{dt} \otimes 1 + 1 \otimes L(-1)$. For $v \in V$, let $v(m) = v \otimes t^m + D\mathbb{C}[t, t^{-1}] \otimes V \in \hat{V}$. Then $\hat{V}$ can be given the structure of a $\mathbb{Z}$-graded Lie algebra as follows: Define the degree of $v(m)$ to be $\mathrm{wt}\, v - m - 1$ for homogeneous $v \in V$, and define the Lie bracket on $\hat{V}$ by

$$[u(j), v(k)] = \sum_{i=0}^{\infty} \binom{j}{i} (u_i v)(j + k - i), \tag{8}$$

for $u, v \in V$, $j, k \in \mathbb{Z}$. Denote the homogeneous subspace of degree m by $\hat{V}(m)$. In particular, the degree 0 space of $\hat{V}$, denoted by $\hat{V}(0)$, is a Lie subalgebra.

Denote by $\mathcal{U}(\hat{V})$ the universal enveloping algebra of the Lie algebra $\hat{V}$. Then $\mathcal{U}(\hat{V})$ has a natural $\mathbb{Z}$-grading induced from $\hat{V}$, and we denote by $\mathcal{U}(\hat{V})_k$ the degree k space with respect to this grading, for $k \in \mathbb{Z}$.

We can regard $A_n(V)$ as a Lie algebra via the bracket $[u, v] = u *_n v - v *_n u$, and then the map $v(\mathrm{wt}\, v - 1) \mapsto v + O_n(V)$ is a well-defined Lie algebra epimorphism from $\hat{V}(0)$ onto $A_n(V)$.

Let U be an $A_n(V)$-module. Since $A_n(V)$ is naturally a Lie algebra homomorphic image of $\hat{V}(0)$, we can lift U to a module for the Lie algebra $\hat{V}(0)$, and then to a module for $P_n = \bigoplus_{p>n} \hat{V}(-p) \oplus \hat{V}(0) = \bigoplus_{p<-n} \hat{V}(p) \oplus \hat{V}(0)$ by letting $\hat{V}(-p)$ act trivially for $p \neq 0$. Define

$$M_n(U) = \mathrm{Ind}_{P_n}^{\hat{V}}(U) = \mathcal{U}(\hat{V}) \otimes_{\mathcal{U}(P_n)} U.$$

We impose a grading on $M_n(U)$ with respect to U, n, and the $\mathbb{Z}$-grading on $\mathcal{U}(\hat{V})$, by letting U be degree n, and letting $M_n(U)(i)$ be the $\mathbb{Z}$-graded subspace of $M_n(U)$ induced from $\hat{V}$, i.e., $M_n(U)(i) = \mathcal{U}(\hat{V})_{i-n}U$.

For $v \in V$, define $Y_{M_n(U)}(v, x) \in (\mathrm{End}(M_n(U)))((x))$ by

$$Y_{M_n(U)}(v, x) = \sum_{m \in \mathbb{Z}} v(m) x^{-m-1}. \tag{9}$$

Let W_A be the subspace of $M_n(U)$ spanned linearly by the coefficients of

$$\begin{aligned}(x_0 + x_2)^{\mathrm{wt}\, v+n} Y_{M_n(U)}(v, x_0 + x_2) Y_{M_n(U)}(w, x_2)u \\ -(x_2 + x_0)^{\mathrm{wt}\, v+n} Y_{M_n(U)}(Y(v, x_0)w, x_2)u\end{aligned} \tag{10}$$

for $v, w \in V$, with v homogeneous, and $u \in U$. Set

$$\overline{M}_n(U) = M_n(U)/\mathcal{U}(\hat{V})W_A.$$

It is shown in [5] that if U is an $A_n(V)$-module that does not factor through $A_{n-1}(V)$, then $\overline{M}_n(U) = \bigoplus_{i \in \mathbb{N}} \overline{M}_n(U)(i)$ is an $\mathbb{N}$-gradable V-module with $\overline{M}_n(U)(0) \neq 0$, and as an $A_n(V)$-module, $\overline{M}_n(U)(n) \cong U$. Note that the condition that U

itself does not factor though $A_{n-1}(V)$ is indeed a necessary and sufficient condition for $\overline{M}_n(U)(0) \neq 0$ to hold.

It is also observed in [5] that $\overline{M}_n(U)$ satisfies the following universal property: For any weak V-module M and any $A_n(V)$-module homomorphism $\phi : U \longrightarrow \Omega_n(M)$, there exists a unique weak V-module homomorphism $\Phi : \overline{M}_n(U) \longrightarrow M$, such that $\Phi \circ \iota = \phi$ where ι is the natural injection of U into $\overline{M}_n(U)$. This follows from the fact that $\overline{M}_n(U)$ is generated by U as a weak V-module, again with the possible need of a grading shift.

Let $U^* = \mathrm{Hom}(U, \mathbb{C})$. As in the construction in [5], we can extend U^* to $M_n(U)$ by first an induction to $M_n(U)(n)$ and then by letting U^* annihilate $\bigoplus_{i \neq n} M_n(U)(i)$. In particular, we have that elements of $M_n(U)(n) = \mathcal{U}(\hat{V})_0 U$ are spanned by elements of the form

$$o_{p_1}(a_1) \cdots o_{p_s}(a_s) U$$

where $s \in \mathbb{N}$, $p_1 \geq \cdots \geq p_s$, $p_1 + \cdots + p_s = 0$, $p_i \neq 0$, $p_s \geq -n$, $a_i \in V$ and $o_{p_i}(a_i) = (a_i)(\mathrm{wt}\, a_i - 1 - p_i)$. Then inducting on s by using Remark 3.3 in [5] to reduce from length s vectors to length $s-1$ vectors, we have a well-defined action of U^* on $M_n(U)(n)$.

Set

$$J = \{v \in M_n(U) \mid \langle u', xv \rangle = 0 \text{ for all } u' \in U^*, x \in \mathcal{U}(\hat{V})\}$$

and

$$L_n(U) = M_n(U)/J.$$

Remark 3 It is shown in [5], Propositions 4.3, 4.6 and 4.7, that if U does not factor through $A_{n-1}(V)$, then $L_n(U)$ is a well-defined $\mathbb{N}$-gradable V-module with $L_n(U)(0) \neq 0$; in particular, it is shown that $\mathcal{U}(\hat{V}) W_A \subset J$, for W_A the subspace of $M_n(U)$ spanned by the coefficients of (10), i.e., giving the associativity relations for the weak vertex operators on $M_n(U)$.

We have the following theorem from [3]:

Theorem 1 ([3]) *For $n \in \mathbb{N}$, let U be a nonzero $A_n(V)$-module such that if $n > 0$, then U does not factor through $A_{n-1}(V)$. Then $L_n(U)$ is an $\mathbb{N}$-gradable V-module with $L_n(U)(0) \neq 0$. If we assume further that there is no nonzero submodule of U that factors through $A_{n-1}(V)$, then $\Omega_n/\Omega_{n-1}(L_n(U)) \cong U$.*

One of the main reasons we are interested in Theorem 1 is what it implies for the question of when modules for the higher level Zhu algebras give rise to indecomposable non simple modules for V not seen by the lower level Zhu algebras.

To this end, we recall the following two corollaries to Theorem 1 from [3]:

Corollary 1 ([3]) *Suppose that for some fixed $n \in \mathbb{Z}_+$, $A_n(V)$ has a direct sum decomposition $A_n(V) \cong A_{n-1}(V) \oplus A_n'(V)$, for $A_n'(V)$ a direct sum complement to $A_{n-1}(V)$, and let U be an $A_n(V)$-module. If U is trivial as an $A_{n-1}(V)$-module, then $\Omega_n/\Omega_{n-1}(L_n(U)) \cong U$.*

An example of this setting is given by the level one Zhu algebra for the Heisenberg vertex operator algebra as constructed in Sect. 4 of this paper.

Corollary 2 ([3]) *Let $n \in \mathbb{Z}_+$ be fixed, and let U be a nonzero indecomposable $A_n(V)$-module such that there is no nonzero submodule of U that can factor through $A_{n-1}(V)$. Then $L_n(U)$ is a nonzero indecomposable $\mathbb{N}$-gradable V-module generated by its degree n subspace, $L_n(U)(n) \cong U$, and satisfying*

$$\Omega_n / \Omega_{n-1}(L_n(U)) \cong L_n(U)(n) \cong U$$

as $A_n(V)$-modules.

Furthermore if U is a simple $A_n(V)$-module, then $L_n(U)$ is a simple V-module as well.

Definition 2 For $n \in \mathbb{N}$, denote by $\mathcal{A}_{n,n-1}$ the category of $A_n(V)$-modules such that there are no nonzero submodules that can factor through $A_{n-1}(V)$.

Definition 3 For $n \in \mathbb{N}$, denote by $\mathcal{V}_n$ the category of weak V-modules whose objects W satisfy: W is $\mathbb{N}$-gradable with $W(0) \neq 0$; and W is generated by $W(n)$.

With these definitions, we have that Theorem 1 shows that $\Omega_n / \Omega_{n-1} \circ L_n$ is the identity functor on the category $\mathcal{A}_{n,n-1}(V)$, and Corollary 2 shows that L_n sends indecomposable objects in $\mathcal{A}_{n,n-1}$ to indecomposable objects in $\mathcal{V}_n$. We have the following Theorem from [3] (see also [5]):

Theorem 2 ([3]) *L_n and Ω_n / Ω_{n-1} are equivalences when restricted to the full subcategories of completely reducible $A_n(V)$-modules whose irreducible components do not factor through $A_{n-1}(V)$ and completely reducible $\mathbb{N}$-gradable V-modules that are generated by their degree n subspace (or equivalently, that have a nonzero degree n subspace), respectively. In particular, L_n and Ω_n / Ω_{n-1} induce naturally inverse bijections on the isomorphism classes of simple objects in the category $\mathcal{A}_{n,n-1}$ and isomorphism classes of simple objects in $\mathcal{V}_n$.*

We also recall the following result which gives a stronger statement than Corollary 2:

Proposition 1 ([3]) *Let U be an indecomposable $A_n(V)$-module that does not factor through $A_{n-1}(V)$. Then $L_n(U)$ is an indecomposable $\mathbb{N}$-gradable V-module generated by its degree n subspace.*

Furthermore, if U is finite dimensional, then $L_n(U)$ is an indecomposable $\mathbb{N}$-gradable generalized V-module.

Regarding the question: When is $L_n(\Omega_n / \Omega_{n-1}(W)) \cong W$ for W an $\mathbb{N}$-gradable V-module? It is clear that we must at least have that W is an object in the category $\mathcal{V}_n$, i.e., W must be generated by its degree n subspace $W(n)$. However, in general we have that

$$\bigoplus_{k=0}^{n} W(k) \subset \Omega_n(W)$$

with equality holding if, for instance, W is simple. In [4] and [3], we give an example, that of the Virasoro vertex operator algebra, to show that in the indecomposable case equality will not necessarily hold. But we do have the following sufficient criteria, where below we denote the cyclic submodule of W generated by $w \in W$ by Vw:

Theorem 3 ([3]) *Let W be an $\mathbb{N}$-gradable V-module that is generated by $W(n)$ such that $\Omega_j(W) = \bigoplus_{k=0}^{j} W(k)$, for $j = n$ and $n-1$. Then $L_n(\Omega_n/\Omega_{n-1}(W))$ is naturally isomorphic to a quotient of W.*

Furthermore, suppose W also satisfies the property that for any $w \in W$, $w = 0$ if and only if $Vw \cap W(n) = 0$. Then

$$L_n(\Omega_n/\Omega_{n-1}(W)) \cong W.$$

It is these criteria stated in this Theorem, i.e., those that are necessary for the property $L_n(\Omega_n/\Omega_{n-1}(W)) \cong W$ to hold, that are realized for the example constructed in this paper — namely for $n = 1$ and where V is the Heisenberg vertex operator algebra.

We have the following corollary from [3]:

Corollary 3 ([3]) *Let $\mathcal{A}_{n,n-1}^{Res}$ denote the subcategory of objects U in $\mathcal{A}_{n,n-1}$ that satisfy $\Omega_j(L_n(U)) = \bigoplus_{k=0}^{j} L_n(U)(k)$ for $j = n$ and $n-1$, and let $\mathcal{V}_n^{Res}$ denote the subcategory of objects W in $\mathcal{V}_n$ that satisfy: $\Omega_j(W) = \bigoplus_{k=0}^{j} W(k)$ for $j = n$ and $n-1$; For any $w \in W$, $w = 0$ if and only if $Vw \cap W(n) = 0$; and $W(n)$ has no nonzero $A_n(V)$-submodule that is an $A_{n-1}(V)$-module.*

Then the functors Ω_n/Ω_{n-1} and L_n are mutual inverses on the categories $\mathcal{V}_n^{Res}$ and $\mathcal{A}_{n,n-1}^{Res}$, respectively. In particular, the categories $\mathcal{V}_n^{Res}$ and $\mathcal{A}_{n,n-1}^{Res}$ are equivalent.

Furthermore, the subcategory of simple objects in $\mathcal{V}_n^{Res}$ is equivalent to the subcategory of simple objects in $\mathcal{A}_{n,n-1}^{Res}$.

Ultimately, we are more interested in understanding the nature of the types of indecomposable V-modules that can be constructed from various classes of $A_n(V)$-modules through L_n, and in fact the functor Ω_n/Ω_{n-1} is more useful in the indecomposable non simple setting as it gives more information when it is *not* an inverse to L_n. In particular, as we shall see below, the Heisenberg vertex operator algebra is the perfect example to show when the functorial correspondence holds, and that in fact then, there is no new information to be gleaned from the higher level Zhu algebras. We study this issue further in [4] where we construct the level one Zhu algebra for the Virasoro vertex operator algebra which does not have the level zero Zhu algebra as a direct summand, see also [3]. In this case, one can construct through the functor L_1 indecomposable V-modules from indecomposable $A_1(V)$-modules that do not arise from L_0 inducing from a $A_0(V)$-module. In particular, for the Virasoro vertex operator algebra, we see that the higher level Zhu algebras, i.e., $A_n(V)$ for $n \geq 1$, can be used to construct new (i.e. not arising from inducing from an $A_{n-1}(V)$-module) indecomposable nonsimple modules for V with Jordan blocks for the $L(0)$ operator of sizes k through $k+n$, exactly when $A_n(V)$ does not decompose into a direct sum with $A_{n-1}(V)$ (e.g. as in Corollary 1).

3 The Heisenberg Vertex Operator Algebra and Its Irreducible Modules

We denote by $\mathfrak{h}$ a one-dimensional abelian Lie algebra spanned by α with a bilinear form $\langle\cdot,\cdot\rangle$ such that $\langle\alpha,\alpha\rangle = 1$, and by

$$\hat{\mathfrak{h}} = \mathfrak{h} \otimes \mathbb{C}[t, t^{-1}] \oplus \mathbb{C}\mathbf{k}$$

the affinization of $\mathfrak{h}$ with bracket relations

$$[a(m), b(n)] = m\langle a, b\rangle \delta_{m+n,0}\mathbf{k}, \quad a, b \in \mathfrak{h},$$

$$[\mathbf{k}, a(m)] = 0,$$

where we define $a(m) = a \otimes t^m$ for $m \in \mathbb{Z}$ and $a \in \mathfrak{h}$.

Set

$$\hat{\mathfrak{h}}^+ = \mathfrak{h} \otimes t\mathbb{C}[t] \quad \text{and} \quad \hat{\mathfrak{h}}^- = \mathfrak{h} \otimes t^{-1}\mathbb{C}[t^{-1}].$$

Then $\hat{\mathfrak{h}}^+$ and $\hat{\mathfrak{h}}^-$ are abelian subalgebras of $\hat{\mathfrak{h}}$. Consider the induced $\hat{\mathfrak{h}}$-module given by

$$M(1) = U(\hat{\mathfrak{h}}) \otimes_{U(\mathbb{C}[t]\otimes\mathfrak{h}\oplus\mathbb{C}c)} \mathbb{C}\mathbf{1} \simeq S(\hat{\mathfrak{h}}^-) \quad \text{(linearly)},$$

where $U(\cdot)$ and $S(\cdot)$ denote the universal enveloping algebra and symmetric algebra, respectively, $\mathfrak{h} \otimes \mathbb{C}[t]$ acts trivially on $\mathbb{C}\mathbf{1}$ and $\mathbf{k}$ acts as multiplication by 1. Then $M(1)$ is a vertex operator algebra, often called the *vertex operator algebra associated to the rank one Heisenberg*, or the *rank one Heisenberg vertex operator algebra*, or the *one free boson vertex operator algebra*. Here, the Heisenberg Lie algebra in question being precisely $\hat{\mathfrak{h}} \setminus \mathbb{C}\alpha(0)$.

Any element of $M(1)$ can be expressed as a linear combination of elements of the form

$$\alpha(-k_1)\cdots\alpha(-k_j)\mathbf{1}, \quad \text{with} \quad k_1 \geq \cdots \geq k_j \geq 1. \tag{11}$$

The conformal element for $M(1)$ is given by

$$\omega = \frac{1}{2}\alpha(-1)^2\mathbf{1}.$$

However we can also shift the conformal element by

$$\omega_a = \frac{1}{2}\alpha(-1)^2\mathbf{1} + a\alpha(-2)\mathbf{1},$$

for any $a \in \mathbb{C}$, and then $M(1)$ with the conformal element ω_a is a vertex operator algebra with central charge $c = 1 - 12a^2$. We denote this vertex operator algebra with shifted conformal elements by $M_a(1)$. So, for instance $M(1) = M_0(1)$.

Writing $Y(\omega_a, x) = \sum_{k\in\mathbb{Z}} L_a(k)x^{-k-2}$ we have that the representation of the Virasoro algebra corresponding to the vertex operator algebras structure $(M_a(1), Y, \mathbf{1}, \omega_a)$ is given as follows: For n even

$$L_a(-n) = \sum_{k=n/2+1}^{\infty} \alpha(-k)\alpha(-n+k) + \frac{1}{2}\alpha(-n/2)^2 + a(n-1)\alpha(-n), \tag{12}$$

and for n odd, we have

$$L_a(-n) = \sum_{k=(n+1)/2}^{\infty} \alpha(-k)\alpha(-n+k) + a(n-1)\alpha(-n). \tag{13}$$

In Sect. 5, we will discuss the modules for $M_a(1)$ in more detail, but for now we just note that $M_a(1)$ is simple and has infinitely many inequivalent irreducible modules, which can be easily classified (see [7]). In particular, for every highest weight irreducible $M_a(1)$-module W there exists $\lambda \in \mathbb{C}$ such that

$$W \cong M_a(1) \otimes_{\mathbb{C}} \Omega_\lambda,$$

where Ω_λ is the one-dimensional $\mathfrak{h}$-module such that $\alpha(0)$ acts as multiplication by λ. We denote these irreducible $M_a(1)$-modules by $M_a(1, \lambda) = M_a(1) \otimes_{\mathbb{C}} \Omega_\lambda$, so that $M_a(1, 0) = M_a(1)$. If we let $v_\lambda \in \Omega_\lambda$ such that $\Omega_\lambda = \mathbb{C}v_\lambda$, then for instance,

$$L_a(-1)v_\lambda = \alpha(-1)\alpha(0)v_\lambda = \lambda\alpha(-1)v_\lambda, \tag{14}$$

$$L_a(0)v_\lambda = \left(\frac{1}{2}\alpha(0)^2 - a\alpha(0)\right)v_\lambda = \left(\frac{\lambda^2}{2} - a\lambda\right)v_\lambda. \tag{15}$$

Note that $M_a(1, \lambda)$ is admissible with

$$M_a(1, \lambda) = \coprod_{k\in\mathbb{N}} M_a(1, \lambda)_k$$

and $M_a(1, \lambda)_0 \neq 0$ where $M_a(1, \lambda)_k$ is the eigenspace of eigenvectors of weight $k + \frac{\lambda^2}{2} - a\lambda$. And since $M_a(1)$ is simple these $M_a(1, \lambda)$ are faithful $M_a(1)$-modules.

The following is proved in [6]:

Theorem 4 ([6]) *As algebras*

$$A_0(M_a(1)) \cong \mathbb{C}[x, y]/(x^2 - y) \cong \mathbb{C}[x]$$

under the indentifications

$$\alpha(-1)\mathbf{1} + O_0(M(1)) \longleftrightarrow x, \qquad \text{and} \qquad \alpha(-1)^2\mathbf{1} + O_0(M(1)) \longleftrightarrow y.$$

In addition, there is a bijection between isomorphism classes of irreducible admissible $M_a(1)$-modules and irreducible $\mathbb{C}[x]$-modules given by $M_a(1, \lambda) \longleftrightarrow \mathbb{C}[x]/(x - \lambda)$.

4 The Algebra $A_1(M_a(1))$

In this section we construct the level one Zhu algebra for the Heisenberg vertex operator algebra $M_a(1)$ using only the internal vertex operator algebra structure of $M_a(1)$, the level zero Zhu algebra, and some very basic facts about the irreducible $M_a(1)$-modules, namely that $M_a(1)$ has irreducibles with nontrivial degree one spaces.

Let

$$O_1'(M_a(1)) := \{(\alpha(-m-4) + 2\alpha(-m-3) + \alpha(-m-2))v \mid v \in M_a(1),\ m \geq 0\}.$$

For $u, v \in M_a(1)$, we say $u \sim v$ if and only if $u \equiv v \pmod{O_1'(M_a(1))}$.

Then $O_1'(M_a(1)) \subset O_1(M_a(1))$, since from (2) with $k = 0$, we have

$$\Big(\alpha(-m-4) + 2\alpha(-m-3) + \alpha(-m-2)\Big)v = \mathrm{Res}_x\Big(\frac{(1+x)^2}{x^{m+4}} Y(\alpha(-1)\mathbf{1}, x)v\Big) \in O_1(M_a(1)), \tag{16}$$

for every $m \geq 0$, and $v \in M_a(1)$. Thus by recursion, we have that for $m \geq 2$

$$\alpha(-m)v \sim (-1)^{m+1}\left((m-3)\alpha(-2) + (m-2)\alpha(-3)\right)v, \tag{17}$$

or equivalently

$$\alpha(-m)v \sim (-1)^m\alpha(-2)v + (-1)^{m+1}(m-2)\left(\alpha(-2) + \alpha(-3)\right)v. \tag{18}$$

From Eq. (18), we have the following lemma which will be important later in the determination of $A_1(M_a(1))$; in particular, in Theorems 5 and 6.

Lemma 1 *For $v \in M_a(1)$, we have the following identities:*

$$\alpha(-1)^2\mathbf{1} \circ_1 v \sim \big(\alpha(-1) + \alpha(-2)\big)\big(\alpha(-1) + 3\alpha(-2) + 2\alpha(-3)\big)v \tag{19}$$

$$\alpha(-2)^2\mathbf{1} \circ_1 v \sim -2\alpha(-1)\alpha(-2)\mathbf{1} \circ_1 v \sim -4\big(\alpha(-2) + \alpha(-3)\big)^2 v. \tag{20}$$

Thus

$$\big(\alpha(-1)+\alpha(-2)\big)^2 v + 2\big(\alpha(-1)+\alpha(-2)\big)\big(\alpha(-2)+\alpha(-3)\big)v \in O_1(M_a(1)) \quad (21)$$

$$\big(\alpha(-2)+\alpha(-3)\big)^2 v \in O_1(M_a(1)). \quad (22)$$

Furthermore

$$\big(\alpha(-1)+\alpha(-2)\big)^4 v \in O_1(M_a(1)). \quad (23)$$

Proof For $v \in M_a(1)$, let

$$Y^+(v,x) = \sum_{n\le -1} v_n x^{-n-1} \quad \text{and} \quad Y^-(v,x) = \sum_{n\ge 0} v_n x^{-n-1},$$

denote the regular and singular parts of $Y(v,x)$, respectively. Then using the recursion (18), we have that

$$\begin{aligned} Y^+(\alpha(-1)\mathbf{1},x) &= \sum_{n\ge 1}\alpha(-n)x^{n-1} \quad (24)\\ &\sim \alpha(-1) + \sum_{n\ge 2}(-1)^n\alpha(-2)x^{n-1}\\ &\quad + \sum_{n\ge 2}(-1)^{n+1}(n-2)(\alpha(-2)+\alpha(-3))x^{n-1}\\ &= \alpha(-1) + \alpha(-2)\frac{x}{1+x} + (\alpha(-2)+\alpha(-3))\frac{x^2}{(1+x)^2} \end{aligned}$$

and

$$\begin{aligned} Y^+(\alpha(-2)\mathbf{1},x) &= \sum_{n\ge 2}(n-1)\alpha(-n)x^{n-2} \quad (25)\\ &\sim \sum_{n\ge 2}(-1)^n(n-1)\alpha(-2)x^{n-2}\\ &\quad + \sum_{n\ge 2}(-1)^{n+1}(n-1)(n-2)(\alpha(-2)+\alpha(-3))x^{n-2}\\ &= \alpha(-2)\frac{1}{(1+x)^2} + (\alpha(-2)+\alpha(-3))\frac{2x}{(1+x)^3}. \end{aligned}$$

By the definition of $\circ_1$ and using (24) and (25), we have that for any $v \in M_a(1)$, and $i, j \in \mathbb{N}$,

$$\alpha(-1)^i\alpha(-2)^j\mathbf{1} \circ_1 v = \mathrm{Res}_x \frac{(1+x)^{i+2j+1}Y(\alpha(-1)^i\alpha(-2)^j\mathbf{1},x)v}{x^4}$$

$$= \mathrm{Res}_x \frac{(1+x)^{i+2j+1}}{x^4} {}^\circ_\circ \big(Y^-(\alpha(-1)\mathbf{1},x) + Y^+(\alpha(-1)\mathbf{1},x)\big)^i \big(Y^-(\alpha(-2)\mathbf{1},x)$$

$$+ Y^+(\alpha(-2)\mathbf{1},x)\big)^j {}^\circ_\circ v$$

$$
\begin{aligned}
&\sim \mathrm{Res}_x \frac{(1+x)}{x^4}\Bigg((1+x)Y^-(\alpha(-1)\mathbf{1},x) + (1+x)\left(\alpha(-1)+\alpha(-2)\frac{x}{1+x}\right.\\
&\quad \left.\left. + (\alpha(-2)+\alpha(-3))\frac{x^2}{(1+x)^2}\right)\right)^i \Bigg((1+x)^2 Y^-(\alpha(-2)\mathbf{1},x)\\
&\quad + (1+x)^2\left(\alpha(-2)\frac{1}{(1+x)^2} + (\alpha(-2)+\alpha(-3))\frac{2x}{(1+x)^3}\right)\Bigg)^j v\\
&= \mathrm{Res}_x \frac{(1+x)}{x^4}\Bigg((1+x)Y^-(\alpha(-1)\mathbf{1},x) + \alpha(-1)(1+x) + \alpha(-2)x\\
&\quad + (\alpha(-2)+\alpha(-3))\frac{x^2}{1+x}\Bigg)^i \Bigg((1+x)^2 Y^-(\alpha(-2)\mathbf{1},x)\\
&\quad + \alpha(-2) + (\alpha(-2)+\alpha(-3))\frac{2x}{1+x}\Bigg)^j v.
\end{aligned}
$$

Note that $\mathrm{Res}_x x^k Y^-(\alpha(-1)\mathbf{1},x) = \mathrm{Res}_x x^{k+1} Y^-(\alpha(-2)\mathbf{1},x) = 0$, for $k < 0$. Thus the singular terms $Y^-(\alpha(-1)\mathbf{1},x)$ and $Y^-(\alpha(-2)\mathbf{1},x)$ in the expression above do not contribute in the following cases: If $i = 0$ and $j \le 3$; if $j = 0$ and $i \le 2$; or if $i = j = 1$. For these values of i and j, we have

$$
\begin{aligned}
\alpha(-1)^i\alpha(-2)^j\mathbf{1} \circ_1 v \;\sim\; &\mathrm{Res}_x \frac{(1+x)}{x^4}\Bigg(\alpha(-1)(1+x) + \alpha(-2)x\\
&+(\alpha(-2)+\alpha(-3))\frac{x^2}{1+x}\Bigg)^i \left(\alpha(-2) + (\alpha(-2)+\alpha(-3))\frac{2x}{1+x}\right)^j v.
\end{aligned}
\tag{26}
$$

Setting $i = 2$ and $j = 0$ in Eq. (26), we have that

$$
\begin{aligned}
&\alpha(-1)^2\mathbf{1} \circ_1 v\\
&\sim \mathrm{Res}_x \frac{(1+x)}{x^4}\left(\alpha(-1)(1+x) + \alpha(-2)x + (\alpha(-2)+\alpha(-3))\frac{x^2}{1+x}\right)^2 v\\
&= \mathrm{Res}_x \frac{(1+x)}{x^4}\Bigg(\alpha(-1)^2(1+x)^2 + 2\alpha(-1)\alpha(-2)(1+x)x\\
&\quad + 2\alpha(-1)(\alpha(-2)+\alpha(-3))x^2 + \alpha(-2)^2x^2 + 2\alpha(-2)(\alpha(-2)+\alpha(-3))\frac{x^3}{1+x}\\
&\quad + (\alpha(-2)+\alpha(-3))^2\frac{x^4}{(1+x)^2}\Bigg)v\\
&= \Big(\alpha(-1)^2 + 2\alpha(-1)\alpha(-2) + 2\alpha(-1)(\alpha(-2)+\alpha(-3)) + \alpha(-2)^2\\
&\quad + 2\alpha(-2)(\alpha(-2)+\alpha(-3))\Big)v
\end{aligned}
$$

giving Eq. (19). Setting $i = j = 1$ in Eq. (26), we have

$$
\begin{aligned}
&\alpha(-1)\alpha(-2)\mathbf{1} \circ_1 v \\
&\sim \mathrm{Res}_x \frac{(1+x)}{x^4}\left(\alpha(-1)(1+x) + \alpha(-2)x + (\alpha(-2)+\alpha(-3))\frac{x^2}{1+x}\right)\Big(\alpha(-2) \\
&\quad + (\alpha(-2)+\alpha(-3))\frac{2x}{1+x}\Big)v \\
&= \mathrm{Res}_x \frac{(1+x)}{x^4}\Big(\alpha(-1)\alpha(-2)(1+x) + \alpha(-1)(\alpha(-2)+\alpha(-3))2x + \alpha(-2)^2 x \\
&\quad + \alpha(-2)(\alpha(-2)+\alpha(-3))\frac{2x^2}{1+x} + \alpha(-2)(\alpha(-2)+\alpha(-3))\frac{x^2}{1+x} \\
&\quad + (\alpha(-2)+\alpha(-3))^2\frac{2x^3}{(1+x)^2}\Big)v \\
&= 2(\alpha(-2)+\alpha(-3))^2 v,
\end{aligned}
$$

and setting $i = 0$ and $j = 2$ in Eq. (26), we have

$$
\begin{aligned}
\alpha(-2)^2\mathbf{1} \circ_1 v &\sim \mathrm{Res}_x \frac{(1+x)}{x^4}\left(\alpha(-2) + (\alpha(-2)+\alpha(-3))\frac{2x}{1+x}\right)^2 v \\
&= \mathrm{Res}_x \frac{(1+x)}{x^4}\Big(\alpha(-2)^2 + \alpha(-2)(\alpha(-2)+\alpha(-3))\frac{4x}{1+x} \\
&\quad + (\alpha(-2)+\alpha(-3))^2\frac{4x^2}{(1+x)^2}\Big)v \\
&= \mathrm{Res}_x \frac{4}{x^2(1+x)}(\alpha(-2)+\alpha(-3))^2 \\
&= -4(\alpha(-2)+\alpha(-3))^2
\end{aligned}
$$

giving Eq. 20.

Finally, we observe that

$$
\begin{aligned}
&\alpha(-1)^2\mathbf{1} \circ_1 \left((\alpha(-1)+\alpha(-2))^2 - 2\,(\alpha(-1)+\alpha(-2))\,(\alpha(-2)+\alpha(-3))\right) v \\
&\quad -\alpha(-2)^2\mathbf{1} \circ_1 (\alpha(-1)+\alpha(-2))^2\, v \\
&\sim \left((\alpha(-1)+\alpha(-2))^2 + 2\,(\alpha(-1)+\alpha(-2))\,(\alpha(-2)+\alpha(-3))\right) \\
&\quad \cdot \left((\alpha(-1)+\alpha(-2))^2 - 2\,(\alpha(-1)+\alpha(-2))\,(\alpha(-2)+\alpha(-3))\right) v \\
&\quad + 4\,(\alpha(-2)+\alpha(-3))^2\,(\alpha(-1)+\alpha(-2))^2\, v \\
&= (\alpha(-1)+\alpha(-2))^4\, v,
\end{aligned}
$$

which is by definition in $O_1(M_a(1))$. □

Remark 4 For $u, v \in M_a(1)$, we will use the notation that $u \approx v$ if and only if $u \equiv v \pmod{(L_a(-1)+L_a(0))M_a(1)}$. And if $u \sim v \approx w$, or if more broadly u is equivalent to w modulo $O_1(M_a(1))$, we will write $u \equiv w$. In addition, since by (13) we have $L_a(-1) = L_0(-1)$ and by (12) we have $L_a(0)v = L_0(0)v$ for all $v \in M_a(1)$, we will write $(L(-1) + L(0))v$ for $(L_a(-1) + L_a(0))v$, for $v \in M_a(1)$.

From (11) and (17), we have that $A_1(M_a(1))$ is generated by elements of the form

$$\alpha(-1)^i \alpha(-2)^j \alpha(-3)^k \mathbf{1}, \quad \text{for } i, j, k \in \mathbb{N}.$$

We will eventually show, that in fact $A_1(M_a(1))$ is generated by $\alpha(-1)\mathbf{1}$ and ω_a, or equivalently, generated by $\alpha(-1)\mathbf{1}$ and $\alpha(-1)^2\mathbf{1}$. To show this, we need several preliminary results which we will now give.

Proposition 2 *The algebra $A_1(M_a(1))$ is generated by elements of the form*

$$\alpha(-1)^i \alpha(-2)^j \mathbf{1}$$

for $i, j \in \mathbb{N}$.

Proof Using (12) and (13), and then (16), we have that for $i, j, k \in \mathbb{N}$,

$$\begin{aligned}
0 &\approx (L(-1) + L(0))\alpha(-1)^i \alpha(-2)^j \alpha(-3)^k \mathbf{1} \\
&= i\alpha(-1)^{i-1} \alpha(-2)^{j+1} \alpha(-3)^k \mathbf{1} + 2j\alpha(-1)^i \alpha(-2)^{j-1} \alpha(-3)^{k+1} \mathbf{1} \\
&\quad +3k\alpha(-1)^i \alpha(-2)^j \alpha(-3)^{k-1} \alpha(-4)\mathbf{1} + (i + 2j + 3k)\alpha(-1)^i \alpha(-2)^j \alpha(-3)^k \mathbf{1} \\
&\sim i\alpha(-1)^{i-1} \alpha(-2)^{j+1} \alpha(-3)^k \mathbf{1} + 2j\alpha(-1)^i \alpha(-2)^{j-1} \alpha(-3)^{k+1} \mathbf{1} \\
&\quad -3k\alpha(-1)^i \alpha(-2)^{j+1} \alpha(-3)^{k-1} \mathbf{1} + (i + 2j - 3k)\alpha(-1)^i \alpha(-2)^j \alpha(-3)^k \mathbf{1}.
\end{aligned}$$

Thus letting $j \mapsto j + 1$, and $k \mapsto k - 1$ in the equation above, we have that for $i, j \in \mathbb{N}$ and $k \in \mathbb{Z}_+$,

$$\begin{aligned}
\alpha(-1)^i \alpha(-2)^j \alpha(-3)^k \mathbf{1} \equiv \frac{1}{2(j+1)} \Big(&-i\alpha(-1)^{i-1} \alpha(-2)^{j+2} \alpha(-3)^{k-1} \mathbf{1} \\
&+3(k-1)\alpha(-1)^i \alpha(-2)^{j+2} \alpha(-3)^{k-2} \mathbf{1} \\
&- (i + 2j - 3k + 5)\alpha(-1)^i \alpha(-2)^{j+1} \alpha(-3)^{k-1} \mathbf{1} \Big).
\end{aligned} \tag{27}$$

It follows by induction on k, that $A_1(M_a(1))$ is generated by $\alpha(-1)^i \alpha(-2)^j \mathbf{1}$, for $i, j \in \mathbb{N}$. □

Remark 5 Equation (27) implies that any expression of the form $\alpha(-1)^i \alpha(-2)^j \alpha(-3)^k \mathbf{1}$ is equivalent modulo $O_1(M(1))$ to sums of terms that are still of total degree $i + j + k$ but are of degree i or lower in $\alpha(-1)$, and of degree strictly lower than k in $\alpha(-3)$.

To prove that in fact, $A_1(M_a(1))$ is generated by $\alpha(-1)\mathbf{1}$ and ω_a, we only need to show that $\alpha(-1)^i\alpha(-2)^j\mathbf{1}$ is generated by elements $\alpha(-1)\mathbf{1}$ and ω_a.

Lemma 2 *For $i \geq 2$ and $j \in \mathbb{N}$, we have in $A_1(M_a(1))$*

$$\alpha(-1)^i\mathbf{1} *_1 \alpha(-2)^j\mathbf{1} \equiv (1-i^2)\alpha(-1)^i\alpha(-2)^j\mathbf{1} - \frac{i(i-1)j}{j+1}\alpha(-1)^{i-2}\alpha(-2)^{j+2}\mathbf{1} + i\Big(\frac{i+2j+1}{j+1} - 2i - 1\Big)\alpha(-1)^{i-1}\alpha(-2)^{j+1}\mathbf{1} + r(i,j), \tag{28}$$

where $r(i,j)$ is a linear combination of elements of the form $\alpha(-1)^k\alpha(-2)^l\mathbf{1}$, for $k,l \in \mathbb{N}$, with $k+l < i+j-1$, and $k < i$. Furthermore, for $j \in \mathbb{N}$, we also have $r(2,j) \equiv 0$ and

$$\alpha(-1)\mathbf{1} *_1 \alpha(-2)^j\mathbf{1} \equiv -\alpha(-2)^{j+1}\mathbf{1}. \tag{29}$$

Proof From the definition of $*_1$, we have

$$\begin{aligned} &\alpha(-1)^i\mathbf{1} *_1 \alpha(-2)^j\mathbf{1} \\ &= \mathrm{Res}_x(1+x)^{i+1}\left(\frac{1}{x^2} - 2\frac{1}{x^3}\right) Y(\alpha(-1)^i\mathbf{1},x)\alpha(-2)^j\mathbf{1} \\ &= \mathrm{Res}_x \sum_{m=0}^{i+1}\binom{i+1}{m}\left(x^{m-2} - 2x^{m-3}\right) {}^\circ_\circ\alpha(x)\alpha(x)\cdots\alpha(x){}^\circ_\circ\alpha(-2)^j\mathbf{1} \\ &= \mathrm{Res}_x \sum_{m=0}^{i+1}\binom{i+1}{m}\left(x^{m-2} - 2x^{m-3}\right) \sum_{k_1,k_2,\ldots,k_i\in\mathbb{Z}} {}^\circ_\circ\alpha(k_1)\alpha(k_2)\cdots\alpha(k_i){}^\circ_\circ \\ &\quad x^{-k_1-k_2-\cdots-k_i-i}\alpha(-2)^j\mathbf{1}. \end{aligned} \tag{30}$$

Thus letting $i = 1$ in (30), and using Eq. (27), we have

$$\begin{aligned} &\alpha(-1)\mathbf{1} *_1 \alpha(-2)^j\mathbf{1} \\ &= \mathrm{Res}_x \sum_{k\in\mathbb{Z}}\left(x^{-k-1} - 3x^{-k-3} - 2x^{-k-4}\right)\alpha(k)\alpha(-2)^j\mathbf{1} \\ &= (-3\alpha(-2) - 2\alpha(-3))\,\alpha(-2)^j\mathbf{1} \\ &= -3\alpha(-2)^{j+1}\mathbf{1} - 2\alpha(-2)^j\alpha(-3)\mathbf{1} \\ &= -3\alpha(-2)^{j+1}\mathbf{1} - \frac{1}{j+1}(-(2j+2))\alpha(-2)^{j+1}\mathbf{1}, \end{aligned}$$

giving (29).

For $i > 1$, we have from (30) that

$$\begin{aligned}
&\alpha(-1)^i\mathbf{1} *_1 \alpha(-2)^j\mathbf{1}\\
&= \Bigg(\sum_{k_1+\cdots+k_i=-i-1} {}^\circ_\circ\alpha(k_1)\cdots\alpha(k_i){}^\circ_\circ + (i+1)\sum_{k_1+\cdots+k_i=-i} {}^\circ_\circ\alpha(k_1)\cdots\alpha(k_i){}^\circ_\circ\\
&\quad -2\sum_{k_1+\cdots+k_i=-i-2} {}^\circ_\circ\alpha(k_1)\cdots\alpha(k_i){}^\circ_\circ - 2(i+1)\sum_{k_1+\cdots+k_i=-i-1} {}^\circ_\circ\alpha(k_1)\cdots\alpha(k_i){}^\circ_\circ\\
&\quad -(i+1)i\sum_{k_1+\cdots+k_i=-i} {}^\circ_\circ\alpha(k_1)\cdots\alpha(k_i){}^\circ_\circ + r'(i,j)\Bigg)\alpha(-2)^j\mathbf{1}\\
&= \Bigg(i\alpha(-1)^{i-1}\alpha(-2) + (i+1)\alpha(-1)^i - 2\left(i\alpha(-1)^{i-1}\alpha(-3) + \binom{i}{2}\alpha(-1)^{i-2}\alpha(-2)^2\right)\\
&\quad -2(i+1)i\alpha(-1)^{i-1}\alpha(-2) - (i+1)i\alpha(-1)^i + r'(i,j)\Bigg)\alpha(-2)^j\mathbf{1}\\
&= \Big((1-i^2)\alpha(-1)^i - i(2i+1)\alpha(-1)^{i-1}\alpha(-2) - 2i\alpha(-1)^{i-1}\alpha(-3)\\
&\quad -i(i-1)\alpha(-1)^{i-2}\alpha(-2)^2 + r'(i,j)\Big)\alpha(-2)^j\mathbf{1}\\
&= (1-i^2)\alpha(-1)^i\alpha(-2)^j\mathbf{1} - i(2i+1)\alpha(-1)^{i-1}\alpha(-2)^{j+1}\mathbf{1} - 2i\alpha(-1)^{i-1}\alpha(-2)^j\alpha(-3)\mathbf{1}\\
&\quad -i(i-1)\alpha(-1)^{i-2}\alpha(-2)^{j+2}\mathbf{1} + r'(i,j)\alpha(-2)^j\mathbf{1}
\end{aligned}$$

where $r'(i,j)\alpha(-2)^j\mathbf{1}$ involves terms of the form

$$\alpha(-1)^{i-i'}\alpha(-m_1)\alpha(-m_2)\cdots\alpha(-m_p)\alpha(2)^q\alpha(-2)^j\mathbf{1},$$

with $m_s \geq 2$, for $s = 1,\ldots p$, and $i', p, q \in \mathbb{Z}_+$, and $p+q = i'$, which is a constant times

$$\alpha(-1)^{i-i'}\alpha(-m_1)\alpha(-m_2)\cdots\alpha(-m_p)\alpha(-2)^{j-q}\mathbf{1}.$$

Thus using (17), we have that $r'(i,j)\alpha(-2)^j\mathbf{1}$ involves terms of the form

$$\alpha(-1)^{i-i'}\alpha(-2)^{j-q}\sum_{t=0}^{p} c_t\alpha(-2)^t\alpha(-3)^{p-t}\mathbf{1}$$

with $i', p, q \in \mathbb{Z}_+$ and $p+q = i'$, and for some constants $c_t \in \mathbb{C}$, and with total degree in $\alpha(-1)$, $\alpha(-2)$, and $\alpha(-3)$ equal to $i - i' + j - q + p = i + j - 2q$ for $q > 0$. Using Eq. (27), this is equivalent to a linear combination of elements of the form

$$\alpha(-1)^k\alpha(-2)^l\mathbf{1}$$

with $k \leq i - i'$, and $k + l = i + j - 2q$, for $i', q \in \mathbb{Z}_+$. That is we have that $r'(i,j)\alpha(-2)^j\mathbf{1}$ which we will now call $r(i,j)$, consists of elements of total degree less than $i + j - 1$, and with degree in $\alpha(-1)$ less than i.

Then using Eq. (27) with $k = 1$, we have that

$$\begin{aligned}
&-2i\alpha(-1)^{i-1}\alpha(-2)^j\alpha(-3)\mathbf{1}\\
&\equiv \frac{-i}{j+1}\Big(-(i-1)\alpha(-1)^{i-2}\alpha(-2)^{j+2}\mathbf{1} - (i-1+2j+2)\alpha(-1)^{i-1}\alpha(-2)^{j+1}\mathbf{1}\Big)\\
&= \frac{i}{j+1}\Big((i-1)\alpha(-1)^{i-2}\alpha(-2)^{j+2}\mathbf{1} + (i+2j+1)\alpha(-1)^{i-1}\alpha(-2)^{j+1}\mathbf{1}\Big).
\end{aligned}$$

Therefore

$$\begin{aligned}
&\alpha(-1)^i\mathbf{1} *_1 \alpha(-2)^j\mathbf{1}\\
&\equiv (1-i^2)\alpha(-1)^i\alpha(-2)^j\mathbf{1} - i(2i+1)\alpha(-1)^{i-1}\alpha(-2)^{j+1}\mathbf{1}\\
&\quad -i(i-1)\alpha(-1)^{i-2}\alpha(-2)^{j+2}\mathbf{1}\\
&\quad +\frac{i}{j+1}\Big((i-1)\alpha(-1)^{i-2}\alpha(-2)^{j+2}\mathbf{1} + (i+2j+1)\alpha(-1)^{i-1}\alpha(-2)^{j+1}\mathbf{1}\Big) + r(i,j)
\end{aligned}$$

which gives (28).

It remains to prove that $r(2, j) \equiv 0$ in $A_1(M_a(1))$. For this, we note that from (30), and then using (16) repeatedly, we have that

$$\begin{aligned}
r(2,j) &= 2\left(-2\alpha(-6) - 5\alpha(-5) - 3\alpha(-4) + \alpha(-3) + \alpha(-2)\right)\alpha(2)\alpha(-2)^j\mathbf{1}\\
&\sim 2\left(-\alpha(-5) - \alpha(-4) + \alpha(-3) + \alpha(-2)\right)\alpha(2)\alpha(-2)^j\mathbf{1}\\
&\sim 2\left(\alpha(-4) + 2\alpha(-3) + \alpha(-2)\right)\alpha(2)\alpha(-2)^j\mathbf{1}\\
&\sim 0.
\end{aligned}$$

□

Theorem 5 *The algebra $A_1(M_a(1))$ is generated by $\alpha(-1)\mathbf{1}$ and $\alpha(-1)^2\mathbf{1}$, or equivalently by $\alpha(-1)\mathbf{1}$ and ω_a.*

Proof We need only show that $\alpha(-1)^i\alpha(-2)^j\mathbf{1}$ is generated by elements $\alpha(-1)\mathbf{1}$ and $\alpha(-1)^2\mathbf{1}$.

We order the elements $\alpha(-1)^i\alpha(-2)^j\mathbf{1}$ as follows:

$$\alpha(-1)^i\alpha(-2)^j\mathbf{1} < \alpha(-1)^{i'}\alpha(-2)^{j'}\mathbf{1} \iff \begin{cases} i+j < i'+j' \text{ or} \\ i+j = i'+j' \text{ and } i < i'. \end{cases} \tag{31}$$

We proceed by induction on this order. By Lemma 2, we have that for $i > 1$

$$\begin{aligned}
&(1-i^2)\alpha(-1)^i\alpha(-2)^j\mathbf{1}\\
&\equiv \alpha(-1)^i\mathbf{1} *_1 \alpha(-2)^j\mathbf{1} + \frac{i(i-1)j}{j+1}\alpha(-1)^{i-2}\alpha(-2)^{j+2}\mathbf{1}\\
&\quad - i\Big(\frac{i+2j+1}{j+1} - 2i - 1\Big)\alpha(-1)^{i-1}\alpha(-2)^{j+1}\mathbf{1} - r(i,j)
\end{aligned} \tag{32}$$

with $r(i, j)$ consisting of terms of total degree less than $i + j - 1$ in $\alpha(-1)$ and $\alpha(-2)$ and of degree less than i in $\alpha(-1)$. Thus $\alpha(-1)^i\alpha(-2)^j\mathbf{1}$ for $i > 1$ is generated by elements of lower total order, and hence generated by $\alpha(-1)\mathbf{1}$ and $\alpha(-1)^2\mathbf{1}$ by induction on this ordering, once we have proved the base cases. That is, it remains to show that $\alpha(-1)^i\mathbf{1}$, $\alpha(-2)^j\mathbf{1}$ and $\alpha(-1)\alpha(-2)^j\mathbf{1}$ are generated by $\alpha(-1)\mathbf{1}$ and $\alpha(-1)^2\mathbf{1}$ for $i, j > 0$.

By (29) in Lemma 2, we have $\alpha(-2)^j\mathbf{1} \equiv -\alpha(-1)\mathbf{1} *_1 \alpha(-2)^{j-1}\mathbf{1}$, for $j > 0$, which implies

$$\alpha(-2)^j\mathbf{1} \equiv (-1)^j(\alpha(-1)\mathbf{1})^j. \tag{33}$$

Thus the claim holds for $\alpha(-2)^j\mathbf{1}$ for $j \in \mathbb{Z}_+$.

Next we note that by (19) with $v = \alpha(-2)^j\mathbf{1}$, we have

$$\begin{aligned} 0 &\sim \big(\alpha(-1) + \alpha(-2)\big)\big(\alpha(-1) + 3\alpha(-2) + 2\alpha(-3)\big)\alpha(-2)^j\mathbf{1} \\ &= \alpha(-1)^2\alpha(-2)^j\mathbf{1} + 4\alpha(-1)\alpha(-2)^{j+1}\mathbf{1} + 3\alpha(-2)^{j+2}\mathbf{1} + 2\alpha(-1)\alpha(-2)^j\alpha(-3)\mathbf{1} \\ &\quad + 2\alpha(-2)^{j+1}\alpha(-3)\mathbf{1}. \end{aligned} \tag{34}$$

From (27), we have that

$$\alpha(-2)^j\alpha(-3)\mathbf{1} \equiv -\alpha(-2)^{j+1}\mathbf{1},$$

and

$$2\alpha(-1)\alpha(-2)^j\alpha(-3)\mathbf{1} \equiv -\frac{2j+3}{j+1}\alpha(-1)\alpha(-2)^{j+1}\mathbf{1} - \frac{1}{j+1}\alpha(-2)^{j+2}\mathbf{1}.$$

Substituting the previous two relations into (34), we obtain

$$\alpha(-1)^2\alpha(-2)^j\mathbf{1} + \frac{2j+1}{j+1}\alpha(-1)\alpha(-2)^{j+1}\mathbf{1} + \frac{j}{j+1}\alpha(-2)^{j+2}\mathbf{1} \equiv 0. \tag{35}$$

From Lemma 2, with $i = 2$ in Eq. (28) along with the fact that $r(2, j) = 0$, we have that

$$\begin{aligned} \alpha(-1)^2\mathbf{1} *_1 \alpha(-2)^j\mathbf{1} \equiv -3\alpha(-1)^2\alpha(-2)^j\mathbf{1} - \frac{2j}{j+1}\alpha(-2)^{j+2}\mathbf{1} \\ + 2\left(\frac{-3j-2}{j+1}\right)\alpha(-1)\alpha(-2)^{j+1}\mathbf{1}. \end{aligned} \tag{36}$$

Adding Eqs. (36) and three times (35), we have

$$\alpha(-1)^2\mathbf{1} *_1 \alpha(-2)^j\mathbf{1} \equiv \frac{j}{j+1}\alpha(-2)^{j+2}\mathbf{1} - \frac{1}{j+1}\alpha(-1)\alpha(-2)^{j+1}\mathbf{1}$$

giving

$$\alpha(-1)\alpha(-2)^j\mathbf{1} \equiv -j\alpha(-1)^2\mathbf{1} *_1 \alpha(-2)^{j-1}\mathbf{1} + (j-1)\alpha(-2)^{j+1}\mathbf{1}. \tag{37}$$

This along with our previous determination for $\alpha(-2)^k\mathbf{1}$, shows that $\alpha(-1)\alpha(-2)^j\mathbf{1}$ is generated by $\alpha(-1)\mathbf{1}$ and $\alpha(-1)^2\mathbf{1}$.

It remains to show that $\alpha^i(-1)\mathbf{1}$ for $i > 2$ is generated by $\alpha(-1)\mathbf{1}$ and $\alpha(-1)^2\mathbf{1}$. But since $L(-1)\alpha(-1)^i\mathbf{1} = i\alpha(-1)^{i-1}\alpha(-2)\mathbf{1}$, we have that

$$\alpha(-1)^i\mathbf{1} \approx -\alpha(-1)^{i-1}\alpha(-2)\mathbf{1}. \tag{38}$$

Thus by Eq. (32), induction on i, and our previous determination that $\alpha(-2)^k\mathbf{1}$ is generated by $\alpha(-1)\mathbf{1}$ and $\alpha(-1)^2\mathbf{1}$, we have that $\alpha(-1)^i\mathbf{1}$ is generated by $\alpha(-1)\mathbf{1}$ and $\alpha(-1)^2\mathbf{1}$.

Since $\omega_a = \frac{1}{2}\alpha(-1)^2\mathbf{1} + a\alpha(-2)\mathbf{1}$, by (38), we can replace the generators $\alpha(-1)\mathbf{1}$ and $\alpha(-1)^2\mathbf{1}$ with generators $\alpha(-1)\mathbf{1}$ and ω_a. □

We now prove the following theorem:

Theorem 6 *Let I be the ideal generated by the polynomial $(x^2 - y)(x^2 - y + 2)$ in $\mathbb{C}[x, y]$. Then we have the following isomorphisms of algebras*

$$A_1(M_a(1)) \cong \mathbb{C}[x, y]/I \cong \mathbb{C}[x, y]/(x^2 - y) \oplus \mathbb{C}[x, y]/(x^2 - y + 2) \tag{39}$$

$$\cong A_0(M_a(1)) \oplus \mathbb{C}[x, y]/(x^2 - y + 2) \cong \mathbb{C}[x] \oplus \mathbb{C}[x] \tag{40}$$

under the identifications

$$\alpha(-1)\mathbf{1} + O_1(M_a(1)) \longleftrightarrow x + I, \qquad \alpha(-1)^2\mathbf{1} + O_1(M_a(1)) \longleftrightarrow y + I.$$

Equivalently,

$$A_1(M_a(1)) \cong \mathbb{C}[\alpha(-1)\mathbf{1}, \omega_a]/J \tag{41}$$

with

$$J = \left(\left((\alpha(-1)\mathbf{1})^2 - 2\omega_a - 2a\alpha(-1)\mathbf{1}\right)\left((\alpha(-1)\mathbf{1})^2 - 2\omega_a - 2a\alpha(-1)\mathbf{1} + 2\right)\right). \tag{42}$$

Proof We have shown that $A_1(M_a(1))$ is generated by $\alpha(-1)\mathbf{1} + O_1(M_a(1))$ and $\alpha(-1)^2\mathbf{1} + O_1(M_a(1))$. We also note, that as mentioned in Sect. 2, the element $\omega_a + O_1(M_a(1))$, which in this case is $\frac{1}{2}\alpha(-1)^2\mathbf{1} + a\alpha(-2)\mathbf{1} + O_1(M_a(1)) = \frac{1}{2}\alpha(-1)^2\mathbf{1} - a\alpha(-1)\mathbf{1} + O_1(M_a(1))$, is in the center of $A_1(M_a(1))$. Thus we have a surjective algebra homomorphism

$$\begin{aligned} \varphi : \mathbb{C}[x, y] &\longrightarrow A_1(M_a(1)) \\ x &\mapsto \alpha(-1)\mathbf{1} + O_1(M_a(1)) \\ y &\mapsto \alpha(-1)^2\mathbf{1} + O_1(M_a(1)). \end{aligned} \tag{43}$$

Let $p(x, y) = (x^2 - y)(x^2 - y + 2)$. We will first show that $I = (p(x, y)) \subset \ker \varphi$. We do this by showing that the element

$$\varphi(p(x,y)) = (\alpha(-1)\mathbf{1})^4 - 2(\alpha(-1)\mathbf{1})^2 *_1 \alpha(-1)^2\mathbf{1} + (\alpha(-1)^2\mathbf{1})^2 \\ + 2(\alpha(-1)\mathbf{1})^2 - 2\alpha(-1)^2\mathbf{1} \tag{44}$$

lies in $O_1(M_a(1))$. We first rewrite each of the first four terms in (44) using previous results.

For the first term in Eq. (44), using Eq. (33), we have

$$(\alpha(-1)\mathbf{1})^4 \equiv \alpha(-2)^4\mathbf{1}. \tag{45}$$

For the second term in Eq. (44), we note that by the definition of $*_1$ and direct calculation, we have that

$$\alpha(-1)\mathbf{1} *_1 v = \big(-3\alpha(-2) - 2\alpha(-3)\big)v \tag{46}$$

for all $v \in M_a(1)$. Thus for the second term in (44), we have

$$-2(\alpha(-1)\mathbf{1})^2 *_1 \alpha(-1)^2\mathbf{1} = -2\big(-3\alpha(-2) - 2\alpha(-3)\big)^2\alpha(-1)^2\mathbf{1}. \tag{47}$$

Lemma 1, in particular Eqs. (20), and (22) which gives $(\alpha(-2) + \alpha(-3))^2 v \sim 0$ in $O_1(M_a(1))$, implies that, for any $v \in V$,

$$\alpha(-3)^2 v \sim -(\alpha(-2)^2 + 2\alpha(-2)\alpha(-3))v.$$

Substituting this into Eq. (47), and then using Eq. (27), gives that the second term in (44) is given by

$$\begin{aligned} &-2(\alpha(-1)\mathbf{1})^2 *_1 \alpha(-1)^2\mathbf{1} \\ &\equiv -10\alpha(-1)^2\alpha(-2)^2\mathbf{1} - 8\alpha(-1)^2\alpha(-2)\alpha(-3)\mathbf{1} \\ &\equiv -10\alpha(-1)^2\alpha(-2)^2\mathbf{1} - 2\Big(-2\alpha(-1)\alpha(-2)^3\mathbf{1} - 6\alpha(-1)^2\alpha(-2)^2\mathbf{1}\Big) \\ &= 2\alpha(-1)^2\alpha(-2)^2\mathbf{1} + 4\alpha(-1)\alpha(-2)^3\mathbf{1}. \end{aligned} \tag{48}$$

For the third term in (44), by direct calculation and using (16) and (27), we have that

$$\begin{aligned} &(\alpha(-1)^2\mathbf{1})^2 = \alpha(-1)^2\mathbf{1} *_1 \alpha(-1)^2\mathbf{1} \\ &= 2\big(-2\alpha(-5) - 5\alpha(-4) - 3\alpha(-3) + \alpha(-2) + \alpha(-1)\big)\alpha(1)\alpha(-1)^2\mathbf{1} \\ &\quad + \big(-4\alpha(-3)\alpha(-1) - 2\alpha(-2)^2 - 10\alpha(-2)\alpha(-1) - 3\alpha(-1)^2\big)\alpha(-1)^2\mathbf{1} \\ &\sim 2\big(-\alpha(-4) - \alpha(-3) + \alpha(-2) + \alpha(-1)\big)2\alpha(-1)\mathbf{1} \end{aligned}$$

$$
\begin{aligned}
&+\big(-4\alpha(-3)\alpha(-1)-2\alpha(-2)^2-10\alpha(-2)\alpha(-1)-3\alpha(-1)^2\big)\alpha(-1)^2\mathbf{1}\\
&\sim 2\big(\alpha(-3)+2\alpha(-2)+\alpha(-1)\big)2\alpha(-1)\mathbf{1}+\Big(-4\alpha(-3)\alpha(-1)-2\alpha(-2)^2\\
&\quad -10\alpha(-2)\alpha(-1)-3\alpha(-1)^2\Big)\alpha(-1)^2\mathbf{1}\\
&=\Big(-3\alpha(-1)^2-10\alpha(-1)\alpha(-2)-4\alpha(-1)\alpha(-3)-2\alpha(-2)^2\Big)\alpha(-1)^2\mathbf{1}\\
&\quad +4\big(\alpha(-3)+2\alpha(-2)+\alpha(-1)\big)\alpha(-1)\mathbf{1}\\
&\equiv -3\alpha(-1)^4\mathbf{1}-10\alpha(-1)^3\alpha(-2)\mathbf{1}-4\Big(\frac{1}{2}\Big(-3\alpha(-1)^2\alpha(-2)^2\mathbf{1}-5\alpha(-1)^3\alpha(-2)\mathbf{1}\Big)\Big)\\
&\quad -2\alpha(-1)^2\alpha(-2)^2\mathbf{1}+4\big(\alpha(-3)+2\alpha(-2)+\alpha(-1)\big)\alpha(-1)\mathbf{1}\\
&=-3\alpha(-1)^4\mathbf{1}+4\alpha(-1)^2\alpha(-2)^2\mathbf{1}+4\big(\alpha(-3)+2\alpha(-2)+\alpha(-1)\big)\alpha(-1)\mathbf{1}. \qquad (49)
\end{aligned}
$$

For the fourth term in (44), using Eq. (46), we have

$$
2(\alpha(-1)\mathbf{1})^2=-2\big(3\alpha(-2)+2\alpha(-3)\big)\alpha(-1)\mathbf{1}. \qquad (50)
$$

Substituting the Eqs. (45), (48)–(50) into the expression (44), and using Eqs. (38) and (33), we have

$$
\begin{aligned}
&\varphi(p(x,y))\\
&=(\alpha(-1)\mathbf{1})^4-2(\alpha(-1)\mathbf{1})^2\alpha(-1)^2\mathbf{1}+(\alpha(-1)^2\mathbf{1})^2+2(\alpha(-1)\mathbf{1})^2-2\alpha(-1)^2\mathbf{1}\\
&\equiv \alpha(-2)^4\mathbf{1}+4\alpha(-2)^3\alpha(-1)\mathbf{1}+6\alpha(-2)^2\alpha(-1)^2\mathbf{1}-3\alpha(-1)^4\mathbf{1}\\
&\quad +2(\alpha(-1)^2\mathbf{1}+\alpha(-1)\alpha(-2)\mathbf{1})\\
&=\Big(\alpha(-2)^4\mathbf{1}+4\alpha(-2)^3\alpha(-1)\mathbf{1}+6\alpha(-2)^2\alpha(-1)^2\mathbf{1}+4\alpha(-2)\alpha(-1)^3\mathbf{1}+\alpha(-1)^4\mathbf{1}\\
&\quad -4\alpha(-2)\alpha(-1)^3\mathbf{1}-\alpha(-1)^4\mathbf{1}\Big)-3\alpha(-1)^4\mathbf{1}+2(\alpha(-1)^2\mathbf{1}+\alpha(-1)\alpha(-2)\mathbf{1})\\
&=(\alpha(-2)+\alpha(-1))^4\mathbf{1}-4\alpha(-2)\alpha(-1)^3\mathbf{1}-\alpha(-1)^4\mathbf{1}-3\alpha(-1)^4\mathbf{1}\\
&\quad +2(\alpha(-1)^2\mathbf{1}+\alpha(-1)\alpha(-2)\mathbf{1})\\
&\approx (\alpha(-2)+\alpha(-1))^4\mathbf{1}.
\end{aligned}
$$

Thus by Lemma 1, we have that $\varphi(p(x,y))\in O_1(M_a(1))$, and therefore $I=((x^2-y)(x^2-y+2))\subset \ker\varphi$.

Note that letting $I_1=(x^2-y)$ and $I_2=(x^2-y+2)$, we have that $I_1+I_2=\mathbb{C}[x,y]$, and $I=I_1I_2=I_1\cap I_2$, i.e., both x^2-y and x^2-y+2 are irreducible, and the ideals I_1 and I_2 these polynomials generate, respectively, are coprime. Therefore

$$
\begin{aligned}
\mathbb{C}[x,y]/I&=(I_1+I_2)/(I_1\cap I_2)\cong (I_1+I_2)/I_1\oplus(I_1+I_2)/I_2\\
&=\mathbb{C}[x,y]/(x^2-y)\oplus\mathbb{C}[x,y]/(x^2-y+2)\\
&\cong \mathbb{C}[x]\oplus\mathbb{C}[x].
\end{aligned}
$$

We now show that in fact $\ker\varphi = I$. Suppose not, i.e., suppose $I \subsetneq \ker\varphi$. Then $\ker\varphi$ must be one of the following three ideals of $\mathbb{C}[x, y]$ containing I: $I_1 = (x^2 - y)$, $I_2 = (x^2 - y + 2)$ or $I_1 + I_2 = \mathbb{C}[x, y]$ itself since $x^2 - y$ and $x^2 - y + 2$ are irreducible and coprime. However, we have from Theorem 4 that $A_0(M_a(1)) \cong \mathbb{C}[x, y]/(x^2 - y)$ under the algebra homomorphism given by $\alpha(-1)\mathbf{1} + O_0(M_a(1)) \mapsto x + I_1$ and $\alpha(-1)^2\mathbf{1} + O_0(M_a(1)) \mapsto y + I_1$. Furthermore we have that $A_0(M_a(1))$ is a homomorphic image of $A_1(M_a(1))$ via (3), implying that $\ker\varphi$ cannot be I_2 or $I_1 + I_2 = \mathbb{C}[x, y]$.

From [3] (see also [5]), we have that there is a bijection between the isomorphism classes of irreducible $A_1(M_a(1))$-modules which cannot factor through $A_0(M_a(1)))$ and irreducible admissible $M_a(1)$-modules generated by their degree one subspace. Since the irreducible $M_a(1)$ modules $M_a(1, \lambda)$ for $\lambda \in \mathbb{C}$ are generated by their weight one subspaces, this implies that $A_1(M_a(1)) \neq A_0(M_a(1))$, and thus $\ker\varphi \neq I_1$. □

As a consequence of Theorem 6, we have the following Corollary:

Corollary 4 *The Heisenberg vertex operator algebra $M_a(1)$ gives a realization of the setting of Corollary 2.6 for the level one Zhu algebra. Namely, that $A_1(M_a(1)) \cong A_0(M_a(1)) \oplus A_1'(M_a(1))$. Thus for any $A_1(M_a(1))$-module U that is trivial as an $A_0(M_a(1))$-module, we have that*

$$\Omega_1/\Omega_0(L_1(U)) \cong U.$$

In Sect. 4 below, we give the actual module structure for all indecomposable $M_a(1)$-modules to further illustrate this phenomenon of Corollary 4.

Remark 6 We were motivated to explore $(p(x, y)) = ((x^2 - y)(x^2 - y + 2))$ as the kernel of the surjective algebra homomorphism $\varphi : \mathbb{C}[x, y] \longrightarrow A_1(M_a(1))$ due to the action of $\varphi(p(x, y))$ on the irreducible modules for $M_a(1)$. In particular in exploring possibilities for $\ker\varphi$, we observed that for $M_a(1, \lambda) = M_a(1) \otimes \mathbb{C}v_\lambda$, we have

$$\begin{aligned} o(\varphi(x^2 - y))v_\lambda &= \left(\alpha(0)^2 - 2L_a(0) - 2a\alpha(0)\right) v_\lambda \\ &= \left(\lambda^2 - (\lambda^2 - 2a\lambda) - 2a\lambda\right) v_\lambda \\ &= 0, \qquad (51) \\ o(\varphi(x^2 - y + 2))\alpha(-1)v_\lambda &= \left(\alpha(0)^2 - 2L_a(0) - 2a\alpha(0) + 2\right) \alpha(-1)v_\lambda \\ &= \left(\lambda^2 - (2 + \lambda^2 - 2a\lambda) - 2a\lambda + 2\right) \alpha(-1)v_\lambda \qquad (52) \\ &= 0. \end{aligned}$$

Thus $\varphi((x^2 - y)(x^2 - y + 2))$ acts as zero on the degree zero and one spaces for all irreducible modules for $A_1(M_a(1))$.

5 Indecomposable Modules for $M_a(1)$ and Inducing From $A_0(M_a(1))$ Versus $A_n(M_a(1))$ for $n > 0$

The irreducible modules for $M_a(1)$ were briefly introduced in Sect. 3. However, the indecomposable non simple generalized modules have been completely determined as well, e.g., see [8]. In particular, we have

Proposition 3 *([8]) Let W be an indecomposable generalized $M_a(1)$-module. Then as an $\hat{\mathfrak{h}}$-module*

$$W \cong M_a(1) \otimes \Omega(W) \tag{53}$$

where $\Omega(W) = \{w \in W \mid \alpha(n)w = 0 \text{ for all } n > 0\}$ is the vacuum space.

Remark 7 Note that in terms of the functors Ω_n for $n \in \mathbb{N}$, if W is an indecomposable $M_a(1)$-module, then

$$\mathbf{1} \otimes \Omega(W) = \Omega_0(W).$$

The indecomposable modules for $A_0(M_a(1)) \cong \mathbb{C}[x, y]/(y - x^2) \cong \mathbb{C}[x]$ are given by

$$U_0(\lambda, k) = \mathbb{C}[x, y]/((y - x^2), (x - \lambda)^k) \cong \mathbb{C}[x]/(x - \lambda)^k \tag{54}$$

for $\lambda \in \mathbb{C}$ and $k \in \mathbb{Z}_+$, and

$$L_0(U_0(\lambda, k)) \cong M_a(1) \otimes_{\mathbb{C}} \Omega(\lambda, k),$$

where $\Omega(\lambda, k)$ is a k-dimensional vacuum space such that $\alpha(0)$ acts with Jordan form given by

$$\begin{bmatrix} \lambda & 1 & 0 & \cdots & 0 & 0 \\ 0 & \lambda & 1 & \cdots & 0 & 0 \\ 0 & 0 & \lambda & \cdots & 0 & 0 \\ \vdots & \vdots & \vdots & \ddots & \vdots & \vdots \\ 0 & 0 & 0 & \cdots & \lambda & 1 \\ 0 & 0 & 0 & \cdots & 0 & \lambda \end{bmatrix}. \tag{55}$$

Note then that the zero mode of ω_a which is given by

$$L_a(0) = \sum_{m \in \mathbb{Z}_+} \alpha(-m)\alpha(m) + \frac{1}{2}\alpha(0)^2 - a\alpha(0) \tag{56}$$

acts on $\Omega(\lambda, k)$ such that the only eigenvalue is $\frac{1}{2}\lambda^2 - a\lambda$ (which is the lowest conformal weight of $M_a(1) \otimes \Omega(\lambda, k)$), and $L_a(0) - (\frac{1}{2}\lambda^2 - a\lambda)Id_k$ with respect to a Jordan basis for $\alpha(0)$ acting on $\Omega(\lambda, k)$ is given by

$$\begin{bmatrix} 0 & \lambda - a & \frac{1}{2} - a & 0 & \cdots & 0 & 0 \\ 0 & 0 & \lambda - a & \frac{1}{2} - a & \cdots & 0 & 0 \\ 0 & 0 & 0 & \lambda - a & \cdots & 0 & 0 \\ \vdots & \vdots & \vdots & \ddots & \vdots & \vdots \\ 0 & 0 & 0 & 0 & \cdots & \lambda - a & \frac{1}{2} - a \\ 0 & 0 & 0 & 0 & \cdots & 0 & \lambda - a \\ 0 & 0 & 0 & 0 & \cdots & 0 & 0 \end{bmatrix}. \tag{57}$$

Also note that $L(0)$ is diagonalizable if and only if: (i) $k = 1$ which corresponds to the case when $M_a(1) \otimes \Omega(\lambda, k)$ is irreducible; (ii) $k = 2$ and $\lambda = a$; or (iii) $k > 2$ and $\lambda = a = \frac{1}{2}$.

These $M_a(1) \otimes \Omega(\lambda, k)$ exhaust all the indecomposable generalized $M_a(1)$-modules and the $\mathbb{N}$-grading of $M_a(1) \otimes \Omega(\lambda, k)$ is explicitly given by

$$M_a(1) \otimes \Omega(\lambda, k) = \coprod_{m \in \mathbb{N}} M_a(1)_m \otimes \Omega(\lambda, k)$$

where $M_a(1)_m$ is the weight m space of the vertex operator algebra $M_a(1)$ and thus $M_a(1)_m \otimes \Omega(\lambda, k)$ is the space of generalized eigenvectors of weight $m + \frac{1}{2}\lambda^2 - a\lambda$ with respect to $L(0)$. Therefore, the generalized graded dimension of $M_a(1) \otimes \Omega(\lambda, k)$ is given by

$$\begin{aligned} \mathrm{gdim}_q\, M_a(1) \otimes \Omega(\lambda, k) &= q^{-1/24} \sum_{m \in \mathbb{N}} (k \dim M_a(1)_m)\, q^{m + \frac{1}{2}\lambda^2 - a\lambda} \\ &= q^{\frac{1}{2}\lambda^2 - a\lambda - 1/24} k \sum_{m \in \mathbb{N}} (\dim M_a(1)_m)\, q^m \\ &= q^{\frac{1}{2}\lambda^2 - a\lambda} k\, \eta(q)^{-1} \end{aligned} \tag{58}$$

where $\eta(q)$ is the Dedekind η-function.

The indecomposable modules for $A_1(M_a(1)) \cong \mathbb{C}[x, y]/((y - x^2)(y - x^2 - 2)) \cong A_0(M_a(1)) \oplus \mathbb{C}[x]$ are given by either the indecomposable modules $U_0(\lambda, k)$ for $A_0(M_a(1))$ as given in (54), or by

$$U_1(\lambda, k) = \mathbb{C}[x, y]/((y - x^2 - 2), (x - \lambda)^k) \cong \mathbb{C}[x]/(x - \lambda)^k \tag{59}$$

for $\lambda \in \mathbb{C}$ and $k \in \mathbb{Z}_+$. In the latter case,

$$U_1(\lambda, k) \cong \alpha(-1)\mathbf{1} \otimes \Omega(\lambda, k) \quad \text{and} \quad L_1(U_1(\lambda, k)) \cong M_a(1) \otimes \Omega(\lambda, k).$$

Therefore, there are no new modules obtained via inducing from a module for $A_1(M_a(1))$ versus from $A_0(M_a(1))$.

If we allow for inducing by L_1 for any indecomposable $A_1(M_a(1))$ module (including those that factor through $A_0(M_a(1))$ so that $L_1(U)(0)$ might be zero), then the possible cases for $\Omega_1/\Omega_0(L_1(U))$ for U an indecomposable $A_1(M_a(1))$-module are

$$U = U_0(\lambda, k), \quad L_1(U)(0) = 0, \quad \text{and} \quad \Omega_1/\Omega_0(L_1(U)) \cong U_1(\lambda, k) \not\cong U$$

or

$$U = U_1(\lambda, k) \quad L_1(U)(0) = \Omega(\lambda, k) \neq 0, \quad \text{and} \quad \Omega_1/\Omega_0(L_1(U)) \cong U.$$

Note however that in the case of $U = U_0(\lambda, k)$, the $M_a(1)$-module $L_1(U)$ is in fact $M_a(1) \otimes \Omega(\lambda, k)$ but the grading as an $\mathbb{N}$-gradable module is shifted up one. Thus by regrading to obtain an $\mathbb{N}$-gradable $M_a(1)$-module in the sense of Definition 1, this module is again just $M_a(1) \otimes \Omega(\lambda, k)$, and the level one Zhu algebra gives no new information about the indecomposable $M_a(1)$-modules not already given by the level zero Zhu algebra.

This example illustrates Corollary 2.6 (see Corollary 4): Any indecomposable U module for $A_1(M_a(1))$ which does not factor through $A_0(M_a(1))$, will satisfy

$$U \cong \Omega_1/\Omega_0(L_1(U)),$$

without any requirement that no nonzero submodule of U factor through $A_0(M_a(1))$, and as in Corollary 3, we have that $\Omega_j(M_a(1) \otimes \Omega(\lambda, k)) = \bigoplus_{m=0}^{j} M_a(1)_m \otimes \Omega(\lambda, k)$.

Acknowledgements The authors thank Darlayne Addabbo and Kiyo Nagatomo for reading a draft of this paper and making comments, suggestions, and corrections. The first author is the recipient of a Simons Foundation Collaboration Grant 282095, and greatly appreciates their support.

References

1. Adamović, D., Milas, A.: On the triplet vertex algebra $W(p)$. Adv. Math. **217**, 2664–2699 (2008)
2. Adamovic, D., Milas, A.: The structure of Zhu's algebras for certain W-algebras. Adv. Math. **227**, 2425–2456 (2011)
3. Barron, K., Vander Werf, N., Yang, J.: Higher level Zhu algebras and modules for vertex operator algebras. J. Pure Appl. Alg. **223**, 3295–3317 (2019)
4. Barron, K., Vander Werf, N., Yang, J.: The level one Zhu algebra for the vertex operator algebra associated to the Virasoro algebra and implications. In: Krauel, M., Tuite, M., Yamskulna, G. (eds.) Vertex Operator Algebras, Number Theory, and Related Topics (To appear). Contemporary Mathematics. American Mathematical Society, Providence
5. Dong, C., Li, H., Mason, G.: Vertex operator algebras and associative algebras. J. Alg. **206**, 67–98 (1998)
6. Frenkel, I., Zhu, Y.-C.: Vertex operator algebras associated to representations of affine and Virasoro algebras. Duke Math. J. **66**, 123–168 (1992)
7. Lepowsky, J., Li, H.: Introduction to Vertex Operator Algebras and Their Representations, Progress in Math, vol. 227. Birkhäuser, Boston (2003)

8. Milas, A.: Logarithmic intertwining operators and vertex operators. Commun. in Math. Phys. **277**, 497–529 (2008)
9. Miyamoto, M.: Modular invariance of vertex operator algebras satisfying C_2-cofiniteness. Duke Math. J. **122**, 51–91 (2004)
10. Nagatomo, K., Tsuchiya, A.: The triplet vertex operator algebra $\mathcal{W}(p)$ and the restricted quantum group at root of unity. Exploring New Structures and Natural Constructions in Mathematical Physics. Advanced Studies in Pure Mathematics, vol. 61, p. 149. Mathematical Society of Japan, Tokyo (2011)
11. Tsuchiya, A., Wood, S.: The tensor structure on the representation category of the $\mathcal{W}_p$ triplet algebra. J. Phys. A: Math. Theor. **46**, 445203 (2013)
12. Zhu, Y.-C.: Modular invariance of characters of vertex operator algebras. J. Amer. Math. Soc. **9**, 237–302 (1996)

Quasi-particle Bases of Principal Subspaces of Affine Lie Algebras

Marijana Butorac

Abstract This note is a survey of recent results on the construction of combinatorial bases of principal subspaces of generalized Verma module $N(k\Lambda_0)$ and standard module $L(k\Lambda_0)$ appearing in [5–7]. By using these bases, we obtain characters of principal subspaces.

Keywords Affine Lie algebras · Vertex operator algebras · Principal subspaces · Combinatorial bases

1 Introduction

Let $\mathfrak{g}$ be a complex simple Lie algebra of type X_l with a triangular decomposition $\mathfrak{g} = \mathfrak{n}_- \oplus \mathfrak{h} \oplus \mathfrak{n}_+$, where the nilpotent subalgebra $\mathfrak{n}_+$ of $\mathfrak{g}$ is spanned with one-dimensional subalgebras corresponding to the positive roots. Denote by $\mathcal{L}(\mathfrak{n}_+) = \mathfrak{n}_+ \otimes \mathbb{C}[t, t^{-1}]$ a subalgebra of an affine Kac–Moody Lie algebra $\tilde{\mathfrak{g}}$ of type $X_l^{(1)}$. Let V be a highest $\tilde{\mathfrak{g}}$-module with highest weight vector v_V. The principal subspace W_V of V is defined as

$$W_V = U\left(\mathcal{L}(\mathfrak{n}_+)\right) \cdot v_V.$$

In our papers [5–7] we constructed combinatorial bases of principal subspaces of generalized Verma module $N(k\Lambda_0)$ and of its irreducible quotient $L(k\Lambda_0)$ for the cases of affine Lie algebras of type $B_l^{(1)}$, $C_l^{(1)}$ and $G_2^{(1)}$ for all positive integer levels k.

The approach of finding an explicit basis for standard (i.e., integrable highest weight) $\tilde{\mathfrak{g}}$-module by using vertex operators originates from the work of J. Lepowsky and R. L. Wilson in [38] and from the work of Lepowsky and M. Primc in [37], and it was done by constructing bases of monomial vectors parameterized by partitions of integers, which led to a Lie theoretic proof of the Rogers–Ramanujan identities by Lepowsky and Wilson. In 1994. B. L. Feigin and A. V. Stoyanovsky in [23]

M. Butorac (✉)
Department of Mathematics, University of Rijeka, Radmile Matejčić 2, 51000 Rijeka, Croatia
e-mail: mbutorac@math.uniri.hr

D. Adamović and P. Papi (eds.), *Affine, Vertex and W-algebras*,
Springer INdAM Series 37, https://doi.org/10.1007/978-3-030-32906-8_4

introduced principal subspace of standard modules of affine Lie algebras of type $A_1^{(1)}$ and $A_2^{(1)}$. Principal subspace turned out to be a useful concept since Feigin and Stoyanovsky showed that the characters of principal subspaces are related to Rogers–Ramanujan-type identities.

Feigin and Stoyanovsky obtained combinatorial bases of principal subspaces, which have an interesting physical interpretation, as quasi-particles, whose energies comply the difference-two condition (see in particular [18, 23, 25]). Shortly after paper of Feigin and Stoyanovsky, G. Georgiev constructed bases for principal subspaces of certain standard modules for affine Lie algebras of types $A_l^{(1)}$. He showed that principal subspaces have bases whose elements are quasi-particle monomials acting on the highest weight vector. From obtained combinatorial bases one can easily write down the Rogers–Ramanujan type character formulas and his approach easily extends to all algebras of ADE type. In the proof of linear independence of quasi-particle bases Georgiev used intertwining operators constructed in [19], which were also used in the work of S. Capparelli, Lepowsky and A. Milas in [16, 17] (and later in the work of C. Calinescu, Lepowsky and Milas in [11–13]) to obtain recursions satisfied by the characters of principal subspaces, but without explicit use of combinatorial bases.

Many other authors have been further studied combinatorial properties of principal subspaces (along with variants of these spaces), in particular see [2–4, 8–10, 14, 15, 21, 27–29, 31–35, 41–44, 46–50] and others.

In our papers we extended Georgiev's approach to the cases of affine Lie algebras of type $B_l^{(1)}$, $C_l^{(1)}$ and $G_2^{(1)}$. As a consequence, from obtained bases we proved three series of new fermionic characters for principal subspaces of $L(k\Lambda_0)$, and also three series of generalizations of Euler–Cauchy identity obtained from the characters of principal subspaces of $N(k\Lambda_0)$. Here we give an overview of the basic ideas and results from these papers.

2 Principal Subspaces

Let $\mathfrak{g}$ be a complex simple Lie algebra of type X_l with a triangular decomposition $\mathfrak{g} = \mathfrak{n}_- \oplus \mathfrak{h} \oplus \mathfrak{n}_+$, where the nilpotent subalgebra $\mathfrak{n}_+$ of $\mathfrak{g}$ is spanned with one-dimensional subalgebras $\mathfrak{n}_\alpha = \mathbb{C}x_\alpha$ for every positive root $\alpha \in R_+$. Denote by Π the set of simple roots, by θ the highest root and by ω_i, $i = 1, \ldots, l$ the fundamental weights. We identify $\mathfrak{h}$ and $\mathfrak{h}^*$ via the invariant bilinear symmetric form $\langle \cdot, \cdot \rangle$ normalized in a such way that $\langle \alpha, \alpha \rangle = 2$ for every long root (cf. [26]).

Affine Lie algebra associated to $\mathfrak{g}$ is

$$\widetilde{\mathfrak{g}} = \mathfrak{g} \otimes \mathbb{C}[t, t^{-1}] \oplus \mathbb{C}c \oplus \mathbb{C}d,$$

c denoting the canonical central element and d the degree element and with commutation relations given in the usual way (cf. [30]). Of particular interest will be the following subalgebras of $\widetilde{\mathfrak{g}}$:

$$\mathcal{L}(\mathfrak{n}_+) = \mathfrak{n}_+ \otimes \mathbb{C}[t, t^{-1}], \quad \mathcal{L}(\mathfrak{n}_\alpha) = \mathfrak{n}_\alpha \otimes \mathbb{C}[t, t^{-1}],$$

and

$$\mathcal{L}(\mathfrak{n}_+)_{\geq 0} = \mathfrak{n}_+ \otimes \mathbb{C}[t], \quad \mathcal{L}(\mathfrak{n}_+)_{<0} = \mathfrak{n}_+ \otimes t^{-1}\mathbb{C}[t^{-1}].$$

We use the notation $L(\Lambda_i)$ to denote an irreducible integrable highest weight $\widetilde{\mathfrak{g}}$-module (i.e., standard module) with highest weight vector $v_{L(\Lambda_i)}$, where Λ_i, $i = i, \ldots, l$ are fundamental weights of $\widetilde{\mathfrak{g}}$. We consider generalized Verma module $N(k\Lambda_0)$ with highest weight vector $v_{N(k\Lambda_0)}$, and its irreducible quotient $L(k\Lambda_0)$ with highest weight vector $v_{L(k\Lambda_0)}$, where k is fixed positive integral level. For $x \in \mathfrak{g}$ we use the notation $x(m)$ to denote the action of $x \otimes t^m$ on highest weight module. Following [23, 25], we refer to $x_{\alpha_i}(m)$, $\alpha_i \in \Pi$, as quasi-particle of charge 1, color i and energy $-m$.

The generalized Verma module $N(k\Lambda_0)$ has a structure of vertex operator algebra (see [36, 39, 40]), where v_N is the vacuum vector. For $x \in \mathfrak{g}$

$$Y(x(-1)v_{N(k\Lambda_0)}, z) = x(z) = \sum_{m \in \mathbb{Z}} x(m) z^{-m-1}$$

is a vertex operator associated with the vector $x(-1)v_{N(k\Lambda_0)}$. On $L(k\Lambda_0)$ we have the structure of a simple vertex operator algebra and all the level k standard modules are modules for this vertex operator algebra (cf. [36, 40]).

Now, let V be $N(k\Lambda_0)$ or $L(k\Lambda_0)$ with highest weight vector v_V. We define the principal subspace W_V of V as

$$W_V = U\left(\mathcal{L}(\mathfrak{n}_+)\right) \cdot v_V.$$

Note that the map

$$f : U(\mathcal{L}(\mathfrak{n}_+)_{<0}) \to W_{N(k\Lambda_0)}, \quad f(b) = b v_{N(k\Lambda_0)} \tag{1}$$

is an isomorphism of $\mathcal{L}(\mathfrak{n}_+)_{<0}$-modules and by using Poincaré–Birkhoff–Witt theorem we can find basis of principal subspace $W_{N(k\Lambda_0)}$. In the next sections we describe the construction of bases in terms of certain coefficients of vertex operators corresponding to vectors $x_{\alpha_i}(-1)^r v_V$, where $r \geq 1$ and $\alpha_i \in \Pi$. From now on we will assume that $\mathfrak{g}$ is a simple Lie algebra of type B_l $(l \geq 2)$, C_l $(l \geq 3)$ or G_2. In these cases we make the standard choice of simple roots, which we now recall. In the case of B_l, we let

$$\Pi = \{\alpha_1 = \epsilon_1 - \epsilon_2, \ldots, \alpha_{l-1} = \epsilon_{l-1} - \epsilon_l, \alpha_l = \epsilon_l\},$$

where $\{\epsilon_i\}$ denotes the usual orthonormal basis of the $\mathbb{R}^l$. In the case of C_l, we have the following base of the root system:

$$\Pi = \left\{\alpha_1 = \frac{1}{\sqrt{2}}(\epsilon_1 - \epsilon_2), \ldots, \alpha_{l-1} = \frac{1}{\sqrt{2}}(\epsilon_{l-1} - \epsilon_l), \alpha_l = \sqrt{2}\epsilon_l\right\}.$$

In the case of G_2 we have

$$\Pi = \left\{\alpha_1 = \frac{1}{\sqrt{3}}(-2\epsilon_1 + \epsilon_2 + \epsilon_3), \alpha_2 = \frac{1}{\sqrt{3}}(\epsilon_1 - \epsilon_2)\right\},$$

where $\epsilon_1, \epsilon_2, \epsilon_3$ are vectors of the standard basis of $\mathbb{R}^3$.

If we choose a special subspace U of $U(\mathcal{L}(\mathfrak{n}_+))$:

$$U = U(\mathcal{L}(\mathfrak{n}_{\alpha_l})) \cdots U(\mathcal{L}(\mathfrak{n}_{\alpha_1})) \quad \text{in the case of } B_l^{(1)},$$

$$U = U(\mathcal{L}(\mathfrak{n}_{\alpha_1})) \cdots U(\mathcal{L}(\mathfrak{n}_{\alpha_l})) \quad \text{in the case of } C_l^{(1)},$$

and

$$U = U(\mathcal{L}(\mathfrak{n}_{\alpha_2}))U(\mathcal{L}(\mathfrak{n}_{\alpha_1})) \quad \text{in the case of } G_2^{(1)},$$

then using the same arguments as Georgiev (see Lemma 3.1 in [25]), we can show that W_V is generated by operators in U acting on the highest weight vector v_V.

3 Quasi-particle Bases of Principal Subspaces

In the description of our bases, we use quasi-particles as in [25].

3.1 *Definition of Quasi-particles*

Fix a simple root α_i $(1 \leq i \leq l)$ of $\mathfrak{g}$, positive integer r and integer m. We define a quasi-particles $x_{r\alpha_i}(m)$ of color i, charge r and energy $-m$ as

$$x_{r\alpha_i}(m) = \mathrm{Res}_z \left\{z^{m+r-1} x_{r\alpha_i}(z)\right\}, \tag{2}$$

where $x_{r\alpha_i}(z) = x_{\alpha_i}(z)^r$ denotes the vertex operator associated with the vector $x_{\alpha_i}(-1)^r v_V$. We call $x_{r\alpha_i}(z)$ a generating function of quasi-particle of color i and charge r.

3.2 *Quasi-particle Monomials*

We denote a product of quasi-particles of color i by

$$b(\alpha_i) = x_{n_{r_i^{(1)},i}\alpha_i}(m_{r_i^{(1)},i}) \cdots x_{n_{1,i}\alpha_i}(m_{1,i}).$$

Since quasi-particles of the same color commute, we arrange quasi-particles in monomial $b(\alpha_i)$ so that the charges $n_{p,i}$ and values $m_{p,i}$ (in the case of quasi-particles with the same charges), for $1 \leq p \leq r_i^{(1)}$, form a decreasing sequence of integers from right to left. We say that a monochromatic quasi-particle monomial is of color-type r_i, charge-type

$$\left(n_{r_i^{(1)},i}, \ldots, n_{1,i}\right) \tag{3}$$

where

$$0 \leq n_{r_i^{(1)},i} \leq \cdots \leq n_{1,i},$$

and dual-charge-type

$$\left(r_i^{(1)}, r_i^{(2)}, \ldots, r_i^{(s)}\right), \tag{4}$$

where

$$r_i^{(1)} \geq r_i^{(2)} \geq \cdots \geq r_i^{(s)} \geq 0 \text{ and } s \geq 1,$$

if (3) and (4) are conjugate partitions of r_i (cf. [5–7, 25]).

We extend this dictionary to the case of polychromatic monomials, that is, for monomial

$$b = b(\alpha_l) \cdots b(\alpha_1)$$

$$= x_{n_{r_l^{(1)},l}\alpha_l}(m_{r_l^{(1)},l}) \cdots x_{n_{1,l}\alpha_l}(m_{1,l}) \cdots x_{n_{r_1^{(1)},1}\alpha_1}(m_{r_1^{(1)},1}) \cdots x_{n_{1,1}\alpha_1}(m_{1,1})$$

we will say it is of color-type

$$(r_l, \ldots, r_1),$$

of charge-type

$$\mathfrak{R}' = \left(n_{r_l^{(1)},l}, \ldots, n_{1,l}; \ldots; n_{r_1^{(1)},1}, \ldots, n_{1,1}\right),$$

where

$$0 \leq n_{r_i^{(1)},i} \leq \cdots \leq n_{1,i},$$

and dual-charge-type

$$\mathfrak{R} = \left(r_l^{(1)}, \ldots, r_l^{(s_l)}; \ldots; r_1^{(1)}, \ldots, r_1^{(s_1)}\right),$$

where

$$r_i^{(1)} \geq r_i^{(2)} \geq \cdots \geq r_i^{(s_i)} \geq 0$$

and where for every color i, $1 \leq i \leq l$ we have

$$r_i = \sum_{p=1}^{r_i^{(1)}} n_{p,i} = \sum_{t=1}^{s_i} r_i^{(t)} \text{ and } s_i \in \mathbb{N}.$$

We can visualize charge-type and dual-charge type of polychromatic monomials $b(\alpha_2)b(\alpha_1)$ in graphic presentation, as in the following example.

Example

For monomial

$$x_{\alpha_2}(m_{4,2})x_{2\alpha_2}(m_{3,2})x_{3\alpha_2}(m_{2,2})x_{4\alpha_2}(m_{1,2})x_{2\alpha_1}(m_{4,1})x_{3\alpha_1}(m_{3,1})x_{4\alpha_1}(m_{2,1})x_{4\alpha_1}(m_{1,1})$$

of color-type $(r_2, r_1) = (9, 13)$, of charge-type $\mathfrak{R}' = (1, 2, 2, 4; 2, 3, 4, 4)$ and dual-charge-type $\mathfrak{R} = (4, 3, 1, 1; 4, 4, 3, 2)$ we have the graphic presentation as given in Fig. 1, where each quasi-particle of charge r is presented by a column of height r.

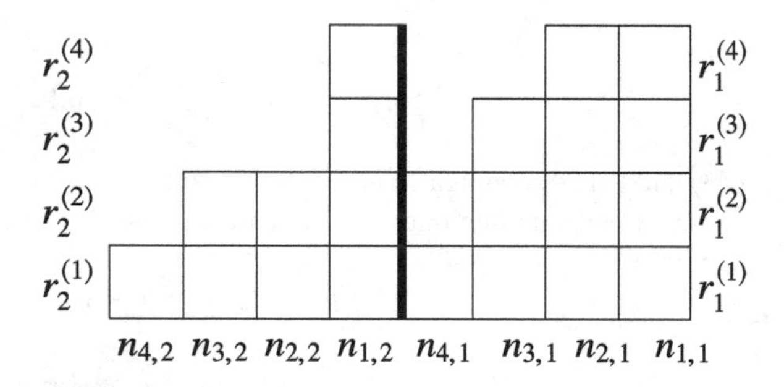

Fig. 1 Graphic presentation

We compare two monomials b and $\overline{b}$ by comparing first their charge-types $\mathcal{R}'$ and $\overline{\mathcal{R}'}$, where $\overline{\mathcal{R}'} = \left(\overline{n}_{\overline{r}_l^{(1)},l}, \ldots, \overline{n}_{1,1}\right)$, so that we compare their charges from right to left, i.e. we write $\mathcal{R}' < \overline{\mathcal{R}'}$ if there is $u \in \mathbb{N}$, such that $n_{1,i} = \overline{n}_{1,i}, n_{2,i} = \overline{n}_{2,i}, \ldots, n_{u-1,i} = \overline{n}_{u-1,i}$, and $u = \overline{r}_i^{(1)} + 1$ or $n_{u,i} < \overline{n}_{u,i}$, starting from color $i = 1$, and then their sequences of energies $\left(m_{r_l^{(1)},l}, \ldots, m_{1,1}\right)$ and $\left(\overline{m}_{\overline{r}_l^{(1)},l}, \ldots, \overline{m}_{1,1}\right)$ (in a similar way as charge-types, again starting from color $i = 1$). So, we will write $b < \overline{b}$ if $\mathcal{R}' < \overline{\mathcal{R}'}$ or, $\mathcal{R}' = \overline{\mathcal{R}'}$ and $\left(m_{r_l^{(1)},l}, \ldots, m_{1,1}\right) < \left(\overline{m}_{\overline{r}_l^{(1)},l}, \ldots, \overline{m}_{1,1}\right)$.

3.3 *Quasi-particle Bases of Principal Subspaces*

In reducing the set $U v_V$ to the spanning set we use relations among quasi-particles. First note that from (2) follows

$$x_{r\alpha_i}(z)v_V \in W_V\,[[z]]\,, \tag{5}$$

and

$$x_{r\alpha_i}(m)v_V = 0, \quad \text{for } m > -r. \tag{6}$$

On a standard module $L(k\Lambda_0)$ we have vertex operator algebra relations

$$x_{(tk+1)\alpha_i}(z) = 0, \tag{7}$$

where $t = 1$ if α_i is a long root, $t = 2$ if α_i is a short root and $\mathfrak{g}$ is of type B_l or C_l and $t = 3$ if α_i is a short root and $\mathfrak{g}$ is of type G_2 (see [36], [39] or [40]).

We use relations among quasi-particles of the same color:

Lemma 1 ([29]) *For fixed $M, j \in \mathbb{Z}$ and $1 \leq n \leq n'$ a sequence of $2n$ monomials from the set*

$$A = \{x_{n\alpha_i}(j)x_{n'\alpha_i}(M-j), x_{n\alpha_i}(j+1)x_{n'\alpha_i}(M-j-1), \ldots$$
$$\ldots, x_{n\alpha_i}(j+2n-1)x_{n'\alpha_i}(M-j-2n+1)\}$$

can be expressed as a linear combination of monomials from the set

$$\left\{x_{n\alpha_i}(m)x_{n'\alpha_i}(m') : m+m' = M\right\} \setminus A$$

and monomials which have as a factor quasi-particle $x_{(n'+1)\alpha_i}(j')$, $j' \in \mathbb{Z}$.

These relations were also proved in [22, 23, 25], and we use these relations in reducing spanning set Uv_V for the following two cases. First, $2n$ monomials with charge-type (n, n'), where $n < n'$,

$$x_{n\alpha_i}(m)x_{n'\alpha_i}(m')v_V, \ldots, x_{n\alpha_i}(m-2n+1)x_{n'\alpha_i}(m'+2n-1)v_V,$$

we express as a (finite) linear combination of monomial vectors

$$x_{n\alpha_i}(j)x_{n'\alpha_i}(j')v_V \text{ such that } j \leq m-2n \text{ and } j' \geq m'+2n$$

and monomial vectors with a factor quasi-particle $x_{(n'+1)\alpha_i}(j_1)$, $j_1 \in \mathbb{Z}$. In the case when $n = n'$, we express $2n$ monomials

$$x_{n\alpha_i}(m)x_{n\alpha_i}(m') \text{ with } m'-2n < m \leq m'$$

as a linear combination of monomials

$$x_{n\alpha_i}(j)x_{n\alpha_i}(j') \text{ with } j \leq j'-2n$$

and monomials with quasi-particle $x_{(n+1)\alpha_i}(j_1)$, $j_1 \in \mathbb{Z}$.

Next, we consider relations among quasi-particles of different colors:

Lemma 2 ([5–7])

1. *In the case of* $B_l^{(1)}$ *for fixed color* i, $1 \leq i \leq l-2$ *we have*

$$(z_1 - z_2)^{min\{n_{i+1},n_i\}} x_{n_{i+1}\alpha_{i+1}}(z_1) x_{n_i\alpha_i}(z_2) = (z_1 - z_2)^{min\{n_{i+1},n_i\}} x_{n_i\alpha_i}(z_2) x_{n_{i+1}\alpha_{i+1}}(z_1),$$

and for colors $l-1$ *and* l *we have*

$$(z_1 - z_2)^{min\{2n_{l-1},n_l\}} x_{n_{l-1}\alpha_{l-1}}(z_1) x_{n_l\alpha_l}(z_2) = (z_1 - z_2)^{min\{2n_{l-1},n_l\}} x_{n_l\alpha_l}(z_2) x_{n_{l-1}\alpha_{l-1}}(z_1).$$

2. *In the case of* $C_l^{(1)}$ *for fixed color* i, $1 \leq i \leq l-2$, *we have*

$$(z_1 - z_2)^{min\{n_{i+1},n_i\}} x_{n_{i+1}\alpha_{i+1}}(z_1) x_{n_i\alpha_i}(z_2) = (z_1 - z_2)^{min\{n_{i+1},n_i\}} x_{n_i\alpha_i}(z_2) x_{n_{i+1}\alpha_{i+1}}(z_1),$$

and for colors $l-1$ *and* l *we have*

$$(z_1 - z_2)^{min\{n_{l-1},2n_l\}} x_{n_{l-1}\alpha_{l-1}}(z_1) x_{n_l\alpha_l}(z_2) = (z_1 - z_2)^{min\{n_{l-1},2n_l\}} x_{n_l\alpha_l}(z_2) x_{n_{l-1}\alpha_{l-1}}(z_1).$$

3. *In the case of* $G_2^{(1)}$, *we have*

$$(z_1 - z_2)^{min\{3n_1,n_2\}} x_{n_1\alpha_1}(z_1) x_{n_2\alpha_2}(z_2) = (z_1 - z_2)^{min\{3n_1,n_2\}} x_{n_2\alpha_2}(z_2) x_{n_1\alpha_1}(z_1).$$

Using described relations, we can define the following sets:

1. for the case of $B_l^{(1)}$:

$$B_{W_{L(k\Lambda_0)}} = \bigcup_{\substack{n_{r_1^{(1)},1} \leq \cdots \leq n_{1,1} \leq k \\ \cdots \\ n_{r_{l-1}^{(1)},l-1} \leq \cdots \leq n_{1,l-1} \leq k \\ n_{r_l^{(1)},l} \leq \cdots \leq n_{1,l} \leq 2k}} \left(\text{or, equivalently,} \quad \bigcup_{\substack{r_1^{(1)} \geq \cdots \geq r_1^{(k)} \geq 0 \\ \cdots \\ r_{l-1}^{(1)} \geq \cdots \geq r_{l-1}^{(k)} \geq 0 \\ r_l^{(1)} \geq \cdots \geq r_l^{(2k)} \geq 0}} \right)$$

$$\{b = b(\alpha_l) \cdots b(\alpha_1)$$

$$= x_{n_{r_l^{(1)},l}\alpha_l}(m_{r_l^{(1)},l}) \cdots x_{n_{1,l}\alpha_l}(m_{1,l}) \cdots x_{n_{r_1^{(1)},1}\alpha_1}(m_{r_1^{(1)},1}) \cdots x_{n_{1,1}\alpha_1}(m_{1,1}) :$$

$$\left|\begin{array}{l} m_{p,i} \leq -n_{p,i} + \sum_{q=1}^{r_{i-1}^{(1)}} \min\left\{n_{q,i-1}, n_{p,i}\right\} - \sum_{p>p'>0} 2\min\{n_{p,i}, n_{p',i}\}, \\ \qquad 1 \leq p \leq r_i^{(1)}, 1 \leq i \leq l-1; \\ m_{p+1,i} \leq m_{p,i} - 2n_{p,i} \text{ if } n_{p+1,i} = n_{p,i},\ 1 \leq p \leq r_i^{(1)} - 1, \\ \qquad 1 \leq i \leq l-1; \\ m_{p,l} \leq -n_{p,l} + \sum_{q=1}^{r_{l-1}^{(1)}} \min\left\{2n_{q,l-1}, n_{p,l}\right\} - \sum_{p>p'>0} 2\min\{n_{p,l}, n_{p',l}\}, \\ \qquad 1 \leq p \leq r_l^{(1)}; \\ m_{p+1,l} \leq m_{p,l} - 2n_{p,l} \text{ if } n_{p,l} = n_{p+1,l},\ 1 \leq p \leq r_l^{(1)} - 1 \end{array}\right\},$$

2. for the case of $C_l^{(1)}$:

$$B_{W_{L(k\Lambda_0)}} = \bigcup_{\substack{n_{r_1^{(1)},1} \leq \cdots \leq n_{1,1} \leq 2k \\ \cdots \\ n_{r_{l-1}^{(1)},l-1} \leq \cdots \leq n_{1,l-1} \leq 2k \\ n_{r_l^{(1)},l} \leq \cdots \leq n_{1,l} \leq k}} \left(\text{or, equivalently, } \bigcup_{\substack{r_1^{(1)} \geq \cdots \geq r_1^{(2k)} \geq 0 \\ \cdots \\ r_{l-1}^{(1)} \geq \cdots \geq r_{l-1}^{(2k)} \geq 0 \\ r_l^{(1)} \geq \cdots \geq r_l^{(k)} \geq 0}}\right)$$

$$\{b = b(\alpha_1) \cdots b(\alpha_l)$$

$$= x_{n_{r_1^{(1)},1}\alpha_1}(m_{r_1^{(1)},1}) \cdots x_{n_{1,1}\alpha_1}(m_{1,1}) \cdots x_{n_{r_l^{(1)},l}\alpha_l}(m_{r_l^{(1)},l}) \cdots x_{n_{1,l}\alpha_l}(m_{1,l}) :$$

$$\left|\begin{array}{l} m_{p,l} \leq -n_{p,l} - \sum_{p>p'>0} 2\min\{n_{p,l}, n_{p',l}\},\ 1 \leq p \leq r_l^{(1)}; \\ m_{p+1,l} \leq m_{p,l} - 2n_{p,l} \text{ if } n_{p,l} = n_{p+1,l},\ 1 \leq p \leq r_l^{(1)} - 1; \\ m_{p,l-1} \leq -n_{p,l-1} + \sum_{q=1}^{r_l^{(1)}} \min\left\{2n_{q,l}, n_{p,l-1}\right\} \\ \qquad - \sum_{p>p'>0} 2\min\{n_{p,l}, n_{p',l}\}, 1 \leq p \leq r_l^{(1)}; \\ m_{p+1,l-1} \leq m_{p,l-1} - 2n_{p,l-1} \text{ if } n_{p+1,l-1} = n_{p,l-1},\ 1 \leq p \leq r_{l-1}^{(1)} - 1; \\ m_{p,i} \leq -n_{p,i} + \sum_{q=1}^{r_{i+1}^{(1)}} \min\left\{n_{q,i+1}, n_{p,i}\right\} - \sum_{p>p'>0} 2\min\{n_{p,i}, n_{p',i}\}, \\ \qquad 1 \leq p \leq r_i^{(1)},\ 1 \leq i \leq l-2; \\ m_{p+1,i} \leq m_{p,i} - 2n_{p,i} \text{ if } n_{p+1,i} = n_{p,i},\ 1 \leq p \leq r_i^{(1)} - 1,\ 1 \leq i \leq l-2 \end{array}\right\},$$

3. for the case of $G_2^{(1)}$:

$$B_{W_{L(k\Lambda_0)}} = \bigcup_{\substack{n_{r_1^{(1)},1} \leq \cdots \leq n_{1,1} \leq k \\ n_{r_2^{(1)},2} \leq \cdots \leq n_{1,2} \leq 3k}} \left(\text{or, equivalently, } \bigcup_{\substack{r_1^{(1)} \geq r_1^{(2)} \geq \cdots \geq r_1^{(k)} \geq 0 \\ r_2^{(1)} \geq r_2^{(2)} \geq \cdots \geq r_2^{(3k)} \geq 0}}\right)$$

$$\{b = b(\alpha_2) b(\alpha_1)$$

$$= x_{n_{r_2^{(1)},2}\alpha_2}(m_{r_2^{(1)},2}) \cdots x_{n_{1,2}\alpha_2}(m_{1,2}) x_{n_{r_1^{(1)},1}\alpha_1}(m_{r_1^{(1)},1}) \cdots x_{n_{1,1}\alpha_1}(m_{1,1}) :$$

$$\left| \begin{array}{l} m_{p,1} \leq -n_{p,1} - \sum_{p>p'>0} 2\min\{n_{p,1}, n_{p',1}\},\ 1 \leq p \leq r_1^{(1)}; \\ m_{p+1,1} \leq m_{p,1} - 2n_{p,1} \text{ if } n_{p+1,1} = n_{p,1},\ 1 \leq p \leq r_1^{(1)} - 1; \\ m_{p,2} \leq -n_{p,2} + \sum_{q=1}^{r_1^{(1)}} \min\{3n_{q,1}, n_{p,2}\} - \sum_{p>p'>0} 2\min\{n_{p,2}, n_{p',2}\}, \\ \qquad 1 \leq p \leq r_2^{(1)}; \\ m_{p+1,2} \leq m_{p,2} - 2n_{p,2} \text{ if } n_{p,2} = n_{p+1,2},\ 1 \leq p \leq r_2^{(1)} - 1 \end{array} \right\}.$$

We showed that quasi-particle monomial vectors $bv_{L(k\Lambda_0)}$, with quasi-particle monomials b from given sets, form bases of principal subspaces of level k standard module $L(k\Lambda_0)$. Note that from combinatorial point of view, difference conditions for energies of quasi-particles of colors i, $1 \leq i \leq l-2$ in the case of $B_l^{(1)}$ are identical with the difference conditions for energies of Georgiev's quasi-particles in the case of level k standard $A_{l-1}^{(1)}$-module.

In the case of $C_l^{(1)}$ difference conditions for energies of quasi-particles colored with colors l and $l-1$ are identical with the difference conditions for energies of quasi-particles for $B_2^{(1)}$-module $L(k\Lambda_0)$, and difference conditions for energies of quasi-particles of colors i, $1 \leq i \leq l-2$, are identical with difference conditions for energies of quasi-particles in the case of standard $A_{l-1}^{(1)}$-modules of level $2k$.

In the case of $G_2^{(1)}$ conditions on energies which follow from relations (5), (6) and relations of Lemma 1 are identical to difference conditions in the case of level k standard $B_2^{(1)}$-module in [5].

By using relations from Lemmas 1 and 2 we also define sets $B_{W_{N(k\Lambda_0)}}$, whose monomials (acting on the highest vector) form bases of principal subspaces $W_{N(k\Lambda_0)}$. The main difference with elements from sets $B_{W_{L(k\Lambda_0)}}$ is that quasi-particles in $B_{W_{N(k\Lambda_0)}}$ have no limit on the charge:

1. for the case of $B_l^{(1)}$:

$$B_{W_{N(k\Lambda_0)}} = \bigcup_{\substack{n_{r_1^{(1)},1} \leq \cdots \leq n_{1,1} \\ \cdots \\ n_{r_{l-1}^{(1)},l-1} \leq \cdots \leq n_{1,l-1} \\ n_{r_l^{(1)},l} \leq \cdots \leq n_{1,l}}} \left(\text{or, equivalently,} \bigcup_{\substack{r_1^{(1)} \geq r_1^{(2)} \geq \cdots \geq 0 \\ \cdots \\ r_{l-1}^{(1)} \geq r_{l-1}^{(2)} \geq \cdots \geq 0 \\ r_l^{(1)} \geq r_l^{(2)} \geq \cdots \geq 0}} \right)$$

$$\{b = b(\alpha_l) \cdots b(\alpha_1)$$

$$= x_{n_{r_l^{(1)},l}\alpha_l}(m_{r_l^{(1)},l}) \cdots x_{n_{1,l}\alpha_l}(m_{1,l}) \cdots x_{n_{r_1^{(1)},1}\alpha_1}(m_{r_1^{(1)},1}) \cdots x_{n_{1,1}\alpha_1}(m_{1,1}) :$$

$$\left.\begin{array}{|l} m_{p,i} \leq -n_{p,i} + \sum_{q=1}^{r_{i-1}^{(1)}} \min\left\{n_{q,i-1}, n_{p,i}\right\} - \sum_{p>p'>0} 2\min\{n_{p,i}, n_{p',i}\}, \\ \qquad 1 \leq p \leq r_i^{(1)}, 1 \leq i \leq l-1; \\ m_{p+1,i} \leq m_{p,i} - 2n_{p,i} \text{ if } n_{p+1,i} = n_{p,i},\ 1 \leq p \leq r_i^{(1)} - 1, \\ \qquad 1 \leq i \leq l-1; \\ m_{p,l} \leq -n_{p,l} + \sum_{q=1}^{r_{l-1}^{(1)}} \min\left\{2n_{q,l-1}, n_{p,l}\right\} - \sum_{p>p'>0} 2\min\{n_{p,l}, n_{p',l}\}, \\ \qquad 1 \leq p \leq r_l^{(1)}; \\ m_{p+1,l} \leq m_{p,l} - 2n_{p,l} \text{ if } n_{p,l} = n_{p+1,l},\ 1 \leq p \leq r_l^{(1)} - 1 \end{array}\right\},$$

2. for the case of $C_l^{(1)}$:

$$B_{W_{N(k\Lambda_0)}} = \bigcup_{\substack{n_{r_1^{(1)},1} \leq \cdots \leq n_{1,1} \\ \cdots \\ n_{r_{l-1}^{(1)},l-1} \leq \cdots \leq n_{1,l-1} \\ n_{r_l^{(1)},l} \leq \cdots \leq n_{1,l}}} \left(\text{or, equivalently,} \bigcup_{\substack{r_1^{(1)} \geq r_1^{(2)} \geq \cdots \geq 0 \\ \cdots \\ r_{l-1}^{(1)} \geq r_{l-1}^{(2)} \geq \cdots \geq 0 \\ r_l^{(1)} \geq r_l^{(2)} \geq \cdots \geq 0}} \right)$$

$$\{b = b(\alpha_1) \cdots b(\alpha_l)$$

$$= x_{n_{r_1^{(1)},1}\alpha_1}(m_{r_1^{(1)},1}) \cdots x_{n_{1,1}\alpha_1}(m_{1,1}) \cdots x_{n_{r_l^{(1)},l}\alpha_l}(m_{r_l^{(1)},l}) \cdots x_{n_{1,l}\alpha_l}(m_{1,l}) :$$

$$\left.\begin{array}{|l} m_{p,l} \leq -n_{p,l} - \sum_{p>p'>0} 2\min\{n_{p,l}, n_{p',l}\},\ 1 \leq p \leq r_l^{(1)}; \\ m_{p+1,l} \leq m_{p,l} - 2n_{p,l} \text{ if } n_{p,l} = n_{p+1,l},\ 1 \leq p \leq r_l^{(1)} - 1; \\ m_{p,l-1} \leq -n_{p,l-1} + \sum_{q=1}^{r_l^{(1)}} \min\left\{2n_{q,l}, n_{p,l-1}\right\} \\ \qquad - \sum_{p>p'>0} 2\min\{n_{p,l}, n_{p',l}\},\ 1 \leq p \leq r_l^{(1)}; \\ m_{p+1,l-1} \leq m_{p,l-1} - 2n_{p,l-1} \text{ if } n_{p+1,l-1} = n_{p,l-1},\ 1 \leq p \leq r_{l-1}^{(1)} - 1; \\ m_{p,i} \leq -n_{p,i} + \sum_{q=1}^{r_{i+1}^{(1)}} \min\left\{n_{q,i+1}, n_{p,i}\right\} - \sum_{p>p'>0} 2\min\{n_{p,i}, n_{p',i}\}, \\ \qquad 1 \leq p \leq r_i^{(1)},\ 1 \leq i \leq l-2; \\ m_{p+1,i} \leq m_{p,i} - 2n_{p,i} \text{ if } n_{p+1,i} = n_{p,i},\ 1 \leq p \leq r_i^{(1)} - 1,\ 1 \leq i \leq l-2 \end{array}\right\},$$

3. for the case of $G_2^{(1)}$:

$$B_{W_{N(k\Lambda_0)}} = \bigcup_{\substack{n_{r_1^{(1)},1} \leq \cdots \leq n_{1,1} \\ n_{r_2^{(1)},2} \leq \cdots \leq n_{1,2}}} \left(\text{or, equivalently,} \bigcup_{\substack{r_1^{(1)} \geq r_1^{(2)} \geq \cdots \geq 0 \\ r_2^{(1)} \geq r_2^{(2)} \geq \cdots \geq 0}} \right)$$

$$\{b = b(\alpha_2) b(\alpha_1)$$

$$= x_{n_{r_2^{(1)},2}\alpha_2}(m_{r_2^{(1)},2}) \cdots x_{n_{1,2}\alpha_2}(m_{1,2}) x_{n_{r_1^{(1)},1}\alpha_1}(m_{r_1^{(1)},1}) \cdots x_{n_{1,1}\alpha_1}(m_{1,1}) :$$

$$\left| \begin{array}{l} m_{p,1} \leq -n_{p,1} - \sum_{p>p'>0} 2\min\{n_{p,1}, n_{p',1}\},\ 1 \leq p \leq r_1^{(1)}; \\ m_{p+1,1} \leq m_{p,1} - 2n_{p,1} \text{ if } n_{p+1,1} = n_{p,1},\ 1 \leq p \leq r_1^{(1)} - 1; \\ m_{p,2} \leq -n_{p,2} + \sum_{q=1}^{r_1^{(1)}} \min\left\{3n_{q,1}, n_{p,2}\right\} - \sum_{p>p'>0} 2\min\{n_{p,2}, n_{p',2}\}, \\ \qquad 1 \leq p \leq r_2^{(1)}; \\ m_{p+1,2} \leq m_{p,2} - 2n_{p,2} \text{ if } n_{p,2} = n_{p+1,2},\ 1 \leq p \leq r_2^{(1)} - 1 \end{array} \right\} .$$

Now, we can state the main theorem:

Theorem 1 ([5–7]) *The set $\mathfrak{B}_{W_V} = \{bv_V : b \in B_V\}$ is a basis of the principal subspace W_V.*

The proof that $\mathfrak{B}_{W_V}$ is the spanning set goes by induction on charge-type and total energie of quasi-particle monomials and follows closely [25].

4 Proof of Linear Independence

For the purpose of proving the linear independence of $\mathfrak{B}_{W_V}$, we realize the principal subspace $W_{L(k\Lambda_0)}$ as a subspace of the tensor product of k principal subspaces of standard modules $L(\Lambda_0)$. This enables us to use certain maps defined on level one standard modules. Here we will introduce these maps and we will illustrate how we used them in the proof of linear independence for principal subspace of $L(\Lambda_0)$.

4.1 A Coefficient of an Intertwining Operator

In the case of $B_l^{(1)}$ and $C_l^{(1)}$ we use coefficient A_{ω_i} of intertwining operator $I(\cdot, z)$

$$I(w, z)v = \exp(zL(-1))Y(v, -z)w, \quad w \in L(\Lambda_i),\ v \in L(\Lambda_0)$$

of type

$$\binom{L(\Lambda_i)}{L(\Lambda_i)\ L(\Lambda_0)},$$

(cf. [24]). From [24] we have fusion rule

$$\dim I\binom{L(\Lambda_i)}{L(\Lambda_i)\ L(\Lambda_0)} = 1,$$

where $I\binom{L(\Lambda_i)}{L(\Lambda_i)\ L(\Lambda_0)}$ denotes the vector space of all intertwining operators of type $\binom{L(\Lambda_i)}{L(\Lambda_i)\ L(\Lambda_0)}$. A_{ω_i} is defined by

$$A_{\omega_i} = \mathrm{Res}_z\, z^{-1} I(v_{L(\Lambda_i)}, z),$$

where $i = 1$ in the case of $B_l^{(1)}$ and $i = l$ in the case of $C_l^{(1)}$ so that

$$A_{\omega_i} v_{L(\Lambda_0)} = v_{L(\Lambda_i)}.$$

In the case of $G_2^{(1)}$ we define the coefficient A_{ω_i} of an intertwining operator $x_\theta(z)$ as

$$A_{\omega_i} = \mathrm{Res}_z z^{-1}\, x_\theta(z) = x_\theta(-1),$$

so that

$$A_{\omega_i} v_{L(\Lambda_0)} = x_\theta(-1) v_{L(\Lambda_0)}.$$

These coefficients commute with the action of quasi-particles:

Lemma 3 ([5–7]) *For every $j = 1, \ldots, l$ and fixed $m \in \mathbb{Z}$, we have:*

$$\left[x_{\alpha_j}(m), A_{\omega_i}\right] = 0. \tag{8}$$

4.2 *Simple Current Operator*

In the case of $B_l^{(1)}$ and $C_l^{(1)}$ we also use simple current operator e_{ω_i} on the level 1 standard modules

$$e_{\omega_i} : L(\Lambda_0) \to L(\Lambda_i),$$

associated to $\omega_i \in \mathfrak{h}$, where $i = 1$ in the case of $B_l^{(1)}$ and $i = l$ in the case of $C_l^{(1)}$, such that

$$e_{\omega_i} v_{L(\Lambda_0)} = v_{L(\Lambda_i)}, \tag{9}$$

and such that

$$x_\alpha(m) e_{\omega_i} = e_{\omega_i} x_\alpha(m + \alpha(\omega_i)), \tag{10}$$

for all $\alpha \in R$ and $m \in \mathbb{Z}$ (cf. [20], see also [45]).

4.3 "Weyl Group Translation" Operator

For every root α, we define on the level 1 standard module $L(\Lambda_0)$, the "Weyl group translation" operator e_α by

$$e_\alpha = \exp\, x_{-\alpha}(1) \exp\,(-x_\alpha(-1)) \exp\, x_{-\alpha}(1) \exp\, x_\alpha(0) \exp\,(-x_{-\alpha}(0)) \exp\, x_\alpha(0),$$

(cf. [30]). Then on the standard $\tilde{\mathfrak{g}}$-module $L(\Lambda_0)$ we have

$$e_\alpha v_{L(\Lambda_0)} = -x_\alpha(-1) v_{L(\Lambda_0)}, \tag{11}$$

for every long root α, and

$$x_\beta(j) e_\alpha = e_\alpha x_\beta(j - \beta(\alpha^\vee)), \quad \alpha, \beta \in R, \quad j \in \mathbb{Z}. \tag{12}$$

4.4 The Proof of Linear Independence of the Set $\mathfrak{B}_{W_{L(\Lambda_0)}}$

Assume that we have a relation of linear dependence between elements of $\mathfrak{B}_{W_{L(\Lambda_0)}}$

$$\sum_{a \in A} c_a b_a v_{L(\Lambda_0)} = 0, \tag{13}$$

where A is a finite non-empty set. Furthermore, assume that all b_a have the same color-type. Let $b = b(\alpha_l) \cdots b(\alpha_1) x_{\alpha_1}(-j) v_{L(\Lambda_0)}$ (or $b = b(\alpha_1) \cdots b(\alpha_l) x_{\alpha_l}(-j) v_{L(\Lambda_0)}$, when $\tilde{\mathfrak{g}}$ is of type $C_l^{(1)}$) be the smallest monomial in (13) such that $c_a \neq 0$. On (13), we act with operator A_{ω_i} and commute to the left with operator e_{ω_i}, where e_{ω_i} denotes simple current operator e_{ω_1} in the case of $B_l^{(1)}$, e_{ω_l} in the case of $C_l^{(1)}$, and "Weyl group translation" operator e_θ in the case of $G_2^{(1)}$:

$$0 = A_{\omega_i} \left(\sum_{a \in A} c_a b_a v_{L(\Lambda_0)} \right) = \sum_{a \in A} c_a b_a A_{\omega_i} v_{L(\Lambda_0)}$$

$$= \sum_{a \in A} c_a b_a e_{\omega_i} v_{L(\Lambda_0)} = e_{\omega_i} \left(\sum_{a \in A} c'_a b_a^+ v_{L(\Lambda_0)} \right).$$

After leaving out the invertible operator e_{ω_i} we get

$$\sum_{a \in A} c'_a b_a^+ v_{L(\Lambda_0)} = 0,$$

where $b_a^+ \in B_{W_{L(\Lambda_0)}}$ and $b_a^+ > b_a$. We repeat described procedure until $-j$ becomes -1. All monomial vectors with $x_{\alpha_i}(m_{1,i})$, where $m_{1,i} > -j$ and $i = 1$ in the case of $B_l^{(1)}$ and $G_2^{(1)}$ or $i = l$ in the case of $C_l^{(1)}$ get annihilated at some step of this procedure and therefore we get

$$\sum_a c'_a b_a^+ x_{\alpha_i}(-1) v_{L(\Lambda_0)} = 0,$$

with $b_a^+ x_{\alpha_i}(-1) \in B_{W_{L(\Lambda_0)}}$. Now, from (11) and (12) follows

$$\sum_a c'_a b'_a v_{L(\Lambda_0)} = 0$$

where b'_a are monomials from $B_{W_{L(\Lambda_0)}}$, but without rightmost quasi-particle.

We repeat the described procedure, until we get monomial vectors from the set $\mathfrak{B}_{W_{L(\Lambda_0)}}$ of the form $b(\alpha_l)\cdots b(\alpha_2) v_{L(\Lambda_0)}$ in the case of $B_l^{(1)}$, $b(\alpha_1)\cdots b(\alpha_{l-1}) v_{L(\Lambda_0)}$ in the case of $C_l^{(1)}$ and monomial vectors $b(\alpha_2) v_{L(\Lambda_0)}$ in the case of $G_2^{(1)}$. In the case of $B_l^{(1)}$ these monomial vectors can be realized as elements of $W_{L(\Lambda_0)}$ of the affine Lie algebra of type $B_{l-1}^{(1)}$. Especially, when $l = 2$ with the described procedure we get monomial vectors which can be realized as elements of $W_{L(2\Lambda_0)}$ of the affine Lie algebra of type $A_1^{(1)}$. In the case of $C_l^{(1)}$, monomial vectors of the form $b(\alpha_1)\cdots b(\alpha_{l-1}) v_{L(\Lambda_0)}$ can be realized as elements in $W_{L(2\Lambda_0)}$ of the affine Lie algebra of type $A_{l-1}^{(1)}$ and monomial vectors of the form $b(\alpha_2) v_{L(\Lambda_0)}$ as elements of $W_{L(3\Lambda_0)}$ of the affine Lie algebra of type $A_1^{(1)}$ in the case of $G_2^{(1)}$. For all of these cases we can use Georgiev argument on linear independence from [25], so proceed inductively on charge-type (and on l for the case of $B_l^{(1)}$) we get $c_a = 0$ and the desired theorem follows.

For higher levels k, modeled on a projection from [25], we define a projection $\pi_{\mathcal{R}}$, for a chosen dual-charge-type $\mathcal{R}$, distributing quasi-particles of monomial vectors from the set $\mathfrak{B}_{W_V}$ among factors of the tensor product of $\mathfrak{h}$-weight subspaces of standard modules $L(\Lambda_0)$. The projection enables us to use described maps.

4.5 The Proof of Linear Independence of the Set $\mathfrak{B}_{W_{N(k\Lambda_0)}}$

The main idea in the proof of linear independence of the set $\mathfrak{B}_{W_{N(k\Lambda_0)}}$ is to use the surjective map $f_{k'} : W_{N(k'\Lambda_0)} \to W_{L(k'\Lambda_0)}$ for some $k' \gg 0$. From relations (7) follow $f'_k(\mathfrak{B}_{W_{N(k'\Lambda_0)}}) = \mathfrak{B}_{W_{L(k'\Lambda_0)}}$, so we can repeat the proof of linear independence for $\mathfrak{B}_{W_{L(k'\Lambda_0)}}$. Therefore in

$$\sum_{a \in A} c_a b_a v_{N(k\Lambda_0)} = 0, \tag{14}$$

where all monomials $b_a \in \mathfrak{B}_{W_{N(k\Lambda_0)}}$ are of the same color-type, we find monomial b', which is the largest monomial, such that $c'_a \neq 0$. We also choose a positive integer k' such that for example $n'_{1,i} \leq k'$ for $1 \leq i \leq l-1$ and $\frac{1}{2}n'_{1,l} \leq k'$ in the case of $B_l^{(1)}$ (similar is for the other two cases). Denote by $W_{N(k'\Lambda_0)} = U(\mathcal{L}(\mathfrak{n}_+))v_{N(k'\Lambda_0)}$ the principal subspace of $N\left(k'\Lambda_0\right)$. The automorphism of $\mathcal{L}(\mathfrak{n}_+)$-modules $\rho_{k'} : W \rightarrow W_{N(k'\Lambda_0)}$, defined by

$$\rho_{k'}(bv_{N(k\Lambda_0)}) = bv_{N(k'\Lambda_0)},$$

maps the set $\mathfrak{B}_{W_{N(k\Lambda_0)}}$ to the set that spans $W_{N(k'\Lambda_0)}$, so we have

$$\sum_{a\in A} c_a b_a v_{N(k'\Lambda_0)} = 0.$$

With the surjective map $f_{k'}$, we get

$$\sum_{a\in A} c_a b_a v_{L(k'\Lambda_0)} = 0.$$

5 Characters of Principal Subspaces

We define character ch W_V of W_V as:

$$\text{ch } W_V = \sum_{m,r_1,\ldots,r_l\geq 0} \dim W_{V\,(m,r_1,\ldots,r_l)} q^m y_1^{r_1} \cdots y_l^{r_l},$$

where $q,\ y_1, \ldots y_l$ are formal variables and

$$W_{V\,(m,r_1,\ldots,r_l)} = W_{V\,-m\delta+r_1\alpha_1+\cdots+r_l\alpha_l}$$

is the $\widetilde{\mathfrak{h}}$-weight subspace of weight $-m\delta + r_1\alpha_1 + \cdots + r_l\alpha_l$.

5.1 *Characters of the Principal Subspace* $W_{L(k\Lambda_0)}$

If we write conditions in the definition of the set $\mathfrak{B}_{W_{L(k\Lambda_0)}}$ in terms of $r_i^{(s)}$, then from quasi-particle bases will follow characters of principal subspaces $W_{L(k\Lambda_0)}$:

Theorem 2 ([5–7]) *Using the notation*

$$(q)_r = (1-q)(1-q^2)\cdots(1-q^r)\, for\ r \geq 0$$

we have

1. *in the case of* $B_l^{(1)}$*:*

$$\operatorname{ch} W_{L(k\Lambda_0)} = \sum_{\substack{r_1^{(1)}\geq\cdots\geq r_1^{(k)}\geq 0\\ \cdots \\ r_{l-1}^{(1)}\geq\cdots\geq r_{l-1}^{(k)}\geq 0\\ r_l^{(1)}\geq\cdots\geq r_l^{(2k)}\geq 0}} \frac{q^{\sum_{i=1}^{l-1}\sum_{s=1}^{k} r_i^{(s)^2}+\sum_{s=1}^{2k} r_l^{(s)^2}-\sum_{i=1}^{l-2}\sum_{s=1}^{k} r_i^{(s)}r_{i+1}^{(s)}-\sum_{s=1}^{k} r_{l-1}^{(s)}(r_l^{(2s-1)}+r_l^{(2s)})}}{\prod_{i=1}^{l-1}((q)_{r_i^{(1)}-r_i^{(2)}}\cdots(q)_{r_i^{(k)}})(q)_{r_l^{(1)}-r_l^{(2)}}\cdots(q)_{r_l^{(2k)}}} \prod_{i=1}^{l} y_i^{r_i},$$

where $r_i = \sum_{s=1}^{k} r_i^{(s)}$, $1 \leq i \leq l-1$ *and* $r_l = \sum_{s=1}^{2k} r_l^{(s)}$*;*

2. *in the case of* $C_l^{(1)}$*:*

$$\operatorname{ch} W_{L(k\Lambda_0)} = \sum_{\substack{r_1^{(1)}\geq\cdots\geq r_1^{(2k)}\geq 0\\ \cdots \\ r_{l-1}^{(1)}\geq\cdots\geq r_{l-1}^{(2k)}\geq 0\\ r_l^{(1)}\geq\cdots\geq r_l^{(k)}\geq 0}} \frac{q^{\sum_{i=1}^{l-1}\sum_{s=1}^{2k} r_i^{(s)^2}+\sum_{s=1}^{k} r_l^{(s)^2}-\sum_{i=1}^{l-2}\sum_{s=1}^{2k} r_i^{(s)}r_{i+1}^{(s)}-\sum_{s=1}^{k} r_l^{(s)}(r_{l-1}^{(2s-1)}+r_{l-1}^{(2s)})}}{\prod_{i=1}^{l-1}((q)_{r_i^{(1)}-r_i^{(2)}}\cdots(q)_{r_i^{(2k)}})(q)_{r_l^{(1)}-r_l^{(2)}}\cdots(q)_{r_l^{(k)}}} \prod_{i=1}^{l} y_i^{r_i},$$

where $r_i = \sum_{s=1}^{2k} r_i^{(s)}$, $1 \leq i \leq l-1$ *and* $r_l = \sum_{s=1}^{k} r_l^{(s)}$*;*

3. *in the case of* $G_2^{(1)}$*:*

$$\operatorname{ch} W_{L(k\Lambda_0)} = \sum_{\substack{r_1^{(1)}\geq\cdots\geq r_1^{(k)}\geq 0\\ r_2^{(1)}\geq\cdots\geq r_2^{(3k)}\geq 0}} \frac{q^{\sum_{s=1}^{k} r_1^{(s)^2}+\sum_{s=1}^{3k} r_2^{(s)^2}-\sum_{s=1}^{k} r_1^{(s)}(r_2^{(3s)}+r_2^{(3s-1)}+r_2^{(3s-2)})}}{(q)_{r_1^{(1)}-r_1^{(2)}}\cdots(q)_{r_1^{(k)}}(q)_{r_2^{(1)}-r_2^{(2)}}\cdots(q)_{r_2^{(3k)}}} y_1^{r_1} y_2^{r_2},$$

where $r_1 = \sum_{s=1}^{k} r_1^{(s)}$ *and* $r_2 = \sum_{s=1}^{3k} r_2^{(s)}$.

5.2 *Characters of the Principal Subspace* $W_{N(k\Lambda_0)}$

In a similar way as in the previous case, from quasi-particle bases we obtain character of $W_{N(k\Lambda_0)}$. But also, we obtain character from base of $U(\mathcal{L}(\mathfrak{n}_+)_{<0})$ (see (1)). From these characters follow a generalization of Euler–Cauchy identity from [1].

Theorem 3 ([5–7]) *Using the notation*

$$(a_1; \ldots, a_n; q)_\infty := (a_1; q)_\infty \cdots (a_n; q)_\infty,$$

where

$$(a; q)_\infty = \prod_{i\geq 1}(1 - aq^{i-1}),$$

we have

1. *in the case of* $B_l^{(1)}$:

$$\frac{1}{(qy_1, qy_1y_2, \ldots, qy_1\cdots y_l, qy_1y_2^2\cdots y_l^2, \ldots, qy_1y_2\cdots y_l^2; q)_\infty}$$

$$\frac{1}{(qy_2, qy_2y_3, \ldots, qy_2\cdots y_l, qy_2y_3^2\cdots y_l^2, \ldots, qy_2y_3\cdots y_l^2; q)_\infty}$$

$$\cdots$$

$$\frac{1}{(qy_{l-1}, qy_{l-1}y_l, qy_{l-1}y_l^2, qy_l; q)_\infty}$$

$$= \sum_{\substack{r_1^{(1)}\geq r_1^{(2)}\geq\cdots\geq 0\\ \cdots \\ r_{l-1}^{(1)}\geq r_{l-1}^{(2)}\geq\cdots\geq 0\\ r_l^{(1)}\geq r_l^{(2)}\geq\cdots\geq 0}} \frac{q^{\sum_{i=1}^{l-1}\sum_{s\geq 1} {r_i^{(s)}}^2 + \sum_{s\geq 1} {r_l^{(s)}}^2 - \sum_{i=1}^{l-2}\sum_{s\geq 1} r_i^{(s)} r_{i+1}^{(s)} - \sum_{s\geq 1} r_{l-1}^{(s)}(r_l^{(2s-1)} + r_l^{(2s)})}}{\prod_{i=1}^{l} (q)_{r_i^{(1)}-r_i^{(2)}} (q)_{r_i^{(2)}-r_i^{(3)}} \cdots} \prod_{i=1}^{l} y_i^{r_i}, \tag{15}$$

where $r_i = \sum_{s\geq 1} r_i^{(s)}$, $1 \leq i \leq l$;

2. *in the case of* $C_l^{(1)}$:

$$\frac{1}{(qy_1, qy_1y_2, \ldots, qy_1\cdots y_l, qy_1y_2^2\cdots y_{l-1}^2 y_l, \ldots, qy_1y_2\cdots y_{l-1}^2 y_l, qy_1^2y_2^2\cdots y_{l-1}^2 y_l; q)_\infty}$$

$$\frac{1}{(qy_2, qy_2y_3, \ldots, qy_2\cdots y_l, qy_2y_3^2\cdots y_{l-1}^2 y_l, \ldots, qy_2y_3\cdots y_{l-1}^2 y_l, qy_2^2y_3^2\cdots y_l; q)_\infty}$$

$$\cdots$$

$$\frac{1}{(qy_{l-1}, qy_{l-1}y_l, qy_{l-1}^2 y_l, qy_l; q)_\infty}$$

$$= \sum_{\substack{r_1^{(1)}\geq r_1^{(2)}\geq\cdots\geq 0\\ \cdots \\ r_{l-1}^{(1)}\geq r_{l-1}^{(2)}\geq\cdots\geq 0\\ r_l^{(1)}\geq r_l^{(2)}\geq\cdots\geq 0}} \frac{q^{\sum_{i=1}^{l-1}\sum_{s\geq 1} {r_i^{(s)}}^2 + \sum_{s\geq 1} {r_l^{(s)}}^2 - \sum_{i=1}^{l-2}\sum_{s\geq 1} r_i^{(s)} r_{i+1}^{(s)} - \sum_{s\geq 1} r_l^{(s)}(r_{l-1}^{(2s-1)} + r_{l-1}^{(2s)})}}{\prod_{i=1}^{l} (q)_{r_i^{(1)}-r_i^{(2)}} (q)_{r_i^{(2)}-r_i^{(3)}} \cdots} \prod_{i=1}^{l} y_i^{r_i}, \tag{16}$$

where $r_i = \sum_{s\geq 1} r_i^{(s)}$, $1 \leq i \leq l$;

3. *in the case of* $G_2^{(1)}$:

$$\frac{1}{(qy_1, qy_2, qy_1y_2, qy_1y_2^2, qy_1y_2^3, qy_1^2y_2^3; q)_\infty}$$

$$= \sum_{\substack{r_1^{(1)} \geq r_1^{(2)} \geq \cdots \geq 0 \\ r_2^{(1)} \geq r_2^{(2)} \geq \cdots \geq 0}} \frac{q^{\sum_{s\geq 1} r_1^{(s)^2} + \sum_{s\geq 1} r_2^{(s)^2} - \sum_{s\geq 1} r_1^{(s)}(r_2^{(3s)} + r_2^{(3s-1)} + r_2^{(3s-2)})}}{(q)_{r_1^{(1)}-r_1^{(2)}} \cdots (q)_{r_2^{(1)}-r_2^{(2)}} \cdots} y_1^{r_1} y_2^{r_2}, \qquad (17)$$

where $r_i = \sum_{s\geq 1} r_i^{(s)}$ *and* $i = 1, 2$.

The sum on the right side of (15)–(17) is over all descending infinite sequences of non-negative integers with finite support.

Acknowledgements The author is partially supported by the Croatian Science Foundation under the project 2634 and by the QuantiXLie Centre of Excellence, a project cofinanced by the Croatian Government and European Union through the European Regional Development Fund—the Competitiveness and Cohesion Operational Programme (Grant KK.01.1.1.01.0004).

References

1. Andrews, G.E.: Partitions and Durfee dissection. Am. J. Math. **101**(3), 735–742 (1979)
2. Ardonne, E., Kedem, R., Stone, M.: Fermionic characters and arbitrary highest-weight integrable $\widehat{\mathfrak{sl}}_{r+1}$-modules. Commun. Math. Phys. **264**(2), 427–464 (2006)
3. Baranović, I.: Combinatorial bases of Feigin-Stoyanovsky's type subspaces of level 2 standard modules for $D_4^{(1)}$. Commun. Algebra **39**(3), 1007–1051 (2011)
4. Baranović, I., Primc, M., Trupčević, G.: Bases of Feigin-Stoyanovsky's type subspaces for $C_l^{(1)}$. Ramanujan J. **45**(1), 265–289 (2018)
5. Butorac, M.: Quasi-particle bases of principal subspaces of the affine Lie algebra of type $G_2^{(1)}$. Glas. Mat. Ser. III **52**(1), 79–98 (2017)
6. Butorac, M.: Quasi-particle bases of principal subspaces for the affine Lie algebras of types $B_l^{(1)}$ and $C_l^{(1)}$. Glas. Mat. Ser. III **51**(1), 59–108 (2016)
7. Butorac, M.: Combinatorial bases of principal subspaces for the affine Lie algebra of type $B_2^{(1)}$. J. Pure Appl. Algebra **218**(3), 424–447 (2014)
8. Butorac M., Sadowski, C.: Combinatorial bases of principal subspaces of modules for twisted affine Lie algebras of type $A_{2n+1}^{(2)}$, $D_n^{(2)}$, $E_6^{(2)}$ and $D_4^{(3)}$. Preprint (2018)
9. Calinescu, C.: Intertwining vertex operators and certain representations of $\widehat{sl(n)}$. Commun. Contemp. Math. **10**(1), 47–79 (2008)
10. Calinescu, C.: Principal subspaces of higher-level standard $\widehat{sl(3)}$-modules. J. Pure Appl. Algebra **210**(2), 559–575 (2007)
11. Calinescu, C., Lepowsky, J., Milas, A.: Vertex-algebraic structure of principal subspaces of standard $A_2^{(2)}$-modules, I. Internat. J. Math. **25**(7), 1450063 (2014) 44 pp
12. Calinescu, C., Lepowsky, J., Milas, A.: Vertex-algebraic structure of the principal subspaces of level one modules for the untwisted affine Lie algebras of types A, D, E. J. Algebra **323**(1), 167–192 (2010)
13. Calinescu, C., Lepowsky, J., Milas, A.: Vertex-algebraic structure of the principal subspaces of certain $A_1^{(1)}$-modules, II: higher-level case. J. Pure Appl. Algebra **212**(8), 1928–1950 (2008)

14. Calinescu, C., Lepowsky, J., Milas, A.: Vertex-algebraic structure of the principal subspaces of certain $A_1^{(1)}$-modules, I: level one case. Int. J. Math. **19**(1), 71–92 (2008)
15. Calinescu, C., Milas, A., Penn, M.: Vertex algebraic structure of principal subspaces of basic $A_{2n}^{(2)}$-modules. J. Pure Appl. Algebra **220**(5), 1752–1784 (2016)
16. Capparelli, S., Lepowsky, J., Milas, A.: The Rogers-Selberg recursions, the Gordon-Andrews identities and intertwining operators. Ramanujan J. **12**(3), 379–397 (2006)
17. Capparelli, S., Lepowsky, J., Milas, A.: The Rogers-Ramanujan recursion and intertwining operators. Commun. Contemp. Math. **5**(6), 947–966 (2003)
18. Dasmahapatra, S., Dedem, R., Klassen, T.R., McCoy, B.M., Melzer, E.: Quasi-particles, conformal field theory and q series. Yang-Baxter equations in Paris (1992). Int. J. Mod. Phys. B **7**, 3617–3648 (1993)
19. Dong, C., Lepowsky, J.: Generalized Vertex Algebras and Relative Vertex Operators. Progress in Mathematics, vol. 112. Birkhäuser, Boston (1993)
20. Dong, C., Li, H., Mason, G.: Simple currents and extensions of vertex operator algebras. Commun. Math. Physics **180**(3), 671–707 (1996)
21. Feigin, B., Feigin, E., Jimbo, M., Miwa, T., Mukhin, E.: Principal $\widehat{sl_3}$ subspaces and quantum Toda Hamiltonian. Algebraic analysis and around, pp. 109–166, Advanced Studies in Pure Mathematics, 54. Mathematical Society of Japan, Tokyo, pp. 109–166 (2009)
22. Feigin, E.: The PBW filtration. Represent. Theory **13**, 165–181 (2009)
23. Stoyanovskiĭ, A.V.; Feĭgin, B. L.: Functional models of the representations of current algebras, and semi-infinite Schubert cells. Funktsional. Anal. i Prilozhen. **28**(1), 96, 68–90 (1994); Funct. Anal. Appl. **28**(1), 55–72 (1994); Feĭgin, B.L., Stoyanovskiĭ, A.V.: Quasi-particles models for the representations of Lie algebras and geometry of flag manifold. Preprint (1993) arXiv:hep-th/9308079. Cited 30 Dec 2018
24. Frenkel, I.B., Huang, Y.-Z., Lepowsky, J.: On axiomatic approaches to vertex operator algebras and modules. Mem. Amer. Math. Soc. **104** (1993)
25. Georgiev, G.: Combinatorial constructions of modules for infinite-dimensional Lie algebras, I. Principal subspace. J. Pure Appl. Algebra **112**(3), 247–286 (1996)
26. Humphreys, J.: Introduction to Lie Algebras and Representation Theory. Graduate Texts in Mathematics, vol. 9. Springer, New York (1972)
27. Jerković, M.: Character formulas for Feigin-Stoyanovsky's type subspaces of standard $\widetilde{sl(3, C)}$-modules. Ramanujan J. **27**(3), 357–376 (2012)
28. Jerković, M.: Recurrence relations for characters of affine Lie algebra $A_l^{(1)}$. J. Pure Appl. Algebra **213**(6), 913–926 (2009)
29. Jerković, M., Primc, M.: Quasi-particle fermionic formulas for $(k, 3)$-admissible configurations. Cent. Eur. J. Math. **10**(2), 703–721 (2012)
30. Kac, V.G.: Infinite-Dimensional Lie Algebras, 3rd edn. Cambridge University Press, Cambridge (1990)
31. Kawasetsu, K.: The free generalized vertex algebras and generalized principal subspaces. J. Algebra **444**, 20–51 (2015)
32. Kawasetsu, K.: The intermediate vertex subalgebras of the lattice vertex operator algebras. Lett. Math. Phys. **104**(2), 157–178 (2014)
33. Kožić, S.: Principal subspaces for double Yangian $DY(sl2)$. J. Lie Theory **28**(3), 673–694 (2018)
34. Kožić, S.: Vertex operators and principal subspaces of level one for $U_q(\widehat{\mathfrak{sl}_2})$. J. Algebra **455**, 251–290 (2016)
35. Kožić, S.: Principal subspaces for quantum affine algebra $U_q(A_n^{(1)})$. J. Pure Appl. Algebra **218**(11), 2119–2148 (2014)
36. Lepowsky, J., Li, H.: Introduction to Vertex Operator Algebras and Their Representations. Progress in Mathematics, vol. 227, xiv+318 pp. Birkhäuser Boston, Inc., Boston (2004)
37. Lepowsky, J., Primc, M.: Structure of the Standard Modules for the Affine Lie Algebra $A_1^{(1)}$. Contemporary Mathematics, vol. 46, ix+84 pp. American Mathematical Society, Providence (1985)

38. Lepowsky, J., Wilson, R.L.: The structure of standard modules, I: universal algebras and the Rogers-Ramanujan identities. Invent. Math. **77**, 199–290 (1984)
39. Li, H.: Local systems of vertex operators, vertex superalgebras and modules. J. Pure Appl. Algebra **109**, 143–195 (1996)
40. Meurman, A., Primc, M.: Annihilating fields of standard modules of $\tilde{sl}(2, \mathbb{C})$ and combinatorial identities. Memoirs Amer. Math. Soc. **137**(652) (1999)
41. Milas, A., Penn, M.: Lattice vertex algebras and combinatorial bases: general case and $\mathcal{W}$-algebras. New York J. Math. **18**, 621–650 (2012)
42. Penn, M., Sadowski, C.: Vertex-algebraic structure of principal subspaces of the basic modules for twisted affine Lie algebras of type $A_{2n-1}^{(2)}, D_n^{(2)}, E_6^{(2)}$. J. Algebra **496**, 242–291 (2018)
43. Penn, M., Sadowski, C.: Vertex-algebraic structure of principal subspaces of basic $D_4^{(3)}$-modules. Ramanujan J. **43**(3), 571–617 (2017)
44. Penn, M., Sadowski, C., Webb, G.: Principal subspaces of twisted modules for certain lattice vertex operator algebras. Preprint (2018)
45. Primc, M.: Combinatorial basis of modules for affine Lie algebra $B_2^{(1)}$. Cent. Eur. J. Math. **11**, 197–225 (2013)
46. Sadowski, C.: Presentations of the principal subspaces of the higher-level standard $\widehat{\mathfrak{sl}(3)}$-modules. J. Pure Appl. Algebra **219**(6), 2300–2345 (2015)
47. Sadowski, C.: Principal subspaces of higher-level standard $\widehat{sl(n)}$-modules. Int. J. Math. **26**(08), 1550053 (2015) 35 pp
48. Trupčević, G.: Characters of Feigin-Stoyanovsky's type subspaces of level one modules for affine Lie algebras of types $A_l^{(1)}$ and $D_4^{(1)}$. Glas. Mat. Ser. III **46**(1), 49–70 (2011)
49. Trupčević, G.: Combinatorial bases of Feigin-Stoyanovsky's type subspaces of level 1 standard mod-ules for $\tilde{sl}(l+1, C)$. Commun. Algebra **38**(10), 3913–3940 (2010)
50. Trupčević, G.: Combinatorial bases of Feigin-Stoyanovsky's type subspaces of higher-level standard $\tilde{sl}(l+1, C)$-modules. J. Algebra **322**(10), 3744–3774 (2009)

The Poisson Lie Algebra, Rumin's Complex and Base Change

Alessandro D'Andrea

Abstract Results from the forthcoming papers [4] and [8] are announced. We introduce a *singular current construction*, or *base change*, for pseudoalgebras which may be used to obtain a primitive Lie pseudoalgebra of type H from a suitable one of type K. When applied to representations, it derives the pseudo de Rham complex of type H from that of type K—which is related to Rumin's construction from [15]—both with standard coefficients and with nontrivial Galois coefficients. In the latter case, the construction yields exact complexes of modules for the Poisson linearly compact Lie algebra P_{2N} exhibiting a nontrivial central action.

Keywords Representation theory · Lie algebras and pseudoalgebras · Conformally symplectic geometry · Hopf–Galois extensions

1 Introduction

The notion of (Lie) pseudoalgebra over a (cocommutative) Hopf algebra was introduced in [1] as a generalization of Lie conformal algebras, which have proved useful in dealing with locality of formal distributions and the description of both vertex algebras [6, 7, 12, 14] and Poisson vertex algebras [9]. However, one of their most natural applications is the study of discrete representations over linearly compact Lie algebras, as the annihilation algebra functor may be used to associate with each (commutative, associative, Lie) pseudoalgebra the corresponding linearly compact algebra and representations of the latter can often by lifted to the pseudoalgebraic language [2, 3].

A special role among primitive (i.e., those that cannot be obtained by means of a nontrivial current construction as in [1, Sect. 4.2]) Lie pseudoalgebras is played by those of type H, that correspond to the Hamiltonian family in Cartan's description of

A. D'Andrea (✉)
Dipartimento di Matematica, Università degli Studi di Roma "La Sapienza", P.le Aldo Moro, 5, 00185 Rome, Italy
e-mail: dandrea@mat.uniroma1.it

D. Adamović and P. Papi (eds.), *Affine, Vertex and W-algebras*, Springer INdAM Series 37, https://doi.org/10.1007/978-3-030-32906-8_5

simple infinite-dimensional linearly compact Lie algebras [5]. Indeed, Lie pseudoalgebras $H(\mathfrak{d}, \chi, \omega)$ are the only finite primitive ones over $\mathcal{U}(\mathfrak{d})$, whose annihilation algebra has a non trivial center; this, however, acts trivially on every irreducible pseudoalgebra representation, unless $\chi = 0$ and ω is exact.

Here, we announce results from the forthcoming paper [8] which generalize and clarify [11]: we show how to use the concept of representations with coefficients [11] so as to construct *projective representations* of $L = H(\mathfrak{d}, \chi, \omega)$ that correspond to irreducible discrete representations of the annihilation algebra with a nontrivial central action. We explain how to set up the machinery and construct such modules starting from irreducible representations of a suitably chosen Lie pseudoalgebra $K(\mathfrak{d}, \theta)$ by means of a singular current construction, or *base change*.

Existence of a class of tensor modules possessing nonconstant singular vectors, which was already observed in [2–4], generalizes to the present setting, as can be shown by explicit computation of Tor-spaces. As a byproduct of our techniques, we obtain an alternate more conceptual proof of the non-exactness of the pseudocomplex of de Rham type of the Lie pseudoalgebras $H(\mathfrak{d}, \chi, \omega)$.

The present constructions also sheds light on the similarity of type H and type K structures, especially when reformulated with a pseudoalgebraic language.

2 Primitive Lie Pseudoalgebras of Type H and K

In this paper, definitions and notation concerning pseudoalgebras and Hopf algebras follow those found in [1].

Henceforth, $\mathfrak{d} \neq 0$ will be a finite dimensional Lie algebra and $H = \mathcal{U}(\mathfrak{d})$ its universal enveloping algebra. A simple Lie H-pseudoalgebra L is said to be *primitive* if its annihilation Lie algebra $\mathcal{L} := H^* \otimes_H L$ is (a central extension of) one of the simple linearly compact Lie algebras from Cartan's classification [5].

Here, we are interested in primitive Lie pseudoalgebras of type H and K, i.e., those such that $\mathcal{L}$ is isomorphic to either the Poisson Lie algebra P_n, which centrally extends H_n, or the contact Lie algebra K_n. It is showed in [1] that n must then equal $\dim \mathfrak{d}$, so that primitive Lie pseudoalgebras of type H (resp. K) only exist when $\mathfrak{d}$ is even (resp. odd) dimensional. There are further constraints that prevent, for instance, $\mathfrak{d}$ from being abelian in the K-type case.

Proposition 1 ([1, Sect. 8.5]) *Let L be a Lie H-pseudoalgebra of type H (resp. K). Then $L = He$ is a free H-module of rank* 1 *and*

$$[e * e] = (r + s \otimes 1 - 1 \otimes s) \otimes_H e,$$

where $0 \neq r \in \bigwedge^2 \mathfrak{d}$, $s \in \mathfrak{d}$ *satisfy*

$$[r, \Delta(s)] = 0, \qquad ([r_{12}, r_{13}] + r_{12}s_3) + \textit{ cyclic permutations} = 0,$$

and

- $\operatorname{supp} r = \mathfrak{d}$ *if L is of type H;*

- $s \notin \operatorname{supp} r$ *and* $\operatorname{supp} r + \mathbf{k}s = \mathfrak{d}$ *in type K,*

where $\operatorname{supp} r \subset \mathfrak{d}$ *denotes the subspace supporting* r.

We shall then call $(r, s, \mathfrak{d})$ a *datum of type H (resp. K).*

Example 1 If $(r, s, \mathfrak{d})$ is a datum of type H, then r is non-degenerate and may be used to identify $\mathfrak{d}$ with its dual. Thus r and s translate to a 2-form $\omega \in \bigwedge^2 \mathfrak{d}^*$ and a trace form $\chi \in \mathfrak{d}^*$ respectively, satisfying

$$d\chi = 0, \qquad d\omega + \chi \wedge \omega = 0.$$

This means that ω is a 2-cocycle of $\mathfrak{d}$ with values in the one-dimensional $\mathfrak{d}$-module $\mathbf{k}_\chi$ defined by χ, yielding an abelian extension

$$0 \to \mathbf{k}_\chi \to \mathfrak{d}' \to \mathfrak{d} \to 0.$$

More explicitly, if $\mathfrak{d}' = \mathfrak{d} \oplus \mathbf{k}c$ as vector spaces, then

$$[g, h]' = [g, h] + \omega(g, h)c, \qquad [g, c]' = \chi(g)c$$

extend to a Lie bracket on $\mathfrak{d}'$. If $\chi = 0$, then ω is a 2-cocycle of $\mathfrak{d}$ and $\mathfrak{d}'$ is the corresponding central extension.

By [1, Remark 8.6], if we identify $\mathfrak{d}$ as a subspace of $\mathfrak{d}'$, then $(r, s + c, \mathfrak{d}')$ is a datum of type K. Notice that not all data of type K are obtained by this construction. □

3 Currents and Base Change

3.1 Rings of Coefficients

Let H be a Hopf algebra over $\mathbf{k}$ with comultiplication Δ. A *left* H*-comodule* is a vector space C over $\mathbf{k}$ endowed with a $\mathbf{k}$-linear map

$$\begin{aligned} \Lambda : C &\longrightarrow H \otimes C \\ c &\longmapsto c_{(1)} \otimes c_{(2)} \end{aligned}$$

such that

$$(\Delta \otimes \mathrm{id}_C) \circ \Lambda = (\mathrm{id}_H \otimes \Lambda) \circ \Lambda, \tag{1}$$

$$(\epsilon \otimes \mathrm{id}_C) \circ \Lambda = \mathrm{id}_C. \tag{2}$$

Λ is called the *comodule structure map* of C, or simply *coaction*. Repeated application of Λ defines the *n-fold coaction maps* $\Lambda^n : C \to H^{n-1} \otimes C, n > 1$. We employ the notation

$$\Lambda^2(c) = (\Delta \otimes \mathrm{id}_C) \circ \Lambda(c) = (\mathrm{id}_H \otimes \Lambda) \circ \Lambda(c) = c_{(1)} \otimes c_{(2)} \otimes c_{(3)}, \qquad (3)$$

and similarly $\Lambda^{n-1}(c) = c_{(1)} \otimes \ldots \otimes c_{(n)}$. Notice that this can be misleading, as only the last tensor factor $c_{(n)}$ lies in C, whereas all others are elements of H. The (left) counit axiom (2) rewrites as

$$(\epsilon \otimes \mathrm{id}_C) \circ \Lambda(c) = \epsilon(c_{(1)})c_{(2)} = c. \qquad (4)$$

Once again, the map ϵ can only be applied on elements of H, as no counit is defined on C, hence no right counit axiom may be required to hold.

An H-*comodule algebra* is a unital **k**-algebra D endowed with an algebra homomorphism

$$\begin{aligned} \Lambda : D &\longrightarrow H \otimes D \\ d &\longmapsto d_{(1)} \otimes d_{(2)}, \end{aligned}$$

making D into an H-comodule. From now on we will call an H-comodule algebra D a *ring of coefficients*, or simply *roc*, over H. When D is a roc over H, we will also say that (H, D) is a *roc*.

Lemma 1 *Let $\phi : H_1 \longrightarrow H_2$ be a Hopf algebra homomorphism and D be a roc over H_1 with comodule map $\Lambda_1 : D \to H_1 \otimes D$. Then $\Lambda_2 = (\phi \otimes 1)\Lambda_1$ makes (H_2, D) into a roc.*

Example 2 Every Hopf algebra H is a roc over itself, with comodule structure map given by Δ. □

Example 3 Let H' be a Hopf subalgebra of a Hopf algebra H. Since H' is a roc over itself and the inclusion from H' to H is a Hopf homomorphism from H' to H, then H' is a roc over H. In particular $H' = \mathbf{k} \subset H$ is a roc over H. □

Definition 1 Let (D_1, Λ_1) and (D_2, Λ_2) be rings of coefficients over H_1 and H_2 respectively. Let $\phi : H_1 \to H_2$ be a Hopf algebra homomorphism and $\psi : D_1 \to D_2$ be an algebra homomorphism. The pair $(\phi, \psi) : (H_1, D_1) \longrightarrow (H_2, D_2)$ is a *roc homomorphism* if

$$\Lambda_2(\psi(d)) = (\phi \otimes \psi)\Lambda_1(d), \qquad (5)$$

for every $d \in D_1$.

Example 4 If Λ_1, Λ_2 are as in Lemma 1, then $(\phi, \mathrm{id}_D) : (H_1, D) \mapsto (H_2, D)$ is a roc homomorphism. □

Notice that if $(\phi, \psi) : (H, D) \to (H', D')$ is a roc homomorphism, then $I = \ker \phi$ is a Hopf ideal of H_1, and $J = \ker \psi$ is an algebra ideal of D satisfying $\Lambda(J) \subset I \otimes D + H \otimes J$. Every pair $(I, J) \subset (H, D)$ satisfying the above requirement is an *ideal* of the roc (H, D). Clearly, whenever (I, J) is an ideal of (H, D), there exists a unique roc structure on $(H/I, D/J)$ making the natural projection (π_I, π_J) into a roc homomorphism.

3.2 Lie Pseudoalgebra Representations with Coefficients

Let (H, D) be a roc, L a Lie H-pseudoalgebra, M a left D-module. A (left) *pseudoaction of* L on M with coefficients in D is an $(H \otimes D)$-linear map

$$\begin{array}{rcl} * : L \otimes M & \longrightarrow & (H \otimes D)\otimes_D M \\ a \otimes m & \mapsto & a * m. \end{array}$$

This pseudoaction defines a Lie pseudoalgebra representation (with coefficients) if

$$[a * b] * m = a * (b * m) - (b * (a * m))^{\sigma_{12}}, \tag{6}$$

for any $a, b \in L, m \in M$, where we have extended $*$ to $(H^{\otimes(i+j-1)} \otimes D)$-linear maps

$$* : (H^{\otimes i} \otimes_H L) \otimes ((H^{\otimes(j-1)} \otimes D)\otimes_D M) \longrightarrow (H^{\otimes(i+j-1)} \otimes D)\otimes_D M,$$

by

$$(F\otimes_H a) * (G\otimes_D m) = (F \otimes G)(\Delta^{i-1} \otimes \Lambda^{j-1})(a * m), \tag{7}$$

where $F \in H^{\otimes i}, G \in H^{\otimes(j-1)} \otimes D, a \in L, m \in M, i, j \geq 1$.

Notice that when $D = H$ the above notion of representation coincides with usual Lie pseudoalgebra representations, as considered in [1]. A representation M is *finite* if it is finitely generated as a D-module. It is *irreducible* if the only D-submodules $N \subset M$ satisfying $L * N \subset (H \otimes D) \otimes_D N$ are the trivial ones.

3.3 Base Change on Pseudoalgebras

Let H, H' be Hopf algebras, L a Lie H-pseudoalgebra. Every Hopf algebra homomorphism $\phi : H \longrightarrow H'$ endows H' with a right H-module structure so that we may consider the tensor product $L' = \phi_* L := H' \otimes_H L$, which is a left H'-module. It is not difficult to show that the Lie H-pseudoalgebra structure on L induces a corresponding Lie H'-pseudoalgebra structure on L', whose pseudobracket satisfies

$$[(h'\otimes_H a) * (k'\otimes_H b)] = \sum_i (h'\phi(h^i) \otimes k'\phi(k^i)) \otimes_{H'} (1\otimes_H e_i), \tag{8}$$

if $[a * b] = \sum_i (h^i \otimes k^i)\otimes_H e_i$, $a, b \in L$.

If $\phi : H \to H'$, and $L' = \phi_* L$, we will say that L' is obtained from L by *extension of scalars* or *base change*. Let us see a few examples.

3.3.1 Current Construction

Let $H \subset H'$ be cocommutative Hopf algebras, $\iota : H \to H'$ the inclusion homomorphism. Then ι_* coincides with the current construction $Cur_H^{H'}$ described in [1, Sect. 4.2]. Clearly, if $H = H'$, then $\iota = \mathrm{id}_H$ and ι_* is canonically isomorphic to the identity functor.

3.3.2 Algebra of 0-Modes

Let A be a Lie conformal algebra, viewed as a Lie pseudoalgebra over $H = \mathbf{k}[\partial]$. Then the counit $\varepsilon : H \to \mathbf{k}$ is a Hopf algebra homomorphism, and $\varepsilon_* A$ coincides with the algebra of Fourier 0-modes of A, see [10, 14].

3.3.3 $K(\mathfrak{d}, \theta)$ and $H(\mathfrak{d}, \chi, \omega)$

Let $L = He$ be a primitive Lie pseudoalgebra of type H corresponding to the datum $(r, s, \mathfrak{d})$ and χ be the corresponding 1-form. If we set $\overline{\partial} := \partial - \chi(\partial)$, then the Lie pseudobracket on L may be rewritten as

$$[e * e] = \sum_i (\overline{\partial_i} \otimes \overline{\partial^i}) \otimes_H e.$$

Consider the datum $(r, s, \mathfrak{d}' = \mathfrak{d} \oplus \mathbf{k}c)$ of type K constructed in Example 1. The Lie pseudobracket of the corresponding Lie pseudoalgebra of type K then rewrites as

$$[e' * e'] = \left(\sum_i \bar{\partial}_i \otimes \bar{\partial}^i + c \otimes 1 - 1 \otimes c \right) \otimes_{H'} e'.$$

The canonical projection $\pi : \mathfrak{d}' \twoheadrightarrow \mathfrak{d}'/\mathbf{k} \simeq \mathfrak{d}$ extends to a Hopf algebra homomorphism $\pi : H' \to H$ mapping c to 0. It is then easy to see that $\pi_* L' = L$. Notice that π is not injective, so that π_* cannot, and should not, be understood in terms of the abovementioned standard current construction.

3.4 Base Change on Representations

It is possible to change scalars on both a Lie pseudoalgebra and its representation (with coefficients), once we make sure to employ a roc homomorphism.

Proposition 2 *Let (H, D), (H', D') be rocs, $\Phi = (\phi, \psi) : (H, D) \to (H', D')$ a roc homomorphism, L a Lie pseudoalgebra over H, M a representation of L with coefficients in D. Then there exists a natural pseudoalgebra action of $L' = \phi_* L$ on*

$M' = \psi_* M := D' \otimes_D M$ with coefficients in D' satisfying

$$(h' \otimes_H a) * (d' \otimes_D m) = \sum_i (h'\phi(h^i) \otimes d'\phi(d^i)) \otimes_{D'} (1 \otimes_D e_i), \tag{9}$$

if $a * m = \sum_i (h^i \otimes d^i) \otimes_H e_i$, $a \in L$, $m \in M$.

4 Galois Objects and Projective Representations

The possibility of straightening a pseudoalgebra action on the right makes it possible to generalize many results from [1]. This naturally occurs when the representation takes its coefficients in a *Galois roc*. Representations of a Lie H-pseudoalgebra with coefficients in a Galois roc are called *projective*, in analogy with [13].

4.1 Straightening on the Right

Definition 2 A roc (H, D) is *Galois* if the Galois map

$$\begin{array}{rcl} \beta : D \otimes D & \longrightarrow & H \otimes D \\ d \otimes d' & \mapsto & (1 \otimes d)\Delta(d') \end{array} \tag{10}$$

is a linear isomorphism.

The map β factors via $D \otimes_{D^{\text{co-}H}} D$, where $D^{\text{co-}H} = \{d \in D | \Delta(d) = 1 \otimes d\}$. Therefore, in order for (H, D) to be Galois, one needs $D^{\text{co-}H} = \mathbf{k}$. This implies that $\mathbf{k} \subset D$ is a Hopf–Galois extension, hence D is a Galois object, thus justifying the terminology. Properties of the inverse map β^{-1} are well understood. We use a Sweedler-like notation for β by setting $\beta^{-1}(h \otimes 1) = h^{[1]} \otimes h^{[2]}$.

Proposition 3 *Let $g, h \in H$, $d \in D$. Then:*

$$h^{[2]}{}_{(1)} \otimes h^{[1]} h^{[2]}{}_{(2)} = h \otimes 1 \tag{11}$$

$$h^{[1]} h^{[2]} = \epsilon(h) 1_D \tag{12}$$

$$(gh)^{[1]} \otimes (gh)^{[2]} = h^{[1]} g^{[1]} \otimes g^{[2]} h^{[2]} \tag{13}$$

$$d_{(2)} (d_{(1)})^{[1]} \otimes (d_{(1)})^{[2]} = 1 \otimes d \tag{14}$$

Proof Compute β on $\beta^{-1}(h \otimes 1) = h^{[1]} \otimes h^{[2]}$ in order to obtain (11). Then, applying $\epsilon \otimes \text{id}_D$ gives (12). Equations (13) and (14) are proved by applying β, which is invertible, on both sides. □

Lemma 2 *Let (H, D) be a Galois roc, let M be a right D-module and N be a left D-module. Then the map*

$$\begin{array}{rcl} \tau^R : (H \otimes M) \otimes_D N & \longrightarrow & M \otimes N \\ (h \otimes m) \otimes_D n & \longmapsto & mh^{[1]} \otimes h^{[2]} n, \end{array} \tag{15}$$

is a well defined linear isomorphism.

Proof Using (11), the linear map extending $m \otimes n \mapsto (1 \otimes m) \otimes_D n$ is easily checked to be an explicit inverse to τ^R. Thus, we only need to worry about well definedness of τ^R. However,

$$(h \otimes m) \otimes_D dn = (h \otimes m)\Delta(d) \otimes_D n = (hd_{(1)} \otimes md_{(2)}) \otimes_D n$$

gets mapped to

$$md_{(2)}(hd_{(1)})^{[1]} \otimes (hd_{(1)})^{[2]} n = md_{(2)}(d_{(1)})^{[1]} h^{[1]} \otimes h^{[2]} (d_{(1)})^{[2]} n,$$

due to (13), and this equals $mh^{[1]} \otimes h^{[2]} dn$ thanks to (14). □

Corollary 1 *Let (H, D) be a Galois roc, M be a right D-module, N a left D-module. Then every element $\alpha \in (H \otimes M) \otimes_D N$ can be expressed as a finite sum*

$$\alpha = \sum_i (1 \otimes m^i) \otimes_D n_i, \tag{16}$$

where both $m^i \in M$ and $n_i \in N$ are linearly independent over $\mathbf{k}$, and $\sum_i m^i \otimes n_i \in M \otimes N$ is uniquely determined by α.

We will refer to (16) as a *right-straightened* expression in $(H \otimes M) \otimes_D N$.

4.2 Galois Rocs and Twists: The Weyl Roc

Let H be a cocommutative Hopf algebra, $\sigma : H \otimes H \to \mathbf{k}$ a Hopf 2-cocycle. Then [16] the twisted product $H_\sigma = \mathbf{k} \#_\sigma H$ is a comodule algebra over H, and a Hopf–Galois extension of $\mathbf{k}$. All Galois rocs (H, D), where D satisfies the normal basis condition, are obtained in this way; for instance, when H is pointed, e.g., when it is cocommutative, all Galois objects satisfy the normal basis condition.

We are going to give an alternate construction of this fact in a special case that is of interest to us: let $(r, s, \mathfrak{d})$ be a datum of type H and $(r, s + c, \mathfrak{d}')$ the corresponding datum of type K as from Example 1. Set $H = \mathcal{U}(\mathfrak{d})$, $H' = \mathcal{U}(\mathfrak{d}')$. Then $\mathbf{k}c$ is an abelian ideal of $\mathfrak{d}'$, which is central when $s = 0$—which corresponds to $\chi = 0$. Choose $\lambda \in \mathbf{k}$ and set $I_\lambda = H' \cdot (c - \lambda)$; notice that c, hence $c - \lambda$, is central if and only if $\chi = 0$.

Lemma 3

- *I_0 is a Hopf ideal of H;*
- *If $\chi = 0$, then (I_0, I_λ) is a roc ideal of (H, H) for all $\lambda \in \mathbf{k}$.*

Proof By construction, $[\mathfrak{d}', c] \subset \mathbf{k}c$, hence $cH' \subset H'c$, thus showing that $I_0 = H'c = cH'$ is a two-sided ideal of H'. However, when $\lambda\chi = 0$, $c - \lambda$ is central in H', and all I_λ are also two-sided ideals. As

$$\Delta(c - \lambda) = c \otimes 1 + 1 \otimes (c - \lambda), \qquad S(c) = -c,$$

then

$$\Delta(I_\lambda) \subset I_0 \otimes H' + H' \otimes I_\lambda, \qquad S(I_0) \subset I_0,$$

whence both claims follow immediately. □

For every choice of $\lambda \in \mathbf{k}$, let $\psi^\lambda : H' \longrightarrow H'/I_\lambda := D_\lambda$ be the natural projection, and denote by

$$\Lambda^\lambda : D_\lambda = H'/I_\lambda \to (H' \otimes H')/(I_0 \otimes H' + H' \otimes I_\lambda) = D_0 \otimes D_\lambda$$

the map induced by Δ. Notice that D_0 identifies with H as a Hopf algebra.

Proposition 4

- *The Hopf algebra homomorphism $\phi : H' \to H$ induced by the surjection $\mathfrak{d}' \to \mathfrak{d}$ coincides with $\psi^0 : H' \to D_0 \simeq H$.*
- *The maps Λ^λ make each D_λ into a roc over H, and all pairs (ϕ, ψ^λ), $\lambda \in \mathbf{k}$ are roc homomorphism.*

Remark 1 Since $\Delta(I_\lambda) \subset I_0 \otimes H' + H' \otimes I_\lambda$, the map $\Lambda^\lambda : H'/I_\lambda \to H'/I_0 \otimes H'/I_\lambda$ induced by Δ always defines a comodule structure. However, H'/I_λ carries a compatible algebra structure, hence is a roc, only when I_λ is two-sided. When $\lambda\chi \neq 0$, the 2-sided ideal of H' generated by I_λ coincides with the whole H'.

It is not difficult to show that the rocs (H, D_λ) are all Galois. We have already mentioned that D_0 is isomorphic to H. The algebras D_λ, $\lambda \neq 0$ are all isomorphic as algebras, and also as H-comodules, but not as rocs over H. If $\mathfrak{d}$ is abelian of dimension $n = 2N$, $\chi = 0$ and ω is symplectic, then D_λ, $\lambda \neq 0$ is isomorphic to the Weyl algebra A_N. In all cases, D_λ is a noetherian domain.

5 An Exact Sequence of Projective Tensor Modules of $H(\mathfrak{d}, 0, \omega)$

Finite irreducible representations (with standard coefficients) of primitive Lie pseudoalgebras of type W, S, K have been considered in [2, 3]. In all cases, one is able to locate a class of representations called *tensor modules* and to prove that each finite irreducible representation arises as a quotient of a suitably chosen tensor module. Tensor modules are generically irreducible, but there are exceptions, that can be put together in exact complexes of representations; then the image (or equivalently, the kernel) of each morphism in the complex provides a maximal and irreducible submodule of the relevant tensor module. Tensor modules are always free as H-modules.

These exceptional complexes possess a geometrical interpretation: in types W and S they arise as twists [2, Sect. 4.2] of the *pseudification* of the de Rham complex, whereas in type K they are related to the complex [15] introduced by Rumin in the context of contact manifolds.

Let us review the type K case more closely. Say $(r, s, \mathfrak{d}')$ is a datum of type K corresponding to the Lie pseudoalgebra $K(\mathfrak{d}', \theta')$; here $\mathfrak{d}' = \mathfrak{d} + \mathbf{k}s$ and $\mathfrak{d}$ is the support of r. Tensor modules for $K(\mathfrak{d}', \theta')$ are parametrized by finite dimensional representations of the Lie algebra $\mathfrak{d}' \oplus \mathfrak{csp}\,\mathfrak{d}$ as in [3, Sect. 5.2]. For the sake of simplicity, we will ignore here the action of $\mathfrak{d}'$, which can be recovered by applying a twist; then singular $K(\mathfrak{d}', \theta')$-tensor modules occur when the $\mathfrak{csp}\,\mathfrak{d}$-action restricts to one of the fundamental representations of $\mathfrak{sp}\,\mathfrak{d} \subset \mathfrak{csp}\,\mathfrak{d}$, for a suitable choice of the scalar action.

The corresponding exact complex of singular tensor modules [3, Theorem 6.1] is then

$$0 \to \Omega^0(\mathfrak{d}')/I^0(\mathfrak{d}') \xrightarrow{\mathrm{d}} \cdots \xrightarrow{\mathrm{d}} \Omega^N(\mathfrak{d}')/I^N(\mathfrak{d}') \xrightarrow{\mathrm{d}^R} J^{N+1}(\mathfrak{d}') \xrightarrow{\mathrm{d}} \cdots \xrightarrow{\mathrm{d}} J^{2N+1}(\mathfrak{d}'), \tag{17}$$

where $\Omega^i(\mathfrak{d}')$ are modules of *pseudoforms* and $I^i(\mathfrak{d}')$, $J^i(\mathfrak{d}')$ are suitable submodules related to Rumin's construction. The last d morphism is not surjective, as the complex provides a resolution of a one-dimensional H'-module that we denote by $\mathbf{k}$.

Let now $(r, 0, \mathfrak{d})$ be a datum of type H, and consider the primitive Lie pseudoalgebra $L' = K(\mathfrak{d}', \theta') = H'e'$ of type K associated with the datum of type K constructed in Example 1. We have seen in Sect. 3.3.3 that $L = H(\mathfrak{d}, 0, \omega) = He$ coincides with the base change $\phi_* L'$, where $\phi : H' \to H$ is the Hopf algebra homomorphism induced by mapping the central element $c \in \mathfrak{d}' \subset H'$ to 0.

As c is central in H', it may be specialized to any scalar value; let $\psi^\lambda : H' \to H'/H'(c - \lambda) \simeq D_\lambda$ denote the natural projection, so that $\phi = \psi_0$. We have already seen in Proposition 4 that each pair $(\phi, \psi_\lambda) : (H', H') \to (H, D_\lambda)$ is a roc homomorphism, which we may use to extend scalars on (17) obtaining the following sequence of $H(\mathfrak{d}, 0, \omega) = \phi_* K(\mathfrak{d}', \theta')$-modules

$$\begin{aligned} 0 \to \psi^\lambda_* (\Omega^0(\mathfrak{d}')/I^0(\mathfrak{d}')) \xrightarrow{\psi^\lambda_* \mathrm{d}} \cdots \xrightarrow{\psi^\lambda_* \mathrm{d}} \psi^\lambda_* (\Omega^N(\mathfrak{d}')/I^N(\mathfrak{d}')) \\ \xrightarrow{\psi^\lambda_* \mathrm{d}^R} \psi^\lambda_* J^{N+1}(\mathfrak{d}') \xrightarrow{\psi^\lambda_* \mathrm{d}} \cdots \xrightarrow{\psi^\lambda_* \mathrm{d}} \psi^\lambda_* J^{2N+1}(\mathfrak{d}'), \end{aligned} \tag{18}$$

which is certainly a complex, by functoriality of ψ^λ_*.

Recall that (17) is a projective resolution of the left H'-module $\mathbf{k}$, so that the homology of (18) computes $\mathrm{Tor}^{H'}_{\bullet}(D_\lambda, \mathbf{k})$, which can also be computed by choosing a projective resolution of the right H'-module D_λ and tensoring it by $\mathbf{k}$. This is easily done, by noticing that $c - \lambda$ is a nonzero divisor in H' so that

$$0 \to H' \xrightarrow{\cdot(c-\lambda)} H' \xrightarrow{\psi^\lambda} D_\lambda \to 0$$

is exact. When $\lambda \neq 0$, applying $\otimes_{H'}\mathbf{k}$ to the above resolution yields

$$0 \to \mathbf{k} \xrightarrow{\cdot(-\lambda)} \mathbf{k} \to 0,$$

which is manifestly exact. This shows that $\mathrm{Tor}^{H'}_i(D_\lambda, \mathbf{k}) = 0$ for all $i > 0$, so that (18) is an exact complex of projective $H(\mathfrak{d}, 0, \omega)$-modules when $\lambda \neq 0$; furthermore, the last connecting homomorphism is surjective as $D_\lambda \otimes_{H'} \mathbf{k} = 0$. By using right-straightening as from Sect. 4.1, one may show that projective $H(\mathfrak{d}, 0, \omega)$-modules in (18) and their twists are the only singular ones, so that we have a complete analogy with results from [2, 3].

The exact sequence (18) may be employed to exhibit submodules of reducible projective tensor modules. A classification of their maximal submodules, hence of their irreducible quotients, will be made explicit in [8].

One may proceed similarly when $L = H(\mathfrak{d}, \chi, \omega)$ and $\lambda = 0$, which gives $D_0 \simeq H$ and yields modules with standard coefficients. In this case, one obtains exactness of the sequence (18) everywhere but at the second to last module. This sequence coincides with the pseudo de Rham complex from [4] and one may control the one-dimensional non-exactness of the complex of de Rham type for primitive Lie pseudoalgebras of type H in terms of $\mathrm{Tor}^{H'}_*(H, \mathbf{k})$, thus providing a more conceptual homological explanation.

References

1. Bakalov, B., D'Andrea, A., Kac, V.G.: Theory of finite pseudoalgebras. Adv. Math. **162**, 1–140 (2001)
2. Bakalov, B., D'Andrea, A., Kac, V.G.: Irreducible modules over finite simple Lie pseudoalgebras I. Primitive pseudoalgebras of type W and S. Adv. Math. **204**, 278–346 (2006)
3. Bakalov, B., D'Andrea, A., Kac, V.G.: Irreducible modules over finite simple Lie pseudoalgebras II. Primitive pseudoalgebras of type K. Adv. Math. **232**, 188–237 (2013)
4. Bakalov, B., D'Andrea, A., Kac, V.G.: Irreducible modules over finite simple Lie pseudoalgebras III. Primitive pseudoalgebras of type H, work in progress
5. Cartan, E.: Les groupes de transformation continus, infinis, simples. Ann. Sci. ÉNS **26**, 93–161 (1909)
6. D'Andrea, A.: Finite vertex algebras and nilpotence. J. Pure Appl. Algebr. **212**(4), 669–688 (2008)
7. D'Andrea, A.: A remark on simplicity of vertex and Lie conformal algebras. J. Algebr. **319**(5), 2106–2112 (2008)
8. D'Andrea, A.: Projective modules over pseudoalgebras and base change, work in progress

9. De Sole, A., Kac, V.G.: Finite vs. infinite W-algebras. Jap. J. of Math. **1**(1), 137–261 (2006)
10. D'Andrea, A., Kac, V.G.: Structure theory of finite conformal algebras. Sel. Math. **4**, 377–418 (1998)
11. D'Andrea, A., Marchei, G.: Representations of Lie pseudoalgebras with coefficients. J. Algebr. (2010)
12. D'Andrea, A., Marchei, G.: A root space decomposition for finite vertex algebras. Doc. Math. **17**, 783–805 (2012)
13. De Commer, K.: On projective representations for compact quantum groups. J. Funct. Anal. **260**, 3596–3644 (1998)
14. Kac, V.G.: Vertex Algebras for Beginners. Mathematical Lecture Series, vol. 10, 2nd Edn, 1996. AMS (1998)
15. Rumin, M.: Formes différentielles sur les variétés de contact. J. Diff. Geom. **39**, 281–330 (1994)
16. Sweedler, M.E.: Cohomology of algebras over Hopf algebras. Trans. AMS **133**, 205–239 (1968)

Classical and Quantum $\mathcal{W}$-Algebras and Applications to Hamiltonian Equations

Alberto De Sole

Abstract We start by giving an overview of the four fundamental physical theories, namely classical mechanics, quantum mechanics, classical field theory and quantum field theory, and the corresponding algebraic structures, namely Poisson algebras, associative algebras, Poisson vertex algebras and vertex algebras. We then focus on classical and quantum $\mathcal{W}$-algebras, with a particular emphasis on their application to integrable Hamiltonian PDE.

Keywords Vertex algebras · Classical $\mathcal{W}$-algebras · Hamiltonian reduction · Integrable Hamiltonian systems

1 Introduction

The present paper is based on a minicourse delivered at the winter school "Geometry, Algebra and Combinatorics of Moduli Spaces and Configurations III", held in Dobbiaco on February 18–22, 2019.

In Sect. 2 we review some ideas from classical and quantum Hamiltonian mechanics. We review the main principles of classical Hamiltonian mechanics over an arbitrary Poisson manifold, and describe how its quantization leads to quantum mechanics. We describe the simplest examples of such theories, such as the point particle or the rigid body in absence of force fields. In fact, we decide to adopt a purely algebraic point of view, and extract from a classical mechanical theory its algebraic structure of a Poisson algebra, and from a quantum mechanical theory its algebraic structure of an associative algebra. From such point of view, the classical limit corresponds to taking the associated graded of a filtered associative algebra. The main focus of the first section is on the notion of classical and quantum Hamiltonian reduction of a Poisson manifold. This indeed is then used in the last section for the construction of $\mathcal{W}$-algebras.

A. De Sole (✉)
Department of Mathematics, University of Rome La Sapienza, Roma, Italy
e-mail: desole@mat.uniroma1.it

D. Adamović and P. Papi (eds.), *Affine, Vertex and W-algebras*,
Springer INdAM Series 37, https://doi.org/10.1007/978-3-030-32906-8_6

In Sect. 3 we pass to infinitely many degrees of freedom, and we try to address the natural question as to what are the algebraic structures of a quantum and classical field theory. For chiral fields in a conformal field theory, an answer is provided by vertex algebras and Poisson vertex algebras. The first were first defined by Borcherd [7], and can be described in terms of a collection of translation covariant pairwise local fields on a vector space. Poisson vertex algebras can be obtained from vertex algebras via a classical limit procedure. As such, they can be used to interpret as Hamiltonian equations the most famous integrable PDE's in fluidodynamics, such as KdV or NLS, [6].

Finally, in Sect. 4 we focus our attention on $\mathcal{W}$-algebras. We review the definition of the Poisson manifold of the Slodovy slice, associated to a nilpotent element f in a reductive Lie algebra $\mathfrak{g}$. We describe it as a Hamiltonian reduction of the manifold $\mathfrak{g}^*$ with the Kirillov-Kostant Poisson bracket, and use this description to describe its quantization, the quantum finite $\mathcal{W}$-algebra $\mathcal{W}^{\text{fin}}(\mathfrak{g}, f)$, and its affine analogue, the classical affine $\mathcal{W}$-algebra $\mathcal{W}^{\text{cl}}(\mathfrak{g}, f)$.

In their 1985 seminal paper [32], Drinfeld and Sokolov constructed an integrable Hamiltonian hierarchy of PDE for any simple Lie algebra $\mathfrak{g}$ and its principal nilpotent element f. In this construction they used Kostant's cyclic elements. Since then there have been various attempts to generalize the DS construction to arbitrary nilpotents see e.g. [8, 11, 34, 36, 37]. In a series of joint papers [18, 19] we present the theory of classical $\mathcal{W}$-algebras in a formal Poisson vertex algebra setting, and we then extended the DS methods to any simple Lie algebra $\mathfrak{g}$ and its nilpotent elements f of "semisimple type". About $\frac{1}{2}$ of nilpotents in exceptional $\mathfrak{g}$ are such: 13 out of 20 in E_6, 21 out of 44 in E_7, 27 out of 69 in E_8, 11 out of 15 in F_4, 3 out of 4 in G_2 [33]. But in fact very few such elements of semisimple type are there in classical Lie algebras $\mathfrak{g}$. In [25, 26, 28] we then develop a new method, based on the notions of Adler type operators and generalized quasideterminant, which allow us to construct an integrable hierarchy of Hamiltonian equations for every $\mathcal{W}$-algebra associated to a classical Lie algebra } and its arbitrary nilpotent element f. This construction, as well as several examples, are reviewed in the last Sects. 4.4–4.9.

2 Lecture 1: Classical and Quantum Hamiltonian Mechanics

2.1 Hamiltonian Equations in Classical Mechanics

In the present notes we will be mainly concerned with the role and applications of the concept of *symmetry* in mechanics, from a *purely algebraic* point of view.

Recall that classical mechanics deals with the dynamics of particles, rigid bodies, etc, while classical field theory deals with the dynamics of continuous media (fluids, plasma), and fields such as the electromagnetic field, gravity, etc. The latter can be viewed as an infinite dimensional analogue of the former. We shall deal with the

Hamiltonian formulation of both theories: in the present section we review the basics of classical Hamiltonian mechanics, while in Sect. 2.2 we shall review some basics of classical (Hamiltonian) field theory.

Standard Hamiltonian Equations

In classical mechanics, the state of a physical system is usually described as a point (q, p) in a $2n$-dimensional *space of states* M (usually $q = (q_1, \dots, q_n)$ denotes the positions of all particles of the system, and $p = (p_1, \dots, p_n)$ the conjugate momenta). The dynamics (=time evolution) of the physical system is described in terms of the Hamiltonian function $H(q, p)$, describing the energy of the system in state (q, p). The time evolution $(q(t), p(t))$, $t > 0$, of a system starting in state $(q(0), p(0)) = (q_0, p_0)$, is then (the unique) solution of the *Hamiltonian equations*:

$$\dot{q}_i = \frac{\partial H}{\partial p_i} , \quad \dot{p}_i = -\frac{\partial H}{\partial q_i} . \tag{1}$$

Newton's law of dynamics $F = ma$ in just an instance of such Hamiltonian evolution, when $F = -\nabla U$ is the force field associated to an energy potential $U(q)$, and the Hamiltonian $H = E_k + U$ is the sum of the kinetic energy $E_k = \frac{|p|^2}{2m}$ and the potential energy $U(q)$. The energy $H(q, p)$, as well as all the coordinates q_i and p_i, are physical observables. In general, the *space of physical observables* is the space $C^\infty(M)$ of all smooth functions of the space of states M. It is a commutative associative algebra (with respect to the usual product of functions). Moreover, it is endowed with the *Poisson bracket*

$$\{f, g\} = \sum_{i=1}^{n} \left(\frac{\partial f}{\partial p_i} \frac{\partial g}{\partial q_i} - \frac{\partial f}{\partial q_i} \frac{\partial g}{\partial p_i} \right) . \tag{2}$$

This is a bilinear product $C^\infty(M) \times C^\infty(M) \to C^\infty(M)$ satisfying the axioms of a Lie algebra bracket, i.e. skewsymmetry

$$\{f, g\} = -\{g, f\} \tag{3}$$

and the Jacobi identity

$$\{f, \{g, h\}\} - \{g, \{f, h\}\} = \{\{f, g\}, h\} , \tag{4}$$

and it also satisfy the Leibniz rule

$$\{f, gh\} = \{f, g\}h + g\{f, h\} . \tag{5}$$

The Hamiltonian equations (1) can then be rewritten in terms of the Poisson bracket (2) as follows:

$$\frac{df}{dt} = \{H, f\} , \tag{6}$$

where $f \in C^\infty(M)$ is an arbitrary observable. Indeed, taking $f = q_i$ and $f = p_i$ in Eq. (6) and using the definition (2) of the Poisson bracket, we recover the Hamilton equations (1); conversely, (6) follows from (1) by the chain rule of derivation.

Hamiltonian Evolution on a Poisson Manifold

The form (6) of Hamilton equations allow us to greatly generalize the theory of classical (Hamiltonian) mechanics. In general, the *space of states* of a classical mechanical system is a Poisson manifold M, i.e. a smooth manifold (NOT necessarily even dimensional) endowed with a Poisson bracket $\{\cdot, \cdot\} : C^\infty(M) \times C^\infty(M) \to C^\infty(M)$ on the algebra of smooth functions $C^\infty(M)$, satisfying axioms (3)–(5). Note that, by the Leibniz rule (5), if $x_1, \dots, x_n$ is a set of local coordinates on M, then any Poisson bracket has the form

$$\{f, g\}(x) = \sum_{i,j=1}^{n} \frac{\partial f}{\partial x_i} \frac{\partial g}{\partial x_j} \{x_i, x_j\} . \tag{7}$$

The dynamics of the physical system is associated to a Hamiltonian function $H \in C^\infty(M)$, and it is given by the Hamilton equations (6). In coordinates, a system starting in a state $x \in M$ at time $t = 0$, evolves to the state $\psi(t; x) = x(t) \in M$ at time $t > 0$, solution of the Hamiltonian flow:

$$\dot{x}_i(t) = \{H, x_i\}(\psi(t; x)) . \tag{8}$$

For example, the dynamics of a rigid body about its center of mass in absence of external forces, in terms of its angular momenta $\Pi_i = I_i \Omega_i$, $i = 1, 2, 3$, (Ω_i being the angular velocities and I_i the moments of inerzia):

$$\dot{\Pi}_1 = \frac{I_2 - I_3}{I_2 I_3} \Pi_2 \Pi_3 , \quad \dot{\Pi}_2 = \frac{I_3 - I_1}{I_3 I_1} \Pi_3 \Pi_1 , \quad \dot{\Pi}_3 = \frac{I_1 - I_2}{I_1 I_2} \Pi_1 \Pi_2 , \tag{9}$$

can be written as the Hamiltonian equation (6), with respect to the Poisson bracket

$$\{f, g\}(\Pi) = \Pi \cdot (\nabla f \times \nabla g) = \sum_{i,j,k=1}^{3} \epsilon_{ijk} \Pi_i \frac{\partial f}{\partial \Pi_j} \frac{\partial g}{\partial \Pi_k} \tag{10}$$

(where ϵ_{ijk} is the totally skewsymmetric tensor such that $\epsilon_{123} = 1$) and the Hamiltonian function

$$H(\Pi) = \frac{1}{2} \left(\frac{\Pi_1^2}{I_1} + \frac{\Pi_2^2}{I_2} + \frac{\Pi_3^2}{I_3} \right) . \tag{11}$$

Exercise 1 Check that formula (10) defines a Poisson bracket on $C^\infty(\mathbb{R}^3)$, i.e. it satisfies axioms (3)–(5).

Exercise 2 Check that the Hamiltonian equations for the Poisson bracket (10) and the Hamiltonian function (11) are indeed the equations of the evolution of the rigid body (9).

The Poisson Manifold $\mathfrak{g}^*$ with the Kirillov-Kostant Poisson Structure

A generalization of the Poisson bracket (10) is the Kirillov-Kostant bracket associated to a Lie algebra $\mathfrak{g}$. The underlying Poisson manifold is the dual space $\mathfrak{g}^*$. Then, the Poisson bracket on the algebra of functions $C^\infty(\mathfrak{g}^*)$ is

$$\{f, g\}(\zeta) = \langle \zeta | [d_\zeta f, d_\zeta g] \rangle \,, \tag{12}$$

where $\langle \cdot | \cdot \rangle$ denotes the pairing of $\mathfrak{g}^*$ and $\mathfrak{g}$, and the differential $d_\zeta f \in \mathfrak{g}$ is defined by ($\zeta, \xi \in \mathfrak{g}^*$):

$$\lim_{\varepsilon \to 0} \frac{1}{\varepsilon}(f(\zeta + \varepsilon\xi) - f(\zeta)) = \langle \xi | d_\zeta f \rangle \,.$$

Exercise 3 The Kirillov-Kostant Poisson bracket (12) satisfies axioms (3)–(5).

Exercise 4 The Poisson bracket of the rigid body (10) is a special case of the Kirillov-Kostant Poisson bracket when $\mathfrak{g}$ is the Lie algebra $\mathbb{R}^3$ with the Lie bracket given by the cross product $u \times v$.

The Hamiltonian evolution (6) w.r.t. the Kirillov-Kostant Poisson bracket and the Hamiltonian function $H(\zeta) \in C^\infty(\mathfrak{g}^*)$ has the form

$$\frac{df}{dt}(\zeta) = \langle \zeta | [d_\zeta H, d_\zeta f] \rangle \,. \tag{13}$$

From a purely algebraic point of view, we can replace the algebra of smooth functions $C^\infty(\mathfrak{g}^*)$ by the algebra of *polynomial functions* $\mathbb{C}[\mathfrak{g}^*]$, which is canonically identified with the symmetric algebra $S(\mathfrak{g})$. This identification maps the element

$$P = \sum \text{coeff.} a_1 \dots a_s \ \in \ S(\mathfrak{g})$$

to the polynomial function

$$P(\zeta) = \sum \text{coeff.} \langle \zeta | a_1 \rangle \dots \langle \zeta | a_s \rangle \,.$$

Under this identification, the Kirillov-Kostant Poisson bracket (12) corresponds to the canonical Poisson bracket of the symmetric algebra $S(\mathfrak{g})$: i.e. obtained by extending the Lie algebra bracket of $\mathfrak{g}$ to $S(\mathfrak{g})$ by the Leibniz rules.

Exercise 5 Check that the Kirollov-Kostant Poisson bracket of $C[\mathfrak{g}^*]$ corresponds to the canonical Poisson bracket of $S(\mathfrak{g})$.

If $x_1, \dots, x_n$ is a basis of the Lie algebra $\mathfrak{g}$ (which we view as coordinates on the Poisson manifold $\mathfrak{g}^*$), the Hamiltonian equations (13) for the Hamiltonian function $H \in S(\mathfrak{g})$ become

$$\dot{x}_i = \sum_{j=1}^{n} \frac{\partial H}{\partial x_j} [x_j, x_i] . \tag{14}$$

2.2 Hamiltonian Equations in Classical Field Theory

Classical field theory describes the evolution of continuous media, such as fluids, strings, electromagnetic field, etc. The state of such a system is described by a function $\varphi(x)$, called a *field*, where x varies in a certain manifold S (the space or space-time). For simplicity, in the present notes we shall assume that the space S is one-dimensional. Hence, the *space of states* $\mathcal{M}$ is an ∞-dimensional space of functions $\varphi(x)$ on S, which are assumed to be smooth and rapidly decreasing at the border of S, so that all integrals that we shall write are defined, and we can perform integration by parts without worrying about boundary terms.

The *physical observables* are *local functionals*, i.e. functionals $F : \mathcal{M} \to \mathbb{R}$ of the form:

$$F(\varphi) = \int_S f(\varphi(x), \varphi'(x), \dots, \varphi^{(n)}(x)) dx , \tag{15}$$

where the function f, called the *density function*, is a smooth function depending on the values of φ and finitely many its derivatives at the point x.

In order to give a Hamiltonian formulation of classical field theory, we need to assume that the space of states $\mathcal{M}$ is a *Poisson manifold*. As for classical mechanics, this means that we have a Lie algebra structure on the space of physical observables. In fact, we want this Lie algebra bracket to be a *local Poisson bracket* $\{\cdot\,,\,\cdot\}$ on $\mathcal{F}(\mathcal{M})$ of local functionals, i.e. a bracket $\{\cdot\,,\,\cdot\} : \mathcal{F}(\mathcal{M}) \times \mathcal{F}(\mathcal{M}) \to \mathcal{F}(\mathcal{M})$ of the form (cf. (7))

$$\{F, G\}(\varphi) = \int_S \frac{\delta G}{\delta \varphi(x)} K(\varphi(x), \varphi'(x), \dots, \varphi^{(n)}(x); \partial_x) \frac{\delta F}{\delta \varphi(x)} , \tag{16}$$

where the *Poisson structure* $K(\varphi(x), \varphi'(x), \dots, \varphi^{(n)}(x); \frac{d}{dx})$ is a finite order differential operator depending on φ and finitely many its derivatives at the point x (here ∂_x denotes the total derivative w.r.t. x), and $\frac{\delta F}{\delta \varphi(x)}$ denotes the *variational derivative*, defined by

$$\lim_{\epsilon \to 0} \frac{1}{\epsilon} (F(\varphi + \epsilon \psi) - F(\varphi)) = \int_S \frac{\delta F}{\delta \varphi(x)} \psi(x) dx ,$$

for every test function $\psi(x) \in \mathcal{M}$. By the chain rule and integration by parts, we easily get an explicit formula for the variational derivative:

$$\frac{\delta F}{\delta \varphi(x)} = \sum_{n \in \mathbb{Z}_+} (-\partial_x)^n \left(\frac{\partial f}{\partial \varphi^{(n)}(x)} \right) . \tag{17}$$

Exercise 6 Check formula (16).

An example is the so-called Gardner–Faddeev–Zakharov (GFZ) local Poisson-bracket, given by

$$\{F, G\}(\varphi) = \int_S \frac{\delta G}{\delta \varphi(x)} \partial_x \frac{\delta F}{\delta \varphi(x)} , \tag{18}$$

i.e., it has the form (16) with Poisson structure $K = \partial_x$.

Exercise 7 Check that the GFZ local Poisson bracket (18) is a Lie algebra bracket on the space of local functionals $\mathcal{F}(\mathcal{M})$.

The dynamics (=time evolution) of the physical system is described in terms of the *Hamiltonian functional* $H(\varphi) \in \mathcal{F}(\mathcal{M})$, describing the energy of the system in the state $\varphi(x)$. The time evolution $\varphi(x, t)$, $t > 0$, of a system starting in state $\varphi(x, 0) = \varphi_0(x)$, is the (unique) solution of the *Hamiltonian equations* (cf. (6)):

$$\frac{dF}{dt} = \{H, F\} , \tag{19}$$

where $F(\varphi)$ is a local functional of the function $\varphi(x)$. We can also write an equation for the evolution of the coordinate variable $\varphi(x, t)$. By the definition of variational derivative, we have

$$\frac{dF}{dt} = \int_S \frac{\delta F}{\delta \varphi(x)} \varphi(x, t) ,$$

which combined to (19) and (16), gives

$$\frac{d}{dt} \varphi(x, t) = K(\varphi(x), \varphi'(x), \dots, \varphi^{(n)}(x); \partial_x) \frac{\delta H}{\delta \varphi(x)} . \tag{20}$$

For example, the Hamiltonian equation associated to the GFZ Poisson bracket on $\mathcal{M}$ and the Hamiltonian functional

$$H(\varphi) = \int_S \left(\frac{1}{2} \varphi(x)^3 - \frac{c}{2} \varphi'(x)^2 \right) dx , \tag{21}$$

is the famous *Korteveg-de Vries* (KdV) equation

$$\partial_t \varphi(x, t) = 3\varphi(x, t) \partial_x \varphi(x, t) + c \partial_x^3 \varphi(x, t) , \tag{22}$$

describing the evolution of waves in shallow water.

Exercise 8 Check that the Hamiltonian equation associated to the GFZ GFZ Poisson bracket (18) and the Hamiltonian functional (21) is the KdV equation (22).

2.3 Quantum Mechanics

In quantum mechanics the *state space*, describing all possible states of a physical system, is a Hilbert space $\mathcal{H}$, i.e. a complex vector space V with a positive definite hermitian inner product $\langle \cdot | \cdot \rangle$. The *physical observables* are selfadjoint operators: $A \in \mathrm{End}(\mathcal{H})$, s.t. $A = A^\dagger$. The physical meaning of such an operator is as follows: if the state of the system is an eigenvector $|v\rangle \in \mathcal{H}$ of eigenvalue $\lambda \in \mathbb{R}$, then the result of the measurement of the observable A is λ. In general, the state $|v\rangle$ of the system is linear combination of eigenvectors: $|v\rangle = \sum_i c_i |v_i\rangle$, with $A|v_i\rangle = \lambda_i |v_i\rangle$ and λ_i's distinct. Assuming that the state $|v\rangle$ and the eigenvectors $|v_i\rangle$ are normalized, i.e. are of length 1, then the result of a measurement of the observable A is λ_i with probability $|c_i|^2$.

The *Hamiltonian* operator H describes the energy of the system and defines its dynamics. In the Schroedinger picture, the observables do not evolve, while the states evolves in time according to the Schroedinger equation

$$\frac{d}{dt}|\psi_t\rangle = H|\psi_t\rangle\,. \tag{23}$$

In the equivalent Heisenberg picture, the the state does not evolve, while the observables evolve according to the evolution equation:

$$\frac{d}{dt}A(t) = [H, A(t)]\,, \tag{24}$$

where $[H, A] = HA - AH$ is the commutator. It is apparent that this is the "quantum version" of the Hamiltonian equation (6) of classical mechanics.

Exercise 9 Show that the Schroedinger equation (23) and the Heisenberg equation (24) are equivalent in the following sense: the evolution of the matrix elements $\langle \varphi_t | A | \psi_t \rangle$ in the Schroedinger picture, and the same evolution $\langle \varphi | A(t) | \psi \rangle$ in the Heisenberg picture, coincide.

2.4 Algebraic Structures of Classical and Quantum Mechanics: Poisson Algebras and Associative Algebras

From a purely algebraic point of view, when considering a classical mechanic physical system, we ignore completely the underlying space of states M, and we just retain the algebraic structure of the space of functions $C^\infty(M)$. What we end up with, is the notion of a Poisson algebra.

By definition, a *Poisson algebra* P is a commutative associative algebra, endowed with a Lie algebra bracket $\{\cdot\,,\,\cdot\}$ satisfying the Leibniz rule (5). The Hamiltonian equations (6) are then entirely written in terms of the Poisson algebra structure: the

Hamiltonian function is an element $H \in P$, and the time evolution $\frac{df}{dt}$ of $f \in P$ can be defined as the Poisson bracket $\{H, f\}$.

Example 1 The example to keep in mind is the Poisson algebra $S(\mathfrak{g})$ with the canonical Poisson bracket, obtained by extending the Lie algebra bracket of $\mathfrak{g}$ to $S(\mathfrak{g})$ by the Leibniz rule (cf. Exercise 5).

In passing from classical to quantum mechanics, the physical observables become non commuting objects. Hence, the algebraic structure of a quantum mechanical system is just that of an *associative algebra* A.

The procedure of passing from a quantum mechanical system to a classical Hamiltonian system, known as *classical limit*, can also be described in purely algebraic terms. We assume that the associative algebra A (associated to the quantum system) has an increasing filtration

$$0 = F^{-1}A \subset F^0A \subset F^1A \subset \cdots \subset A\,, \tag{25}$$

such that

$$F^iA \cdot F^jA \subset F^{i+j}A \quad \text{and} \quad [F^iA, F^jA] \subset F^{i+j-1}\,.$$

Then, the associated graded algebra gr $A = \bigoplus_{n\geq 0} \mathrm{gr}^n A$, where $\mathrm{gr}^n A = F^nA/F^{n-1}A$, is naturally a graded Poisson algebra, with the commutative associative product $\mathrm{gr}^i A \cdot \mathrm{gr}^j A \to \mathrm{gr}^{i+j} A$ induced by the associative product of A: for $\bar{a} = a + F^{i-1}A \in \mathrm{gr}^i A$ and $\bar{b} = b + F^{j-1}A \in \mathrm{gr}^j A$,

$$\bar{a}\bar{b} = ab + F^{i+j-1}A \in \mathrm{gr}^{i+j} A\,,$$

and the Poisson bracket $\{\cdot\,,\,\cdot\} : \mathrm{gr}^i A \cdot \mathrm{gr}^j A \to \mathrm{gr}^{i+j-1} A$ induced by the commutator of A:

$$\{\bar{a}, \bar{b}\} = ab - ba + F^{i+j-2}A \in \mathrm{gr}^{i+j-1} A\,.$$

The Poisson algebra gr A is graded in the sense that $\mathrm{gr}^i A \cdot \mathrm{gr}^j A \subset \mathrm{gr}^{i+j} A$, and $\{\mathrm{gr}^i A, \mathrm{gr}^j A\} \subset \mathrm{gr}^{i+j-1} A$.

Exercise 10 Check that gr A is indeed a graded Poisson algebra.

Conversely, given a graded Poisson algebra P, its *quantization* is, by definition, a filtered associative algebra A, such that gr $A = P$. Note that, while the procedure of classical limit is uniquely defined, the quantization is not.

Example 2 A quantization of the Poisson algebra $S(\mathfrak{g})$ of Example 1 is the universal enveloping algebra $U(\mathfrak{g})$ with the usual polynomial filtration. Indeed, by the PBW Theorem, gr $U(\mathfrak{g}) \simeq S(\mathfrak{g})$.

In the next Lecture 3 we will pass from classical and quantum mechanics to classical and quantum field theories and we will describe the algebraic structures underlying those theories.

2.5 Moment Map and Hamiltonian Reduction

The concept of momentum map is a geometric generalization of the usual linear and angular momentum. To introduce its concept, we consider two examples.

Example 3 *The linear momentum* is the moment map for the translation group. For this, consider the space of states of a point particle, $M = \mathbb{R}^6 = \{(p, q)\}$, where $q \in \mathbb{R}^3$ describes the position of the particle, and $p \in \mathbb{R}^3$ describes its linear momentum. It is endowed with the standard Poisson bracket $\{\cdot\,,\,\cdot\}$ given by (2). We have the action of the translation group $G = \mathbb{R}^3$ on M, $G \times M \to M$, mapping $a \in G$ and $(q, p) \in M$ to $(q + a, p)$. This corresponds to an infinitesimal action of the Lie algebra $\mathfrak{g} = \mathbb{R}^3$ on the algebra of functions $C^\infty(M)$, $\mathfrak{g} \times C^\infty(M) \to C^\infty(M)$, mapping $a \in \mathfrak{g}$ and $f(q, p) \in C^\infty(M)$ to

$$\begin{aligned}(a \cdot f)(q, p) &:= \frac{d}{d\epsilon} f(q - \epsilon a, p)\big|_{\epsilon=0} = -\nabla_q f(q, p) \cdot a \\ &= -\{\mu(q, p; a), f(q, p)\}\,, \quad \text{where} \quad \mu(q, p; a) = p \cdot a\,.\end{aligned}$$

We thus say that the action of the translation group G on M is a Hamiltonian action, with the momentum map $\mu : M \to \mathfrak{g}^*$ given by $\mu(q, p) = \langle p | \cdot \rangle \in \mathfrak{g}^*$, associating to the state (q, p) its linear momentum p.

Example 4 *The angolar momentum* is the moment map for the rotation group. For this, consider as in Example 3 the space of states of a point particle, $M = \mathbb{R}^6 = \{(p, q)\}$. The rotation group $G = SO_3(\mathbb{R})$ acts naturally on M, $G \times M \to M$, mapping $g \in G$ and $(q, p) \in M$ to (gq, gp). The corresponding Lie algebra is $\mathfrak{g} = \mathfrak{so}_3(\mathbb{R}) = \{A \in \mathrm{Mat}_{3\times 3}(\mathbb{R}) \mid A^T = -A\}$ with Lie bracket given by the commutator of matrices. It is isomorphic to the Lie algebra $\mathbb{R}^3$ with Lie bracket given by the cross product $[u, v] = u \times v$, an explicit isomorphism being

$$A = \begin{pmatrix} 0 & -c & b \\ c & 0 & -a \\ -b & a & 0 \end{pmatrix} \in \mathfrak{so}_3(\mathbb{R}) \mapsto a = \begin{pmatrix} a \\ b \\ c \end{pmatrix} \in \mathbb{R}^3$$

The action of the rotation group G on M corresponds to the infinitesimal action of the Lie algebra $\mathfrak{g}$ on the algebra of functions $C^\infty(M)$, $\mathfrak{g} \times C^\infty(M) \to C^\infty(M)$, mapping $a \in \mathfrak{g}$ and $f(q, p) \in C^\infty(M)$ to

$$\begin{aligned}&(a \cdot f)(q, p) := \frac{d}{d\epsilon} f(e^{-\epsilon A} q, e^{-\epsilon A} p)\big|_{\epsilon=0} = -\nabla_q f(q, p) \cdot Aq - \nabla_p f(q, p) \cdot Ap \\ &= -\{\mu(q, p; a), f(q, p)\}\,, \quad \text{where} \quad \mu(q, p; a) = (q \times p) \cdot a\,.\end{aligned} \tag{26}$$

We thus say that the action of the rotation group G on M is a Hamiltonian action, with the momentum map $\mu : M \to \mathfrak{g}^*$ given by $\mu(q, p) = \langle q \times p | \cdot \rangle \in \mathfrak{g}^*$. It associates to the state (q, p) its angolar momentum $q \times p$.

Exercise 11 Check Eq. (26).

Generalizing the above two examples, we get the notion of momentum map.

Definition 1 A *Hamiltonian action* of a Lie group G on a Poisson manifold M, is a Lie group action

$$G \times M \to M, \tag{27}$$

endowed with a *momentum map* $\mu : M \to \mathfrak{g}^*$, such that the infinitesimal action corresponding to (27),

$$\mathfrak{g} \times C^\infty(M) \to C^\infty(M)$$

is given by

$$(a \cdot f)(x) := \frac{d}{d\varepsilon} f(e^{-\varepsilon a}x)|_{\varepsilon=0} = -\{\langle \mu(x), a\rangle, f(x)\}. \tag{28}$$

Now, suppose that the Lie group G is a *group of symmetries* for the physical system, i.e. the Hamiltonian $H(x)$ is constant along the orbits of the Lie group G. Then, the famous Noether's Theorem guarantees that the values of the momentum map are constant of motion.

Theorem 1 (Noether Theorem) *Suppose that the Hamiltonian $H(x) \in C^\infty(M)$ is invariant by the Hamiltonian action of the Lie group G on M, i.e.*

$$H(gx) = H(x) \;\; \textit{for all} \;\; x \in M, \; g \in G.$$

Then the values of the momentum map μ are constant of motion:

$$\mu(x(t)) = \mu(x(0)) \;\; \textit{for all times} \;\; t > 0$$

Proof The proof is very simple, so we review it here. For $a \in \mathfrak{g}$, we have

$$\begin{aligned}
&\frac{d}{dt}\langle \mu(x_t), a\rangle = \{H, \langle \mu(\cdot), a\rangle\}(x_t) \\
&= -\{\langle \mu(x), a\rangle, H(x)\}\big|_{x=x(t)} = \frac{d}{d\varepsilon} H(e^{-\varepsilon a}x(t))|_{\varepsilon=0} = 0
\end{aligned}$$

For the first equality we used the Hamiltonian equation (6), while for the third equality we used the definition (28) of the momentum map μ. □

Noether's Theorem is usually used to reduce the number of degrees of freedom of the physical system, by *twice* the dimension of G. The corresponding reduction process goes under the name of *Hamiltonian reduction*. It is defined as follows. We take a point $\xi \in \mathfrak{g}^*$ fixed by the coadjoint action of the Lie group G:

$$(\mathrm{Ad}^* g)(\xi) = \xi \;\; \forall g \in G.$$

We then take the preimage of the point $\xi \in \mathfrak{g}^*$ via the moment map $\mu^{-1}(\xi) \subset M$. This is a submanifold of M, but NOT, in general, a Poisson submanifold. But if we quotient by the action of the Lie group G and take the space of G-orbits, we in fact get a canonically defined Poisson manifold, with Poisson structure induced by that of M. This is, by definition, the *Hamiltonian reduction* of M: by the Hamiltonian action of the Lie group G at the point ξ:

$$\overline{M} = \text{Ham.red.}(M, \xi, G) = \mu^{-1}(\xi)/G \tag{29}$$

We also have the induced Hamiltonian function H on $\overline{M}$ defining the reduced physical system; its number of freedom is: $\dim \overline{M} = \dim M - 2 \dim G$. In fact, after solving the reduced system on $\overline{M}$, it is possible to *reconstruct* the original system on M.

2.6 Hamiltonian Reduction in Algebraic Terms

We would like to describe the Hamiltonian reduction in purely algebraic terms. As we pointed out in Sect. 2.4, the algebraic structure underlying a Poisson manifold M is the Poisson algebra structure of the space of functions $C^\infty(M)$.

Exercise 12 If we have a Hamiltonian action of the Lie group G on M with moment map $\mu : M \to \mathfrak{g}^*$, then the dual map

$$\mu^* : \mathfrak{g} \to C^\infty(M),$$

is a Lie algebra homomorphism, which, by universal property, extends to a Poisson algebra homomorphism

$$\mu^* : S(\mathfrak{g}) \to C^\infty(M).$$

In the Hamiltonian reduction construction (29) we then take the fiber $\mu^{-1}(\xi)$ of a point $\xi \in \mathfrak{g}^*$ fixed by the coadjoint action of G. The corresponding algebra of functions on it can be identified with the quotient

$$C^\infty(M)/I,$$

where I is the ideal of functions vanishing on $\mu^{-1}(\xi)$.

Exercise 13 We have $I = C^\infty(M)\,\text{Span}\{a - \langle \xi, a\rangle \mid a \in \mathfrak{g}\}$. Moreover, if $\xi \in \mathfrak{g}^*$ is invariant by the coadjoint action of G, then the subset

$$J = \{a - \langle \xi, a\rangle \mid a \in \mathfrak{g}\} \subset S(\mathfrak{g}),$$

is invariant by the adjoint action of $\mathfrak{g}$: $\{\mathfrak{g}, J\} = 0$.

Finally, in the Hamiltonian reduction (29) we take the space of G-orbits $\mu^{-1}(\xi)/G$. Passing to the algebra of functions, this corresponds to taking the subalgebra of functions which are constant along the G-orbits. Equivalently, we need to take the functions which are invariant by the Lie algebra action of $\mathfrak{g}$:

$$(C^\infty(M)/C^\infty(M)J)^{\operatorname{ad}\mu^*(\mathfrak{g})}.$$

Exercise 14 A smooth function f on $\mu^{-1}(\xi)$ is constant along the G-orbits if and only if

$$\{\mu^*(a), f\} = 0 \text{ in } C^\infty(M)/C^\infty(M)J \text{ for all } a \in \mathfrak{g}.$$

In conclusion, we arrive at the following general definition of Hamiltonian reduction in purely algebraic terms.

Definition 2 The starting data are a Poisson algebra P, a Lie algebra $\mathfrak{g}$, a Lie algebra homomorphism $\mu^* : \mathfrak{g} \to P$ (extended to a Poisson algebra homomorphism $\mu^* : S(\mathfrak{g}) \to P$), and a set $J \subset S(\mathfrak{g})$ invariant by the adjoint action of $\mathfrak{g}$. The corresponding *Hamiltonian reduction* is:

$$\overline{P} = \text{Ham.red.}(P, \mathfrak{g}, \mu^*, J) = (P/P\mu^*(J))^{\operatorname{ad}\mu^*(\mathfrak{g})} = N/I, \tag{30}$$

where

$$I = P\mu^*(J) \text{ and } N = \big\{f \in P \,\big|\, \{\mu^*(a), f\} \in J \text{ for all } a \in \mathfrak{g}\big\}.$$

Proposition 1 *$N \subset P$ is a Poisson subalgebra and $I \subset N$ is its Poisson algebra ideal. Hence, $\overline{P}$ has a canonical Poisson algebra structure, induced by that of P.*

Exercise 15 Prove Proposition 1.

3 Lecture 2: Vertex Algebras and Poisson Vertex Algebras

Going from a finite to an infinite number of degrees of freedom, we pass from classical and quantum mechanics to *Classical Field Theory* and *Quantum Field Theory respectively*. It is not clear what are, in general, the corresponding algebraic structures. In conformal field theory, the algebraic structures describing chiral fields are known as *vertex algebras*, introduced by Borcherds [7]. Their quasi-classical limits are the *Poisson vertex algebras*, introduced by [12]. They are associated to classical field theory in the same way as Poisson algebras are associated to classical mechanics.

3.1 Review of Quantum Field Theory

In quantum field theory the space of states is a Hilbert space V. There is a fixed vector of minimal energy, called vacuum $|0\rangle \in V$, describing the empty space. The physical observables are operator valued distributions $\Phi(x)$, called quantum fields; they can be thought of as functions of x in the Minkowski space-time M, with values in End V.

According to *Einstein's relativity principle*, no signal can travel at speed higher than the speed of light $c = 1$. Hence, If $x, y \in M$ are at space-like distance, $|x - y|^2 < 0$, then the measures of the observables $\Phi(x)$ and $\Psi(y)$ must be independent. Moreover, according to *Heisenberg uncertainty principle*, if the observable $\Phi(x)$ and $\Psi(y)$ are independent, then they commute: $[\Phi(x), \Psi(y)] = 0$. In this case, they can be simultaneously diagonalized, i.e. they can both be measured simultaneously without uncertainty.

Next, we restrict ourselves to the case of a Minkowski space of dimension $\dim M = 1 + 1$. In this case, combining the Relativity and the Uncertainty principles, we get *locality* of chiral fields. We start by making the chance of variables

$$z = x_0 - x_1\,, \quad \bar{z} = x_0 + x_1 \quad \Rightarrow \quad |x|^2 = z\bar{z}$$

By the Relativity and Uncertainty principles, we get:

$$(z - w)(\bar{z} - \bar{w}) < 0 \Rightarrow [\Phi(z, \bar{z}), \Psi(w, \bar{w})] = 0 \tag{31}$$

A *chiral* field is a quantum field $\Phi(z)$, depending only on z, not on $\bar{z}$. Hence, if we apply Eq. (31) to chiral fields, we get:

$$z - w \neq 0 \quad \Rightarrow \quad [\Phi(z), \Psi(w)] = 0$$

This in turn is equivalent to the *locality principle*:

$$(z - w)^N[\Phi(z), \Psi(w)] = 0 \ \text{ for } \ N \gg 0$$

Usually a quantum (chiral) field $\Phi(z)$ is expanded in its Fourier modes, as a (formal) power series:

$$\Phi(z) \ = \sum_{n \in \mathbb{Z}} \Phi_n z^n\,, \quad \Phi_n \in \text{End}\, V$$

Conventionally, the negative modes Φ_n, $n < 0$ are *annihilation operators* ($v \in V$), i.e. they satisfy the conditions:

$$\Phi_n|0\rangle = 0\,, \ \forall n < 0 \quad \text{and} \quad \Phi_n v = 0\,, \ \forall n \ll 0\,,$$

while the positive modes Φ_n, $n \geq 0$ are the *creation operators*:

$$\Phi_n|0\rangle = |\Phi_n\rangle \neq 0, \ \forall n \geq 0$$

In terms of the powers series $\Phi(z)$, we get that $\Phi(z)|0\rangle \in V[[z]]$ is a Taylor series in z (involving only non-negative powers of z), while, for every $v \in V$, $\Phi(z)v \in V((z))$ is a Laurent series in z (involving only a finite number of negative powers of z).

3.2 Quantum Field Theory and (Pre) Vertex Algebras

The notion of a vertex algebra describes the algebraic structure of chiral fields in a quantum field theory in $1+1$-dimension.

Let V be a vector space, called the *space of states*; let $|0\rangle \in V$ be a vector, called the *vacuum*; and let $T \in \operatorname{End} V$ be an operator, called the *translation operator*. A *quantum field* is a series $\Phi(z) = \sum_{n\in\mathbb{Z}} \Phi_n z^n \in \operatorname{End} V[[z, z^{-1}]]$ such that, for all $v \in V$, $\Phi(z)v \in V((z))$ (or, equivalently, $\Phi_n(v) = 0$ for $n \ll 0$).

Definition 3 A (pre)*vertex algebra* $(V, |0\rangle, T, \mathcal{F})$ is a collection of quantum fields $\mathcal{F} = \{\Phi^\alpha(z)\}_{\alpha\in J}$, complete in the sense that

$$V = \operatorname{Span}\left\{\Phi^{\alpha_1}_{n_1} \dots \Phi^{\alpha_s}_{n_s}|0\rangle\right\}_{s\geq 0, \alpha_\ell \in J, n_\ell \in \mathbb{Z}}, \tag{32}$$

and satisfying the following axioms:

(i) vacuum axiom: $\Phi^\alpha(z)|0\rangle \in V[[z]]$ for all α;
(ii) translation covariance: $[T, \Phi^\alpha(z)] = \partial_z \Phi^\alpha(z)$;
(iii) locality: $(z-w)^N[\Phi^\alpha(z), \Phi^\beta(w)] = 0$ for $N \gg 0$.

Example 5 Let $\mathfrak{g}$ be a semisimple Lie algebra, endowed with a symmetric invariant bilinear form $(\cdot\,, \cdot)$. Recall that the affine Kac–Moody Lie algebra is the space

$$\hat{\mathfrak{g}} = \mathfrak{g}[t^{\pm 1}] \oplus \mathbb{C}K,$$

with the Lie bracket bracket

$$[at^m, bt^n] = [a, b]t^{m+n} + m\delta_{m,-n}(a|b)K, \quad K : \text{ central}.$$

The vacuum module of $\hat{\mathfrak{g}}$ of level k is

$$V^k(\mathfrak{g}) = \operatorname{Ind}_{\hat{\mathfrak{g}}_+} \mathbb{C}_k,$$

where $\mathbb{C}_k = \mathbb{C}$ is the one-dimensional representation of the subalgebra $\hat{\mathfrak{g}}_+ = \mathfrak{g}[t] \oplus \mathbb{C}K$, with the zero action of $\mathfrak{g}[t] = 0$ and K acting by multiplication by k. The corresponding *affine vertex algebra* is defined as follows: the space of states is $V^k(\mathfrak{g})$, the vacuum vector is $|0\rangle = 1$, the translation operator is $T = -\partial_t$, and the collection of local fields is $\mathcal{F} = \{a(z)\}_{a\in\mathfrak{g}}$, where

$$a(z) = \sum_{n\in\mathbb{Z}} (at^n)z^{-n-1}\,.$$

The completeness (32) of $\mathcal{F}$ is automatic, as well as the vacuum axiom (i) and the translation covariance (ii). We need to check the locality axiom (iii). It follows by the following *Operator Product Expansion*:

$$[a(z), b(w)] = [a, b](w)\delta(z - w) + (a|b)K\,\partial_w\delta(z - w) \qquad (33)$$

where

$$\delta(z - w) = \sum_{n\in\mathbb{Z}} z^n w^{-n-1}\,.$$

Exercise 16 Check the OPE (33) for the affine vertex algebra $V^k(\mathfrak{g})$. Check that

$$(z - w)^{n+1}\partial_w^n\delta(z - w) = 0 \quad \text{for all} \quad n \geq 0\,,$$

and use it to deduce that all the fields $\{a(z)\}_{a\in\mathfrak{g}}$ are pairwise local.

Example 6 Recall that the Virasoro Lie algebra is

$$Vir = \oplus_{n\in\mathbb{Z}}\mathbb{C}L_n \oplus \mathbb{C}C$$

with the Lie bracket

$$[L_m, L_n] = L_{m+n} + c\frac{m^3 - m}{12}\delta_{m,-n}C\,, \quad C:\ \text{central}\,.$$

The vacuum module of central charge c is

$$Vir^c = \mathrm{Ind}_{Vir_+}\mathbb{C}_c\,,$$

where $\mathbb{C}_c = \mathbb{C}$ is the one-dimensional representation of the subalgebra $Vir_+ = \oplus_{n\geq -1}\mathbb{C}L_n \oplus \mathbb{C}C$, with $L_n = 0$ for all $n \geq -1$ and $C = c$. The *Virasoro vertex algebra* is defined as follows: the space of states is Vir^c, the vacuum vector is $|0\rangle = 1$, the translation operator is $T = L_0$, and $\mathcal{F} = \{L(z)\}$ consists of the single field

$$L(z) = \sum_n L_n z^{-n-2}\,.$$

As before, the completeness condition (32), the vacuum axiom (i) and the translation covariance (ii) are very simple to check. Locality is a consequence of the following Operator Product Expansion:

$$[L(z), L(w)] = \left(L'(w) + 2L(w)\partial_w + \frac{1}{12}c\partial_w^3\right)\delta(z-w)\,. \tag{34}$$

Exercise 17 Check the OPE (34) for the Virasoro vertex algebra Vir^c. Deduce that the field $L(z)$ is local with itself.

3.3 λ-Bracket Definition of a Vertex Algebra

The algebraic structure of a vertex algebra is encoded in three operations on local quantum fields. The first is the derivative of a quantum field: $\partial_z\Phi(z)$. Given two quantum field, one can define their normally ordered product

$$:\Phi\Psi:(z) \quad = \quad \Phi_+(z)\Psi(z) + \Psi(z)\Phi_-(z)$$

where

$$\Phi_+(z) = \sum_{n<0} \Phi_n z^{-n-1} \quad \text{(creation part)}\ ,$$

and

$$\Phi_-(z) = \sum_{n\geq 0} \Phi_n z^{-n-1} \quad \text{(annihilation part)}\ .$$

Finally, we can define the λ-*bracket* of $\Phi(z)$ and $\Psi(z)$, defined as the Fourier transform of their OPE:

$$[\Phi_\lambda\Psi](w) = \mathrm{Res}_z\, e^{\lambda(z-w)}[\Phi(z), \Psi(w)]$$

Exercise 18 Check that $\mathrm{Res}_z\, e^{\lambda(z-w)}\partial_w^n\delta(z-w) = \lambda^n$.

It follows from Exercise 18 that, if we have the OPE

$$[\Phi(z), \Psi(w)] = \sum_{n=0}^{N} c_n(w)\partial_w^n\delta(z-w)\,,$$

then the corresponding λ-bracket is

$$[\Phi\,{}_\lambda\,\Psi] = \sum_{n=0}^{N} \lambda^n c_n$$

The proof of the following Dong's Lemma can be found, e.g. in [40].

Lemma 1 (Dong's Lemma) *Let $a(z)$, $b(z)$, $c(z)$ be pairwise local quantum fields. Then:*

(a) $\partial_z a(z), c(z)$ *are local,*
(b) $:a(z)b(z):$, $c(z)$ *are local,*
(c) $a(z)_{(n)}b(z)$, $c(z)$ *are local for every* $n \geq 0$, *where* $a(z)_{(n)}b(z)$ *is the coefficient of* λ^n *in* $[a(z)_\lambda b(z)]$.

Dong's Lemma guarantees that, if we start with a local collection of fields $\mathcal{F}$ in a pre-vertex algebra, we can extend it to a larger collection $\widehat{\mathcal{F}}$, still local, closed under all three operations of taking derivatives, normally ordered products and λ-brackets. Such collection can be obtained as

$$\widehat{\mathcal{F}} = \cup_{n\geq 0}\mathcal{F}_n ,$$

where $\mathcal{F}_0 = \mathcal{F}$, and

$$\mathcal{F}_n = \left\{\partial_z^n a(z),\ :a(z)b(z):,\ a(z)_{(n)}b(z)\ \middle|\ a(z), b(z) \in \mathcal{F}_{n-1},\ n \geq 0\right\}.$$

In conclusion, we can always assume that our collection of fields $\mathcal{F}$ is closed with respect to the derivative $\partial = \partial_z$, the normally ordered products $: \cdot \ \ \cdot :$ and the λ-bracket $[\cdot\,_\lambda\, \cdot]$; i.e., for $a, b \in \mathcal{F}$:

$$\partial a \ \in\ \mathcal{F}\,,\quad :ab:\ \in\ \mathcal{F}\,,\quad [a_\lambda b]\ \in\ \mathcal{F}[\lambda]$$

Moreover, one can prove that, under this assumption, the linear map $\mathcal{F} \to V$ mapping $a(z) \mapsto a(z)|0\rangle|_{z=0}$ becomes a bijection, the inverse map being called the *state-field correspondence* $Y : V \to \mathcal{F}$.

In fact, one can study the properties of the maps ∂, $: \cdot \ \ \cdot :$ and $[\cdot\,_\lambda\, \cdot]$ and end up with an equivalent definition of a vertex algebra. This was originally obtained by Bakalov and Kac in [2].

Definition 4 A vertex algebra is a space of states V, with a vacuum vector $|0\rangle \in V$, a translation operator $\partial \in \operatorname{End} V$, a normally ordered product $: ab :\in V$, and a λ-bracket $[a_\lambda b] \in V[\lambda]$, satisfying the following axioms:

(i) vacuum $:a|0\rangle: \ = \ :|0\rangle a: = a$
(ii) translation covariance $\partial(: ab :) = \ :(\partial a)b: + :a(\partial b):$
(iii) sesquilinearity $[\partial a_\lambda b] = -\lambda[a_\lambda b]$, $[a_\lambda \partial b] = (\partial + \lambda)[a_\lambda b]$
(iv) skewsymmetry $[a_\lambda b] = -[b_{-\lambda-\partial} a]$
(v) Jacobi identity $[a_\lambda[b_\mu c]] - [b_\mu[a_\lambda c]] = [[a_\lambda b]_{\lambda+\mu} c]$
(vi) quasi-associativity

$$:(:ab:): - :a(:bc:): \ = \ :\left(\int_0^\partial d\lambda\, a\right)[b_\lambda c]: + :\left(\int_0^\partial d\lambda\, b\right)[a_\lambda c]:$$

(vii) quasi-commutativity

$$:ab: - :ba: \ = \int_{-\partial}^0 d\lambda\, [a_\lambda b]$$

(viii) Wick formula

$$[a_\lambda :bc:] = :[a_\lambda b]c: + :b[a_\lambda c]: + \int_0^\lambda d\mu\, [[a_\lambda b]_\mu c]$$

Axioms (i) and (ii) can be expressed by saying that $(V, |0\rangle, \partial, :\ :)$ is a unital differential algebra. Moreover, the axioms (iii), (iv) and (v), involving only the λ-bracket structure, can be expressed by saying that $(V, \partial, [\cdot\,_\lambda\, \cdot])$ is a *Lie conformal algebra.*

As for Lie algebras and their universal enveloping vertex algebra, start from a Lie conformal algebra R one can construct its universal enveloping vertex algebra $V(R)$. This is stated in the following:

Theorem 2 ([2]) *Given a Lie conformal algebra R, there exists a unique universal enveloping vertex algebra $V(R)$ containing R as a Lie conformal subalgebra, satisfying the following universal property: for every Lie conformal algebra homomorphism $\varphi : R \to W$ from R to a vertex algebra W, there exists a unique vertex algebra homomorphism $\widetilde{\varphi} : V \to W$ extending φ. Moreover, given an ordered basis $\{a_i\}_{i\in J}$ of R, $V(R)$ has a PBW basis consisting of ordered monomials*

$$\left\{ :a_{i_1} a_{i_2} \ldots a_{i_s}: \;\middle|\; 0 \le i_1 \le \cdots \le i_s \right\}$$

Using Theorem 2 we can describe the two Examples 5 and 6 of vertex algebras, in terms of the equivalent notation of Definition/Theorem 4.

Example 7 Given a Lie algebra $\mathfrak{g}$ with a symmetric invariant bilinear form $(\cdot\,,\,\cdot)$, the corresponding *affine LCA* is

$$\mathrm{Cur}_k \mathfrak{g} = (\mathbb{C}[\partial] \otimes \mathfrak{g}) \oplus \mathbb{C}|0\rangle\,,$$

with λ-bracket ($a, b \in \mathfrak{g}$)

$$[a_\lambda b] = [a, b] + \lambda k(a, b)|0\rangle\,, \tag{35}$$

extended (uniquely) to $\mathrm{Cur}_k\mathfrak{g}$ by the sesquilinearity axioms (Note that (35) is the Fourier transform of the OPE (33)). The universal *affine vertex algebra* is $V^k(\mathfrak{g}) = V(\mathrm{Cur}_k\mathfrak{g})$.

Exercise 19 Check that the λ-bracket (35) defines a structure of a Lie conformal algebra on $\mathrm{Cur}_k\mathfrak{g}$.

Example 8 The *Virasoro LCA* of *central charge* $c \in \mathbb{C}$ is

$$R_c = (\mathbb{C}[\partial]L) \oplus \mathbb{C}|0\rangle\,,$$

with λ-bracket

$$[L_\lambda L] = (\partial + 2\lambda)L + \frac{c}{12}\lambda^3|0\rangle\,, \tag{36}$$

extended (uniquely) to R_c by the sesquilinearity axioms (Note that (36) is the Fourier transform of the OPE (34)). The universal *Virasoro vertex algebra* is $Vir^c = V(R_c)$.

Exercise 20 Check that the λ-bracket (36) defines a structure of a Lie conformal algebra on R_c.

3.4 PVA as Classical Limit of VA

We can apply the classical limit procedure described in Sect. 2.4 to get the algebraic structure of classical field theory: this leads to the notion of a *Poisson vertex algebra.*

Assume that the vertex algebra V has an *increasing filtration*

$$0 = F^{-1}V \subset F^0 V \subset F^1 V \subset F^2 A \subset \cdots \subset V$$

such that

$$:F^i V \cdot F^j V: \subset F^{i+j} V \text{ and } [F^i V_\lambda F^j V] \subset F^{i+j-1} V[\lambda]$$

When we can take the associated graded

$$\text{gr } V = \bigoplus_{n\geq 0} \text{gr}^n V \text{ where } \text{gr}^n V = F^n V / F^{n-1} V \,,$$

all the quantum corrections in the axioms of a vertex algebra disappear, and what we get is a (graded) Poisson vertex algebra:

Definition 5 ([12]) A *Poisson vertex algebra* $\mathcal{V}$ is a unital, commutative, associative, differential algebra, with a Lie conformal algebra λ-bracket $\{\cdot_\lambda \cdot\}$, such that the following Leibniz rule holds:

$$\{a_\lambda bc\} = \{a_\lambda b\}c + \{a_\lambda c\}b \,.$$

Exercise 21 Check that if V is a filtered VA, then gr V is a (graded) Poisson vertex algebra.

Conversely, given a graded Poisson vertex algebra $\mathcal{V}$, its *quantization* is, by definition, a filtered vertex algebra V, such that gr $V \simeq \mathcal{V}$. The main example of quantization/classical limit are provided by the universal enveloping VA and PVA of a Lie conformal algebra R:

Theorem 3 *If R is a Lie conformal algebra, then $\mathcal{V}(R) = S(R)$ has a natural structure of a Poisson vertex algebra. Its quantization is the universal enveloping vertex algebra $V(R)$.*

Exercise 22 Prove the above claim.

Again, we can use Theorem 3 to construct the affine and the Virasoro PVA (Examples 10 and 11 below), which are respectively the classical limits of Examples 7 and 8.

Example 9 The *GFZ PVA* is the algebra of differential polynomials in one variable,

$$\mathcal{V} = \mathbb{F}[u, u', u'', \dots],$$

with λ-bracket given by

$$[u_\lambda u] = \lambda,$$

uniquely extended to $\mathcal{V}$ by the sesquilinearity and the Leibniz rules.

Example 10 Given a Lie algebra $\mathfrak{g}$ with a symmetric invariant bilinear form $(\cdot\,,\,\cdot)$, the *affine PVA* is

$$\mathcal{V}(\mathfrak{g}) = S(\mathbb{F}[\partial]\mathfrak{g}),$$

with λ-bracket given by

$$[a_\lambda b] = [a, b] + (a|b)\lambda, \quad a, b \in \mathfrak{g}, \tag{37}$$

uniquely extended to $\mathcal{V}$ by the sesquilinearity and the Leibniz rules. This is the classical limit of the affine vertex algebra $V^k(\mathfrak{g})$.

Example 11 The *Virasoro–Magri PVA* is the algebra of differential polynomials in one variable,

$$\mathcal{V} = \mathbb{F}[L, L', L'', \dots],$$

with λ-bracket given by

$$[L_\lambda L] = (\partial + 2\lambda)L + \frac{c}{12}\lambda^3, \tag{38}$$

uniquely extended to $\mathcal{V}$ by the sesquilinearity and the Leibniz rules. This is the classical limit of the Virasoro vertex algebra Vir^k.

3.5 PVA's and Hamiltonian PDE's

Poisson vertex algebras can be used in the study of Hamiltonian partial differential equations in classical field theory (in the same way as Poisson algebras are used to study Hamiltonian equations in classical mechanics).

The space of "physical observables" is the space of local functionals $\mathcal{V}/\partial\mathcal{V}$. We denote by $\int f$ the image of $f \in \mathcal{V}$ in the quotient space $\mathcal{V}/\partial\mathcal{V}$, and we say that $\int f$ is the *local functional* with *density function* f. The reason for this choice, is that $\mathcal{V}/\partial\mathcal{V}$ is the universal space where integration by parts holds: $\int f'g = -\int fg'$.

Definition 6 Given a Poisson vertex algebra $\mathcal{V}$ with λ-bracket $\{\,_\lambda\,\}$, the *Hamiltonian equation* with Hamiltonian functional $\int h \in \mathcal{V}/\partial\mathcal{V}$ is:

$$\frac{du}{dt} = \{h_\lambda u\}\big|_{\lambda=0} . \tag{39}$$

A *integral of motion* for the Hamiltonian equation (39) is a local functional $\int f \in \mathcal{V}/\partial\mathcal{V}$ such that

$$\{\textstyle\int h, \int f\} = \displaystyle\int \{h_\lambda f\}|_{\lambda=0} = 0 \text{ in } \mathcal{V}/\partial\mathcal{V} .$$

The minimal requirement for *integrability* is to have an infinite sequence $h_0 = h, h_1, h_2, \dots$ of linearly independent integrals of motion in involution:

$$\{\textstyle\int h_m, \int h_n\} = 0 \text{ for all } m, n \in \mathbb{Z}_+ .$$

In this case, we have the *integrable hierarchy* of Hamiltonian equations

$$\frac{du}{dt_n} = \{h_{n\lambda} u\}\big|_{\lambda=0} . \tag{40}$$

Example 12 The most famous example of Hamiltonian equation is the KdV equation (21). It is Hamiltonian w.r.t. the GFZ PVA and the Hamiltonian functional $\int h = \frac{1}{2}\int(u^3 + cuu'')$:

$$\frac{\partial u}{\partial t} = 3uu' + cu''' = \big\{h_\lambda u\big\}|_{\lambda=0}$$

In fact, it is also Hamiltonian in a different way (it is *biHamiltonian*), i.e. w.r.t. the Virasoro–Magri PVA and the Hamiltonian functional $\int h = \frac{1}{2}\int u^2$.

Exercise 23 Check this using λ-bracket computations (and the axioms of Poisson vertex algebra).

A non-commutative analogue of PVA, the so called double PVA, has also been studied, as well as its connection to the theory of integrable PDE's in non-commutative variables [24]. Also, the differential-difference Hamiltonian equations can be described in terms of a related algebraic structure, the so called multiplicative PVA [30, 31].

One of the main techniques to study integrability of a (bi)Hamiltonian PDE is the so-called Lenard-Magri scheme of integrability [45]. It requires the study of the PVA cohomology [5, 15, 16]. The relation between PVA cohomology and VA cohomology is studied in [3, 4]. In the next section we will present a different method, based on Adler type operators.

4 Lecture 3: $\mathcal{W}$-Algebras

4.1 Overwiew on $\mathcal{W}$-Algebras

As discussed in the previous Lectures, there are four fundamental *physical theories*, together with the corresponding fundamental *algebraic structures*, are:

1. Classical Mechanics—Poisson Algebras (PA);
2. Quantum Mechanics—Associative Algebras (AA);
3. Classical Field Theory—Poisson Vertex Algebras (PVA);
4. Quantum Field Theory—Vertex Algebras (VA).

To go from a Quantum Theory to the corresponding Classical Theory, one performs a *classical limit*, which in algebraic terms corresponds to taking the associated graded of a filtered algebra (AA or VA) There is also a procedure for passing from a positive energy VA or PVA to the corresponding AA or PA, by taking the so called Zhu algebra [50].

$\mathcal{W}$-algebras provide a rich family of examples, parametrized by a simple finite dimensional Lie algebra $\mathfrak{g}$ and a nilpotent element $f \in \mathfrak{g}$ (in fact, a nilpotent orbit), appearing in all 4 fundamental aspects. We thus have

1. The classical finite $\mathcal{W}$-algebras $\mathcal{W}^{cl,fin}(\mathfrak{g}, f)$;
2. The finite $\mathcal{W}$-algebras $\mathcal{W}^{\text{fin}}(\mathfrak{g}, f)$;
3. The classical affine $\mathcal{W}$-algebras $\mathcal{W}^{\text{cl}}(\mathfrak{g}, f)$;
4. The quantum affine $\mathcal{W}$-algebras $\mathcal{W}_k(\mathfrak{g}, f)$.

All of these families of algebras were introduced separately and played important roles in different areas of mathematics. Only later it became fully clear the relations between them.

Classical Finite $\mathcal{W}$-Algebras

The classical finite $\mathcal{W}$-algebra $\mathcal{W}^{cl,fin}(\mathfrak{g}, f)$ is a Poisson algebra, which can be viewed as the algebra of functions on the Poisson manifold of the Slodowy slice $\mathcal{S}$. This was first introduced by [47], with in mind application in the theory of singularities, when studying singularities of the coadjoint nilpotent orbits.

Only later Gan and Ginzburg [38] described the Poisson structure of the Slodowy slice $\mathcal{S}$, as the classical limit of finite $\mathcal{W}$-algebras.

Finite $\mathcal{W}$-Algebras

The first appearance of finite $\mathcal{W}$-algebras $\mathcal{W}^{\text{fin}}(\mathfrak{g}, f)$ was in [43], where he defined it for principal nilpotent $f \in \mathfrak{g}$ (in which case it is commutative), and proved that it is isomorphic to the center of the universal enveloping algebra $Z(U(\mathfrak{g}))$. In [44] (a student of Kostant) the construction is extended to the case of even nilpotent element f. Then, for a long time there was no development on these associative algebras, until the 90s, when some physics papers (e.g: [9, 10]) studied some examples of finite $\mathcal{W}$-algebras. Only in [46] there is the general definition of the finite $\mathcal{W}$-algebras

$\mathcal{W}^{\mathrm{fin}}(\mathfrak{g}, f)$ for arbitrary nilpotent f. He related the representation theory of finite $\mathcal{W}$-algebras to the theory of primitive ideals of $\mathfrak{g}$, and used them to prove the Kac–Weisfeiler conjecture.

Quantum Affine $\mathcal{W}$-Algebras

The first appearance of a quantum affine $\mathcal{W}$-algebras was in [49], who discovered the "$\mathcal{W}_3$-algebra" (which is $\mathcal{W}(\mathfrak{sl}_3, f^{pr.})$). He introduced it as a "non-linear" infinite dimensional Lie algebra, extending the Virasoro Lie algebra, as the algebra of symmetries of some CFT. Then, a bunch of other physicists encountered similar other infinite-dimensional Lie algebras with "non-linearities". In [35] they gave a general definition of the $\mathcal{W}$-algebras $\mathcal{W}(\mathfrak{g}, f)$ for principal nilpotent element f, describing it as a vertex algebra. They obtained it as quantization of the Drinfeld–Sokolov construction. Finally, in [41, 42] they gave a general construction of $\mathcal{W}(\mathfrak{g}, f)$ for arbitrary nilpotent element f. Only in [12] the affine $\mathcal{W}$-algebras and the finite $\mathcal{W}$-algebras were related: it is proved that that the Zhu algebra of the affine $\mathcal{W}$-algebra $\mathcal{W}_k(\mathfrak{g}, f)$ is indeed isomorphic to the corresponding finite $\mathcal{W}$-algebra $\mathcal{W}(\mathfrak{g}, f)$. As a consequence, their representation theories are strictly related.

Classical $\mathcal{W}$-Algebras

The classical $\mathcal{W}$-algebras were introduced originally by Drinfeld and Sokolov in [32], for principal nilpotent element f, as Poisson algebras of function on an infinite dimensional Poisson manifold. They used them to study KdV-type integrable bi-Hamiltonian hierarchies of PDE's (which are now known as "Drinfeld Sokolov hierarchies"). Subsequently, in the 90s, there was an extensive literature extending the Drinfeld Sokolv construction to other nilpotent elements, getting "generalized Drinfeld-Sokolv hierarchies" (see e.g. [34, 36, 37]). In [18, 19] the classical $\mathcal{W}$-algebras were then described as Poisson vertex algebras (introduced only in 2006).

4.2 The Slodowy Slice and the Classical Finite $\mathcal{W}$-Algebras

Let $\mathfrak{g}$ be a simple finite dimensional Lie algebra, and let $(\cdot\,|\,\cdot)$ be a non-degenerate symmetric invariant bilinear form on $\mathfrak{g}$. It allows us to identify $\Phi : \mathfrak{g} \xrightarrow{\sim} \mathfrak{g}^*$. Let $f \in \mathfrak{g}$ be a nilpotent element, which, by the Jacobson–Morozov Theorem, we can include it in an $\mathfrak{sl}_2$-triple $(f, h = 2x, e)$. The corresponding *Slodowy slice* $\mathcal{S}$ is defined as follows:

$$\mathcal{S} = \Phi(f + \mathfrak{g}^e) \subset \mathfrak{g}^*$$

We claim that it has a natural structure of a Poisson manifold. This is a consequence of the following two facts:

Exercise 24 If $\xi \in \mathcal{S}$, the symplectic form ω_ξ on the symplectic leave $\mathrm{Ad}^*\, G(\xi)$ restricts to a non-degenerate bilinear form on $T_\xi \mathcal{S} \cap T_\xi\, \mathrm{Ad}^*\, G(\xi) = \mathrm{ad}^*(\mathfrak{g})(\xi) \cap \Phi(\mathfrak{g}^e)$.

Exercise 25 The Slodowy slice $\mathcal{S}$ intersects transversally the symplectic leaf: $T_\xi \mathcal{S} \cap T_\xi \operatorname{Ad}^* G(\xi) = T_\xi \mathfrak{g}^*$ (i.e. $\operatorname{ad}^*(\mathfrak{g})(\xi) \cap \Phi(\mathfrak{g}^e) = \mathfrak{g}^*$).

These two facts guarantee that $\mathcal{S} \subset \mathfrak{g}^*$ is indeed a Poisson submanifold, with Poisson structure is induced by that of $\mathfrak{g}^*$, [48].

In order to quantize the theory, we want describe the Slodowy slice as Hamiltonian reduction of the Poisson manifold $\mathfrak{g}^*$ with the Kirillov-Kostant Poisson structure (12), cf. Sect. 2.5.

Let us first introduce some settings. We have the ad x-eignespace decomposition $\mathfrak{g} = \bigoplus_{i \in \frac{1}{2}\mathbb{Z}} \mathfrak{g}_i$. On $\mathfrak{g}_{\frac{1}{2}}$ we have the non-degenerate skewsymmetric bilinear form

$$\omega(u, v) = (f | [u, v]) ,$$

and we let $\ell \subset \mathfrak{g}_{\frac{1}{2}}$ be a maximal isotropic subspace.

Exercise 26 Check that ω is a non-degenerate skewsymmetric form on $\mathfrak{g}_{\frac{1}{2}}$.

We then let $\mathfrak{n}$ be the nilpotent subalgebra

$$\mathfrak{n} = \ell \oplus \mathfrak{g}_{\geq 1} \subset \mathfrak{g} ,$$

and we let N be the corresponding unipotent Lie group. The Lie group N naturally acts on the Poisson manifold $\mathfrak{g}^*$ via coadjoint action, and this action is in fact Hamiltonian:

Proposition 2 *The coadjoint action of N on the Poisson manifold $\mathfrak{g}^*$ is a Hamiltonian action, with momentum map $\mu : \mathfrak{g}^* \to \mathfrak{n}^*$ given by restriction.*

Exercise 27 Prove Proposition 2.

Exercise 28 The dual of the momentum map $\mu^* : \mathfrak{n} \to \mathfrak{g}$ is the inclusion map.

Next, we let $\chi = (f | \cdot)|_{\mathfrak{n}} \in \mathfrak{n}^*$. It is a character of $\mathfrak{n}$, in the sense that $\chi([\mathfrak{n}, \mathfrak{n}]) = 0$ (by the assumption that ℓ is maximal isotropic).

Exercise 29 Check that $\chi([\mathfrak{n}, \mathfrak{n}]) = 0$. Use this to prove that χ is invariant by the coadjoint action of N.

The pre image of $\chi \in \mathfrak{g}^*$ via the moment map is

$$\mu^{-1}(\chi) = \chi + \Phi(\mathfrak{n}^\perp) = \Phi(f + \mathfrak{n}^\perp) = \Phi(f + [f, \ell] \oplus \mathfrak{g}_{\geq 0}) .$$

According to the general Hamiltonian reduction construction (29), we then have the corresponding reduced Poisson manifold

$$\text{Ham.Red.}(\mathfrak{g}^*, N, \chi) = \mu^{-1}(\chi)/N = \Phi(f + \mathfrak{n}^\perp)/N .$$

Proposition 3 ([38]) *The coadjoint action $N \times (f + \mathfrak{g}^e) \to f + \mathfrak{n}^\perp$ is an isomorphism of affine varieties.*

It thus follows from Proposition 3 that

$$\text{Ham.Red.}(\mathfrak{g}^*, N, \chi) = \Phi(f + \mathfrak{n}^\perp)/N \simeq \Phi(f + \mathfrak{g}^e) = \mathcal{S} .$$

(It is not hard to check that the Poisson structure is the same.)

As a consequence, we can describe the Poisson algebra of (polynomial functions) on $\mathcal{S}$ by the general Hamiltonian reduction procedure (30):

$$\mathcal{W}^{cl,fin}(\mathfrak{g}, f) \left(\text{“} = \mathbb{C}^{pol.}(\mathcal{S})\text{"} \right) = \left(S(\mathfrak{g}) \Big/ S(\mathfrak{g})\{n - (f|n)\}_{n \in \mathfrak{n}} \right)^{\mathfrak{n}} . \qquad (41)$$

4.3 *Quantum Finite $\mathcal{W}$-Algebras as Quantization of the Slodovy Slice*

To define the finite $\mathcal{W}$-algebra, we want to quantize the classical finite $\mathcal{W}$-algebra and, in order to do so, we use its description (41) via Hamiltonian reduction. There is an obvious way to quantize the symmetric algebra $S(\mathfrak{g})$, by considering the universal enveloping algebra $U(\mathfrak{g})$. Then, a natural way to define the finite $\mathcal{W}$-algebra is

$$\mathcal{W}^{\text{fin}}(\mathfrak{g}, f) = \left(U(\mathfrak{g}) \Big/ U(\mathfrak{g})\{n - (f|n)\}_{n \in \mathfrak{n}} \right)^{\text{ad}\,\mathfrak{n}} = \mathcal{N}/\mathcal{I} , \qquad (42)$$

where

$$\mathcal{N} = \left\{ u \in U(\mathfrak{g}) \,\middle|\, \text{ad}\, a(u) \in \mathcal{I} \;\; \forall a \in \mathfrak{n} \right\} ,$$

and

$$\mathcal{I} = U(\mathfrak{g})\{n - (f|n)\}_{n \in \mathfrak{n}} .$$

Exercise 30 The numerator $\mathcal{N}$ is a subalgebra of $U(\mathfrak{g})$, and the denominator is its two-sided ideal. So, the quotient $\mathcal{W}^{\text{fin}}(\mathfrak{g}, f)$ is a well defined associative algebra.

We want to show that the quantum finite $\mathcal{W}$-algebra (42) is indeed a quantization of the classical finite $\mathcal{W}$-algebra (41). In order to do so, we introduce the following *Kazhdan filtration* of the universal enveloping algebra $U(\mathfrak{g})$: for $a \in \mathfrak{g}_i$, we let $\Delta(a) = 1 - i$ (we call this the “conformal weight” of a). Then, we let

$$F_n U(\mathfrak{g}) = \text{Span} \left\{ a_1 \dots a_s \,\middle|\, \Delta(a_1) + \dots + \Delta(a_s) \leq n \right\} .$$

Obviously, we have $F_m U(\mathfrak{g}) \cdot F_n U(\mathfrak{g}) \subset F_{m+n} U(\mathfrak{g})$.

Exercise 31 Show that

$$\Delta([a,b]) = 1-(i+j) = (1-i)+(1-j)-1 = \Delta(a)+\Delta(b)-1\,.$$

As a consequence, $[F_m U(\mathfrak{g}), F_n U(\mathfrak{g})] \subset F_{m+n-1}U(\mathfrak{g})$. Hence, $U(\mathfrak{g})$ is a filtered algebra, and the associated graded carries naturally a structure of a Poisson algebra, which is isomorphic to $S(\mathfrak{g})$.

Since $n-(f|n)$ is "homogeneous" w.r.t. conformal weight, the Kazhdan filtration of $U(\mathfrak{g})$ induces a filtration on $\mathcal{W}^{\mathrm{fin}}(\mathfrak{g},f)$, and we have:

Proposition 4 (GG01) $gr\mathcal{W}^{fin.}(\mathfrak{g},f) \simeq \mathcal{W}^{cl.fin.}(\mathfrak{g},f)$.

Exercise 32 Prove Proposition 4.

4.4 Classical Affine $\mathcal{W}$-Algebras

We are not going to describe the quantum affine $\mathcal{W}$-algebra $\mathcal{W}_k(\mathfrak{g},f)$. This is a vertex algebra, and its definition is obtained via the so called BRST cohomology [35, 41]. The main observations, from our point of view, are the following: $\mathcal{W}_k(\mathfrak{g},f)$ is a conformal vertex algebra, and its Zhu algebra is canonically isomorphic to the quantum finite $\mathcal{W}$-algebra $\mathcal{W}^{\mathrm{fin}}(\mathfrak{g},f)$ [12]; moreover, it has a classical limit, given by the classical affine $\mathcal{W}$-algebra. In the present section we will describe this classical limit.

The set up for the construction of the classical affine $\mathcal{W}$-algebras is the same as for classical finite and quantum finite $\mathcal{W}$-algebras: we fix a finite dimensional simple (or reductive) Lie algebra $\mathfrak{g}$, with a non-degenerate symmetric invariant bilinear form $(\cdot\,|\,\cdot)$, and we let $f\in\mathfrak{g}$ be a nilpotent element. By the Jacobson–Morozov Theorem, we include f in an $\mathfrak{sl}_2$-triple $(f,h=2x,e)\subset\mathfrak{g}$, and we have the corresponding $\operatorname{ad}x$-eigenspace decomposition $\mathfrak{g}=\bigoplus_{i\in\frac12\mathbb{Z}}\mathfrak{g}_i$. On $\mathfrak{g}_{\frac12}$ we have the non-degenerate skewsymmetric bilinear form $\omega(u,v)=(f|[u,v])$, and we let $\ell\subset\mathfrak{g}_{\frac12}$ be a maximal isotropic subspace (w.r.t. ω). We then consider the nilpotent subalgebra $\mathfrak{n}=\ell\oplus\mathfrak{g}_{\geq1}\subset\mathfrak{g}$. We also denote $\chi=(f|\cdot)|_{\mathfrak{n}}\in\mathfrak{n}^*$.

Recall the algebraic Hamiltonian reduction definition (41) of the classical finite $\mathcal{W}$-algebra. We want to consider its "affine analogue". Clearly, the affine analogue of the Poisson algebra $S(\mathfrak{g})$ is the Poisson vertex algebra $\mathcal{V}(\mathfrak{g})=S(\mathbb{C}[\partial]\mathfrak{g})$ from Example 10. The affine analogue of the Lie algebra $\mathfrak{n}$, is the Lie conformal algebra $\mathbb{C}[\partial]\mathfrak{n}$, with λ-bracket $\{a_\lambda b\}=[a,b]$ $(a,b\in\mathfrak{n})$. Note that, since $(\mathfrak{n}|\mathfrak{n})=0$, this is a Lie conformal subalgebra of $\mathcal{V}(\mathfrak{g})$. We then apply the general Hamiltonian reduction procedure (30). We start with the PVA $\mathcal{V}(\mathfrak{g})$. We then quotient it by

$$\mathcal{I}=\langle n-(f|n)\rangle_{n\in\mathfrak{n}}\subset\mathcal{V}(\mathfrak{g})\,,\tag{43}$$

the differential algebra ideal generated by all the elements $n-(f|n)$, i.e. the algebra ideal generated by $n-(f|n)$, n', n'', ..., for all $n\in\mathfrak{n}$. Since $\chi([\mathfrak{n},\mathfrak{n}])=0$, this ideal is invariant by the λ-action of the Lie conformal algebra $\mathbb{C}[\partial]\mathfrak{n}$. It is not a PVA ideal, hence the quotient is not a PVA, but if we take the invariants with respect to the λ-action of $\mathfrak{n}$, we get a well defined PVA. In conclusion, we get the following

Definition 7 ([18]) The *classical $\mathcal{W}$-algebra* is:

$$\mathcal{W}^{\mathrm{cl}}(\mathfrak{g}, f) = \Big(\mathcal{V}(\mathfrak{g}) \big/ \langle n - (f|n)\rangle_{n\in\mathfrak{n}}\Big)^{\mathrm{ad}_\lambda \mathfrak{n}} = \mathcal{N}/\mathcal{I}\,, \tag{44}$$

where $\mathcal{I}$ is as in (43) and

$$\mathcal{N} = \big\{P \in \mathcal{V}(\mathfrak{g}) \,\big|\, \{n_\lambda P\} \in \mathcal{I}\big\}\,. \tag{45}$$

This Hamiltonian reduction procedure for Poisson vertex algebras can also be interpreted as a Dirac reduction [20, 22]. In order to do so, one need to consider a "non-local" generalization of the notion of Poisson vertex algebra [17].

Exercise 33 Show that $\mathcal{N}$ is a PVA subalgebra of $\mathcal{V}(\mathfrak{g})$ and $\mathcal{I}$ is its PVA ideal. Hence, $\mathcal{W}^{cl,fin}(\mathfrak{g}, f)$ has a natural structure of PVA.

As a differential algebra, it is not hard to prove that $\mathcal{W}^{\mathrm{cl}}(\mathfrak{g}, f)$ is isomorphic to an algebra of differential polynomials in finitely many variables:

$$W(\mathfrak{g}, f) \simeq \mathbb{F}\Big[w_i^{(n)} \,\Big|\, \begin{matrix} i = 1,\ldots, \dim(\mathfrak{g}^f) \\ n \in \mathbb{Z}_+ \end{matrix}\Big]\,. \tag{46}$$

The generators w_i are called the Premet's generators [46] and their number equals $\dim(\mathfrak{g}^f)$, the dimension of the centralizer of f in $\mathfrak{g}$. Note that the same result holds for all other types of $\mathcal{W}$-algebras: classical finite, quantum finite and quantum affine.

Some natural problems arise in the context of classical affine $\mathcal{W}$-algebras:

Problem 1 (1) Find explicit representatives of generators $\{w_i\}_{i=1}^{\dim(\mathfrak{g}^f)}$ in $\mathcal{N}$.
(2) Find explicit formulas for the λ-brackets among generators: $\{w_i{}_\lambda w_j\}$.
(3) Construct, if possible, an integrable hierarchy of Hamiltonian equations for the PVA structure of $\mathcal{W}^{\mathrm{cl}}(\mathfrak{g}, f)$.

For example, it is not hard to check that the classical $\mathcal{W}$-algebra for the Lie algebra $\mathfrak{sl}_2$ is isomorphic to the Virasoro–Magri PVA of Example 11: $W(\mathfrak{sl}_2, f) \simeq \mathcal{V}ir$; hence, the corresponding integrable hierarchy is the KdV hierarchy, cf. Example 12.

An explicit, rather involved, formula for λ-brackets among the $\mathcal{W}$-algebra generators was found in [13, 23].

In the next sections we will describe a method which, for a classical simple Lie algebra $\mathfrak{g}$ and an arbitrary nilpotent $f \in \mathfrak{g}$, solves all three problems above at the same time. It is based on the notions of *Adler type operators* [1, 21] and *generalized quasideterminants* [25, 26, 28].

4.5 Lax Equations

Let $L = \partial^n + f_{n-1}\partial^{n-1} + \cdots$ be a pseudodifferential operator. Consider the operators $(L^{\frac{k}{n}})_+$, the differential part of the fractional powers of L, and the corresponding *Lax hierarchy*

$$\frac{dL}{dt_k} = [(L^{\frac{k}{n}})_+, L]\,. \tag{47}$$

We assume that the Lax equations (47) are *self consistent* for all $k \geq 1$. By this we mean that the pseudodifferential operator $[(L^{\frac{k}{n}})_+, L]$ should have the same order as L, and all the differential relations satisfied by the coefficients of $\frac{dL}{dt}$ (deriving by relations among the coefficients of L), should also be automatically satisfied by the same coefficients of $[(L^{\frac{k}{n}})_+, L]$.

Under this assumption, the "Lax Theorem" guarantees that the hierarchy (47) is integrable, and $\int h_k = \int \mathrm{Res}_\partial L^{k/n}$, $k \geq 1$, are integrals of motion in involution.

Exercise 34 The main example of Lax is given by the operator $L(\partial) = \partial^2 + u$. Check that, in this case, $(L^{\frac{3}{2}})_+ = \partial^3 + 2u\partial + u'$, and $[(L^{\frac{3}{2}})_+, L] = u''' + uu'$. Hence, the Lax equation (47) is self consistent and it coincides with the KdV equation.

Exercise 35 Consider the operators $L = \partial^3 + u$. Show that, in this case, the lax equation $\frac{dL}{dt_k} = [(L^{\frac{k}{3}})_+, L]$ for $k = 1$ is $\frac{du}{dt_1} = u'$, which is self consistent, but for $k = 2$ it is

$$\frac{du}{dt} = 2u'\partial + u''\,,$$

which is NOT self consistent.

It is usually very hard to find Lax operators, for which all Lax equations (47) are self consistent. Here is a list of examples:

(i) $L = \partial^2 + u \quad \Rightarrow$ KdV hierarchy
(ii) $L = \partial^3 + u\partial + v \quad \Rightarrow$ Boussinesq hierarchy
(iii) $L = \partial^n + u_1\partial^{n-2} + \cdots + u_{n-1} \quad \Rightarrow$ nth KdV hierarchy
(iv) $L = \partial + u\partial^{-1}v \quad \Rightarrow$ NLS hierarchy
(v) $L = \partial^2 + u + v\partial^{-1}w \quad \Rightarrow$ Yajima–Oikawa hierarchy

In order to solve all three Problems 1, we will construct for each nilpotent element $f \in \mathfrak{g}$, an operator $L(\partial)$ with the following properties:

(1) $L(\partial)$ has coefficients in the $\mathcal{W}$-algebra $W(\mathfrak{gl}_N, f)$, and it allows us to derive all Premet's generators $\{w_i\}_{i=1}^{\dim \mathfrak{g}^f}$;
(2) $L(\partial)$ satisfies a certain *Adler identity* for the λ-brackets, which allows us to derive all λ-brackets among Premet's generators;
(3) all Lax equations (47) associated to $L(\partial)$ are self consistent and they are Hamiltonian with respect to the PVA structure of $\mathcal{W}^{\mathrm{cl}}(\mathfrak{g}, f)$, and the Hamiltonian functional $\int h_k = \int \mathrm{Res}_\partial(L^{\frac{k}{n}})$.

In conclusion, the construction of this operator $L(\partial)$ solves all 3 problems 1 at the same time!

We will describe the construction of $L(\partial)$ in the special case of $\mathfrak{g} = \mathfrak{gl}_N$, following [25, 26], but a similar construction works for all classical Lie algebras $\mathfrak{sl}_N$, $\mathfrak{so}_N$, $\mathfrak{sp}_N$, see [28]. In fact, a similar construction works in the quantum finite case, to provide generators for the quantum finite $\mathcal{W}$-algebra $\mathcal{W}^{\text{fin}}(\mathfrak{g}, f)$, [14, 27, 29].

4.6 First Ingredient: Adler Type Operators

Let $\mathcal{V}$ be a PVA with λ-bracket $\{\cdot\,_\lambda\,\cdot\}$.

Definition 8 An $N \times N$ matrix pseudodifferential operator with coefficients in $\mathcal{V}$, $A(\partial) \in \text{Mat}_{N\times N}\mathcal{V}((\partial^{-1}))$, is of *Adler type* (w.r.t. $\{\cdot\,_\lambda\,\cdot\}$) if the following identity holds:

$$\begin{aligned}\{A_{ij}(z)_\lambda A_{hj}(w)\} = & A_{hj}(w+\lambda+\partial)(z-w-\lambda-\partial)^{-1}(A_{ik})^*(\lambda-z) \\ & - A_{hj}(z)(z-w-\lambda-\partial)^{-1}A_{ik}(w)\,.\end{aligned} \tag{48}$$

Example 13 The "ancestor" Adler type operator is constructed as follows. Consider the affine PVA $\mathcal{V} = \mathcal{V}(\mathfrak{gl}_N)$ (with $\{a_\lambda b\} = [a, b] + (a|b)\lambda$). Then the following matrix is of Adler type

$$E + \partial\mathbb{1} = \begin{pmatrix} e_{11}+\partial & e_{21} & \dots & e_{N1} \\ e_{12} & e_{22}+\partial & \dots & e_{N2} \\ \vdots & & \ddots & \vdots \\ e_{1N} & & \dots & e_{NN}+\partial \end{pmatrix} = \mathbb{1}_N\partial + \sum_{i,j=1}^N e_{ij}E_{ji} \tag{49}$$

Here and further we denote by $E_{ij} \in \text{Mat}_{N\times N}(\mathbb{F})$, $i, j = 1, \dots, N$, the standard basis of the space of $N \times N$-matrices, and by e_{ij} the same basis, viewed inside the PVA: $e_{ij} \in \mathfrak{gl}_N \subset \mathcal{V}(\mathfrak{gl}_N)$.

Exercise 36 Check that the matrix (49) satisfies the Adler identity (48).

The next theorem shows how Adler type operators can be used to construct integrable systems.

Theorem 4 ([25]) *Let $\mathcal{V}$ be a PVA and let $A(\partial) \in \text{Mat}_{N\times N}\mathcal{V}((\partial^{-1}))$ be an operator of Adler type. Assume that the Kth rootv $A(\partial)^{\frac{1}{K}}$ of the pseudodifferential operator $A(\partial)$ exists, where K is a positive integer. Let*

$$\textstyle\int h_n = \int Res_\partial Tr(A(\partial)^{\frac{n}{K}}) \in \mathcal{V}/\partial\mathcal{V}\,,\ n \in \mathbb{Z}_+$$

Then, all the functionals $\int h_n$, $n \geq 0$, are pairwise in involution:

$$\{{\textstyle\int} h_m, {\textstyle\int} h_n\} = 0 \ \forall m, n$$

As a consequence, we have an integrable hierarchy of Hamiltonian equations:

$$\frac{du}{dt_n} = \{{\textstyle\int} h_n, u\}$$

This hierarchy is equivalently written in Lax form:

$$\frac{dA(\partial)}{dt_n} = [(A(\partial)^{\frac{n}{K}})_+, A(\partial)] \ , \quad n \in \mathbb{Z}_+$$

According to Theorem 4, in order to construct integrable systems of Hamiltonian PDE's, we need to construct Adler operators. The question remains on how to construct such Adler operators. So far we only have the "ancestor" Adler operator $\partial\mathbb{1} + E$ from Example 13, which in fact produces only trivial hierarchies.

A way to construct new Adler operators is based on the notion of (generalized) quasideterminants.

4.7 Second Ingredient: (Generalized) Quasideterminants

Definition 9 Let V be an associative algebra and let $A = (a_{ij}) \in \mathrm{Mat}_{N\times N} V$. The (i, j)-quasideterminant of A (see e.g. [39]) is, if it exists,

$$|A|_{ij} = a_{ij} - R_i^j (A^{ij})^{-1} C_j^i \tag{50}$$

where: R_i^j is ith row of A without j-entry; C_j^i is jth column of A without i-entry; and A^{ij} is the matrix A without row i and column j. We are assuming, in the definition (50), that the matrix A^{ij} is invertible. If this is not the case, the quasideterminant $|A|_{ij}$ does not exist.

The following is a simple linear algebra exercise:

Exercise 37 The quasideterminant (50) is also given by

$$|A|_{ij} = \left(\text{entry } (ji) \text{ of } A^{-1}\right)^{-1},$$

provided that both inverses exist.

We shall need the following generalization of Definition 9, see [25].

Definition 10 Let V be an associative algebra and let $A = (a_{ij}) \in \mathrm{Mat}_{N\times N} V$. Let $I \in \mathrm{Mat}_{N\times M} \mathbb{F}$ and $J \in \mathrm{Mat}_{M\times N} \mathbb{F}$ be such that $\mathrm{rk}(JI) = M$. The (I, J)-*generalized quasideterminant* of A is, if it exists,

$$|A|_{IJ} = (JA^{-1}I)^{-1}.$$

The following result gives us a way to construct new Adler type operators.

Theorem 5 ([25]) *If $A(\partial)$ is of Adler type for $\mathcal{V}$, then any its generalized quasideterminant $|A(\partial)|_{I,J}$ is again of Adler type.*

Exercise 38 Prove Theorem 5.

4.8 Lax Type Operator for $\mathcal{W}(\mathfrak{gl}_N, f)$ and Associated Integrable Hamiltonian Hierarchy

Consider the "ancestor" Adler type operator defined in (49): $L(\partial)\partial\mathbb{1}_N + E \in \mathrm{Mat}_{N\times N}\mathcal{V}(\mathfrak{gl}_N)[\partial]$. Let $f \in \mathfrak{gl}_N$ be a nilpotent matrix, associated to the partition $p_1^{r_1}p_2^{r_2}\dots p_s^{r_s}$ of N. We can include f in an $\mathfrak{sl}_2$-triple $(e, h = 2x, f)$. The matrix x is diagonalizable on $\mathbb{F}^N$, with half-integer eigenvalues

$$\mathbb{F}^N = \oplus_{j=-\Delta}^{\Delta} V[j]. \tag{51}$$

The maximal eigenvalue is $\Delta = \frac{p_1-1}{2}$, with multiplicity r_1. Let $J : \mathbb{F}^N \to V[\Delta]$ be the projection and $I : V[\Delta] \hookrightarrow V$ be the inclusion maps, associated to the decomposition (51). If we fix a basis of V compatible with the decomposition (51), then $I \in \mathrm{Mat}_{N\times r_1}\mathbb{F}$, and $J \in \mathrm{Mat}_{r_1\times N}$, and they are of maximal rank. Let $\rho : \mathcal{V}(\mathfrak{g}) \to V(\mathfrak{g})$ be the differential algebra homomorphism defined by:

$$\rho(a) = \pi_{\leq\frac{1}{2}}(a) + (f|a), \quad a \in \mathfrak{g},$$

where $\pi_{\leq\frac{1}{2}} : \mathfrak{gl}_N \twoheadrightarrow \mathfrak{g}_{\leq\frac{1}{2}}$ denotes the projection onto the $\mathrm{ad}\,x$-eigenspaces of eigenvalues $j \leq \frac{1}{2}$, and $(\cdot\,|\,\cdot)$ denotes the trace form of $\mathfrak{gl}_N$.

The "descendant" Lax operator corresponding to the nilpotent element $f \in \mathfrak{gl}_N$ is obtained as the (I, J)-generalized quasideterminant of the image, via ρ, of the ancestor Lax operator:

$$L_f(\partial) = (J(\rho(\partial\mathbb{1}_N + E))^{-1}I))^{-1} \in \mathrm{Mat}_{r_1\times r_2}\mathcal{V}\left(\mathfrak{gl}_{\leq\frac{1}{2}}\right). \tag{52}$$

Theorem 6 ([26]) *The operator $L_f(\partial)$ in (52) is an $r_1 \times r_1$ matrix pseudodifferential operator with leading term ∂^{p_1} and coefficients in the $\mathcal{W}$-algebra $\mathcal{W}^{cl}(\mathfrak{gl}_N, f)$:*

$$L_f(\partial) = \partial^{p_1}\mathbb{1}_{r_1\times r_1} + \cdots \in \mathcal{W}^{cl}(\mathfrak{gl}_N, f)((\partial^{-1})) \otimes \mathrm{End}\,V[\Delta]$$

In fact, $L_f(\partial)$ encodes all the generators $\{w_i\}_{i=1}^{\dim\mathfrak{g}^f}$ of the $\mathcal{W}$-algebra $\mathcal{W}^{cl}(\mathfrak{gl}_N, f)$.

Theorem 7 ([26]) *The operator $L_f(\partial)$ satisfies the Adler identity* (48). *In fact, the Adler identity encodes all λ-brackets among the generators of the $\mathcal{W}$-algebra $\mathcal{W}^{cl}(\mathfrak{g}, f)$.*

According to Theorem 4, we automatically get the corresponding integrable hierarchy of Hamiltonian equations in Lax form:

Theorem 8 ([26]) *The local functionals*

$$\int h_n = \int Res_\partial \operatorname{Tr} L_f(\partial)^{\frac{n}{p_1}} \in \mathcal{W}^{cl}(\mathfrak{g}, f)/\partial W\,,$$

are Hamiltonian functionals in involution:

$$\left\{\int h_m, \int h_n\right\} = 0 \text{ for all } m, n\,.$$

We thus get an integrable hierarchy of Hamiltonian equations for $W(\mathfrak{g}, f)$

$$\frac{du}{dt_n} = \{\textstyle\int h_n, u\}\,,\ n \geq 0\,.$$

This hierarchy can be written in Lax form:

$$\frac{dL_f(\partial)}{dt_n} = [L_f(\partial)_+^{\frac{n}{p_1}}, L_f(\partial)]\,..$$

Remark 1 The above results have been generalized to all classical Lie algebras: $\mathfrak{sl}_N$, $\mathfrak{so}_N$, $\mathfrak{sp}_N$, see [28].

4.9 Examples

Recall that the nilpotent orbits of $\mathfrak{gl}_N$ are parametrized by partitions $N = p_1 + p_2 + \cdots + p_s$, with $p_1 \geq p_2 \geq \cdots \geq p_s$.

Example 14 Partition $2 = 2$. It corresponds to the KdV hierarchy, the simplest equation being:

$$\frac{\partial u}{\partial t} = \frac{\partial^3 u}{\partial x^3} + u\frac{\partial u}{\partial x}\,.$$

The first important discovery in theory of integrable systems was that KdV is integrable.

Example 15 Partition $2 = 1 + 1$. It corresponds to the NLS hierarchy (=AKNS) in two variables u and v, the simplest equation being

$$\begin{cases} \frac{\partial u}{\partial t} = \frac{\partial^2 u}{\partial x^2} + ku^2 v \\ \frac{\partial v}{\partial t} = -\frac{\partial^2 v}{\partial x^2} v - kuv^2 \end{cases} \quad (1964)\,.$$

Example 16 Partition $3 = 3$. It corresponds to the Boussinesq hierarchy, the simplest equation being the Boussinesq equations

$$\begin{cases} \frac{\partial u}{\partial t} = \frac{\partial v}{\partial x} \\ \frac{\partial v}{\partial t} = \frac{\partial^3 u}{\partial x^3} + u\frac{\partial u}{\partial x} \end{cases} \quad (1872)\,.$$

Example 17 Partition $3 = 1 + 1 + 1$. It corresponds to the 3 wave equation.

Example 18 Partition $3 = 2 + 1$. It corresponds to the Yajima–Oikawa hierarchy in three variables u, v, w, the simplest equation describing sonic-Langmuir solitons:

$$\begin{cases} \frac{\partial u}{\partial t} = -\frac{\partial^2 u}{\partial x^2} + uw \\ \frac{\partial v}{\partial t} = \frac{\partial^2 v}{\partial x^2} - vw \\ \frac{\partial w}{\partial t} = \frac{\partial}{\partial x}(uv) \end{cases} \quad (1976)\,.$$

Example 19 Partition $N = N$. It corresponds to the Nth Gelfand–Dickey hierarchy (1975).

Example 20 Partition $N = 2 + 1 + \cdots + 1$. It corresponds to the N–2-component Yajima–Oikawa hierarchy.

Example 21 Partition $N = p + p + \cdots + p$ (r times). It corresponds to the pth $r \times r$-matrix Gelfand-Dickey hierarchy.

Exercise 39 Check (some of) these examples.

References

1. Adler, M.: On a trace functional for formal pseudodifferential operators and the symplectic structure of the Korteweg-de Vries equation. Invent. Math. **50**, 219–248 (1979)
2. Bakalov, B., Kac, V.G.: Field Algebras. IMRN **3**, 123–159 (2003)
3. Bakalov, B., De Sole, A., Heluani, R., Kac, V.G.: An operadic approach to vertex algebra and Poisson vertex algebra cohomology. arXiv:1806.08754
4. Bakalov, B., De Sole, A., Heluani, R., Kac, V.G.: Chiral vs classical operad. arXiv:1812.05972
5. Bakalov, B., De Sole, A., Heluani, R., Kac, V.G.: Computation of cohomology of Lie conformal and Poisson vertex algebras. arXiv:19xxx
6. Barakat, A., De Sole, A., Kac, V.G.: Poisson vertex algebras in the theory of Hamiltonian equations. Jpn. J. Math. **4**(2), 141–252 (2009)
7. Borcherds, R.: Vertex algebras, Kac-Moody algebras and the Monster. Proc. Natl. Acad. Sci. USA **83**, 3068–3071 (1986)
8. Burroughs, N., de Groot, M., Hollowood, T., Miramontes, L.: Generalized Drinfeld-Sokolov hierarchies II: the Hamiltonian structures. Commun. Math. Phys. **153**, 187–215 (1993)

9. de Boer, J., Tjin, T.: Representation theory of finite W algebras. Commun. Math. Phys. **158**, 485–516 (1993)
10. de Boer, J., Tjin, T.: The relation between quantum W algebras and Lie algebras. Commun. Math. Phys. **160**, 317–332 (1994)
11. de Groot, M., Hollowood, T., Miramontes, L.: Generalized Drinfeld-Sokolov hierarchies. Commun. Math. Phys. **145**, 57–84 (1992)
12. De Sole, A., Kac, V.G.: Finite vs. affine W-algebras. Jpn. J. Math. **1**(1), 137–261 (2006)
13. De Sole, A.: On classical finite and affine W-algebras. Springer INdAM Ser. **7**, 51–67 (2014)
14. De Sole, A., Fedele, L., valeri, D.: Generators of the quantum finite W-algebras in type A. arXiv:1806.03233
15. De Sole, A., Kac, V.G.: The variational Poisson cohomology Japan. J. Math. **8**, 1–145 (2013)
16. De Sole, A., Kac, V.G.: Essential variational Poisson cohomology Comm. Math. Phys. **313**(3), 837–864 (2012)
17. De Sole, A., Kac, V.G.: Non-local Poisson structures and applications to the theory of integrable systems Jpn. J. Math. **8**(2), 233–347 (2013)
18. De Sole, A., Kac, V.G., Valeri, D.: Classical W-algebras and generalized Drinfeld-Sokolov bi-Hamiltonian systems within the theory of Poisson vertex algebras. Commun. Math. Phys. **323**(2), 663–711 (2013)
19. De Sole, A., Kac, V.G., Valeri, D.: Classical W-algebras and generalized Drinfeld-Sokolov hierarchies for minimal and short nilpotents. Alberto De Sole, Victor G. Kac, Daniele Valeri Commun. Math. Phys. **331**(2), 623–676 (2014)
20. De Sole, A., Kac, V.G., Valeri, D.: Dirac reduction for Poisson vertex algebras Comm. Math. Phys. **331**(3), 1155–1190 (2014)
21. De Sole, A., Kac, V.G., Valeri, D.: Adler-Gelfand-Dickey approach to classical W-algebras within the theory of Poisson vertex algebras. Int. Math. Res. Not. **21**, 11186–11235 (2015)
22. De Sole, A., Kac, V.G., Valeri, D.: Integrability of Dirac reduced bi-Hamiltonian equations. Trends in Contemporary Mathematics. Springer INdAM Series vol. 8, pp. 13–32 (2014)
23. De Sole, A., Kac, V.G., Valeri, D.: Structure of classical (finite and affine) W-algebras. J. Eur. Math. Soc. **18**(9), 1873–1908 (2016)
24. De Sole, A., Kac, V.G., Valeri, D.: Double Poisson vertex algebras and non-commutative Hamiltonian equations. Adv. Math. **281**, 1025–1099 (2015)
25. De Sole, A., Kac, V.G., Valeri, D.: A new scheme of integrability for (bi)Hamiltonian PDE. Commun. Math. Phys. **347**(2), 449–488 (2016)
26. De Sole, A., Kac, V.G., Valeri, D.: Classical W-algebras for $\mathfrak{gl}_N$ and associated integrable Hamiltonian hierarchies. Commun. Math. Phys. **348**(1), 265–319 (2016)
27. De Sole, A., Kac, V.G., Valeri, D.: Finite W-algebras for $\mathfrak{gl}_N$. Adv. Math. **327**, 173–224 (2018)
28. De Sole, A., Kac, V.G., Valeri, D.: Classical affine W-algebras and the associated integrable Hamiltonian hierarchies for classical Lie algebras. Commun. Math. Phys. **360**(3), 851–918 (2018)
29. De Sole, A., Kac, V.G., Valeri, D.: A Lax type operator for quantum finite W-algebras. arXiv:1707.03669
30. De Sole, A., Kac, V.G., Valeri, D., Wakimoto, M.: Local and non-local multiplicative Poisson vertex algebras and differential-difference equations. arXiv:1809.01735
31. De Sole, A., Kac, V.G., Valeri, D., Wakimoto, M.: Poisson λ-brackets for differential-difference equations. arXiv:1806.05536
32. Drinfeld, V.G., Sokolov, V.V.: Lie algebras and equations of Korteweg-de Vries type. Sov. J. Math. **30**, 1975–2036 (1985)
33. Elashvili, A.G., Kac, V.G., Vinberg, E.B.: Cyclic elements in semisimple Lie algebras. Transf. Groups **18**, 97–130 (2013)
34. Fehér, L., Harnad, J., Marshall, I.: Generalized Drinfeld-Sokolov reductions and KdV type hierarchies. Commun. Math. Phys. **154**(1), 181–214 (1993)
35. Feigin, B.L., Frenkel, E.: Quantization of Drinfeld-Sokolov reduction. Phys. Lett., B **246**, 75–81 (1990)

36. Fernández-Pousa, C., Gallas, M., Miramontes, L., Sánchez Guillén, J.: $\mathcal{W}$-algebras from soliton equations and Heisenberg subalgebras. Ann. Phys. **243**(2), 372–419 (1995)
37. Fernández-Pousa, C., Gallas, M., Miramontes, L., Sánchez Guillén, J.: Integrable systems and $\mathcal{W}$-algebras. VIII J. A. Swieca Summer School on Particles and Fields (Rio de Janeiro, 1995), pp. 475–479
38. Gan, W.L., Ginzburg, V.: Quantization of Slodowy slices. Int. Math. Res. Not. **5**, 243–255 (2002)
39. Gelfand, I.M., Gelfand, S.I., Retakh, V., Wilson, R.L.: Quasideterminants. Adv. Math. **193**(1), 56–141 (2005)
40. Kac, V.G.: Vertex Algebras for Beginners. University Lecture Notes, vol. 10, 2nd edn. AMS, Providence (1996)
41. Kac, V.G., Roan, S.-S., Wakimoto, M.: Quantum reduction for affine superalgebras. Commun. Math. Phys. **241**, 307–342 (2003)
42. Kac, V.G., Wakimoto, M.: Quantum reduction and representation theory of superconformal algebras. Adv. Math. **185**, 400–458 (2004). Corrigendum, Adv. Math. **193**, 453–455 (2005)
43. Kostant, B.: On Whittaker vectors and representation theory. Invent. Math. **48**, 101–184 (1978)
44. Lynch, T.E.: Generalized Whittaker vectors and representation theory. Thesis (Ph.D.) - Massachusetts Institute of Technology (1979)
45. Magri, F.: A simple model of the integrable Hamiltonian equation. J. Math. Phys. **19**(5), 1156–1162 (1978)
46. Premet, A.: Special transverse slices and their enveloping algebras. Adv. Math. **170**, 1–55 (2002)
47. Slodowy, P.: Simple Singularities and Simple Algebraic Groups. Lecture Notes in Mathematics, vol. 815. Springer, Berlin (1980)
48. Vaisman, I.: Lectures on the Geometry of Poisson Manifolds. Progress in Mathematics, vol. 118. Birkhäuser, Basel (1994)
49. Zamolodchikov, A.: Infinite extra symmetries in two-dimensional conformal quantum field theory. Teor. Mat. Fiz **65**(3), 347–359 (1985)
50. Zhu, Y.: Modular invariance of characters of vertex operator algebras. J. AMS **9**, 237–302 (1996)

NGK and HLZ: Fusion for Physicists and Mathematicians

Shashank Kanade and David Ridout

Abstract In this expository note, we compare the fusion product of conformal field theory, as defined by Gaberdiel and used in the Nahm–Gaberdiel–Kausch (NGK) algorithm, with the $P(w)$-tensor product of vertex operator algebra modules, as defined by Huang, Lepowsky and Zhang (HLZ). We explain how the equality of the two "coproducts" derived by NGK is essentially dual to the $P(w)$-compatibility condition of HLZ and how the algorithm of NGK for computing fusion products may be adapted to the setting of HLZ. We provide explicit calculations and instructive examples to illustrate both approaches. This document does not provide precise descriptions of all statements, it is intended more as a gentle starting point for the appreciation of the depth of the theory on both sides.

Keywords Vertex operator algebras · Conformal field theory · Tensor categories · Fusion

1 Before We Begin

> Perhaps fusion always has the quasi-rational features seen for the free bosons. It will be important to investigate Calabi-Yau spaces from this point of view, but the tools for this study have yet to be developed. As a first step, we need a rigorous and convenient definition of the fusion product for generic theories, and better algorithms for its evaluation.
>
> ...Werner Nahm [61]

S. Kanade (✉)
Department of Mathematics, University of Denver, Denver, CO 80208, USA
e-mail: shashank.kanade@du.edu

D. Ridout
School of Mathematics and Statistics, University of Melbourne,
Parkville, VIC 3010, Australia
e-mail: david.ridout@unimelb.edu.au

D. Adamović and P. Papi (eds.), *Affine, Vertex and W-algebras*,
Springer INdAM Series 37, https://doi.org/10.1007/978-3-030-32906-8_7

1.1 Why?

It is a highly non-trivial matter to form tensor products (called fusion products in physics parlance) of modules for a given vertex operator algebra (chiral algebra), not to mention building a braided tensor category out of these modules.

The late 80s and early 90s witnessed an intense period of activity devoted to this problem:

- Feigin and Fuchs described the fusion coefficients of certain conformal field theories as dimensions of spaces of coinvariants [14].
- Moore and Seiberg wrote their highly influential papers [58, 59] on rational conformal field theories,[1] introducing a "coproduct-like" formula for the action of the chiral algebra. However, they incorrectly identified the vector space underlying the fusion product as that underlying the usual tensor product.
- Frenkel and Zhu codified the coinvariant approach for certain classes of modules over quite general vertex operator algebras [19], see also work of Li [52, 53], obtaining a formula for the fusion coefficients in the rational case.
- Gaberdiel extended the coproduct formula of Moore-Seiberg and corrected their work by using locality to (morally) define the fusion product as a quotient of the vector space tensor product [20, 21].
- Kazhdan and Lusztig proved [45–49] major theorems defining fusion rigorously and relating certain tensor categories for affine Lie algebras at non-rational levels to quantum group tensor categories.
- Nahm introduced an algorithmic method to analyse and, in favourable cases (rational conformal field theories in particular), identify fusion products [61].
- Huang and Lepowsky wrote their first series of papers [35–38] on rigorously defining fusion and proving tensor structure theorems for appropriate module categories over rational vertex operator algebras.
- Gaberdiel and Kausch extended Nahm's methods to include logarithmic conformal field theories[2] and implemented them on a computer [25]. The resulting algorithm is now known as the Nahm–Gaberdiel–Kausch fusion algorithm.

We will not discuss this history in any more detail, instead referring the reader to resources such as [39] (for mathematicians) and [9, 23] (for physicists), as well as to the references cited therein.

Our aim in this expository note is to focus exclusively on the physicists' computational approach, as explained by Nahm, Gaberdiel and Kausch (NGK henceforth), and the mathematicians' rigorous approach, as developed by Huang, Lepowsky and Zhang (HLZ for short) in [41]. Here, we concentrate only on the definition of the

[1] A conformal field theory is said to be *rational* if its quantum state space is semisimple (completely reducible), decomposing into a finite direct sum of tensor products of irreducible modules. The name reflects the fact that such theories have rational central charges and conformal weights.

[2] A conformal field theory is said to be *logarithmic* if its quantum state space is not completely reducible. The name reflects the fact that all known examples possess a non-diagonalisable action of the hamiltonian which leads to logarithmic singularities in certain correlation functions.

fusion product and its algorithmic construction; the question of whether it is possible to build a braided tensor category from this fusion product is much much more difficult. Readers interested in the categorical structures underlying conformal field theory can instead turn to papers such as [8, 34, 43].

The approaches of NGK and HLZ start from quite different points of view and use very different language. However, the ingredients are almost identical. The main idea is to somehow construct the fusion product out of the vector space tensor product of modules. Along the way, one has to deal with some nasty convergence issues and to make everything work smoothly, a fix is needed. Physicists were well aware of this difficulty, see [20, 61], and the NGK fusion algorithm avoids these convergence issues by working with various quotients of the fusion product. On the other hand, Huang and Lepowsky successfully tackled this issue rigorously for rational models by passing to the dual space and, together with Zhang, subsequently extended their dual space formalism to cover logarithmic cases.

In this note, we wish to explain two main points. First, the NGK "definition" of the fusion product rests on imposing the condition that two seemingly different coproducts are in fact the same, while the HLZ tensor product is built from functionals in the dual space that are required to satisfy a certain compatibility condition. We will show that these two conditions are essentially the same thing (more precisely, they are duals of one another). Second, we will focus on algorithmic implementations for constructing fusion products. We shall explain this with the aid of a specific example in both the NGK and HLZ formalisms.

Our purpose is to facilitate a dialogue between mathematicians and physicists by presenting a coherent fusion (pun intended) of various ideas. In particular, we would like to reassure mathematicians that the fusion rules that physicists compute with NGK can be rigorously justified (in principle) and also to reassure physicists that the theorems that mathematicians prove with the universal tensor product theory of HLZ are indeed results about fusion. We would however like to remind the reader that this paper should not be relied upon for precise statements of results. While we have done our best not to tell any outright lies, the theory of fusion is notoriously subtle and a full account with all details explained would necessarily take more than the space we have here. As always, the cited literature is the canonical source for these details.

1.2 How?

This note is organised as follows.

In Sect. 2, we first provide a review of the (sometimes wildly different) notation and terminology used by mathematicians and physicists, so as to make it easier for both audiences to follow the rest of the paper. We also take the opportunity to fix some of our own choices in this regard. Our compromises will no doubt outrage many readers, but we take solace in the fact that it is really not possible to please everyone in this respect.

We begin our exposition by first presenting Gaberdiel's original approach to defining fusion products in Sect. 3. Here, we will explain how the fusion product of two modules is morally constructed from their vector space tensor product by imposing a certain equality of two coproducts. These coproducts are derived from locality, so the definition appears natural and uncontroversial. However, we point out concretely that this equality of coproducts does not make sense in general because it involves infinite coefficients when expanded in the usual fashion.

We then move on to the HLZ approach. Just like the case of tensor products of modules over commutative rings, they define fusion via a universal property which we cover in Sect. 4. Here, we introduce and explain the concept of intertwining maps. These maps are a rigorous version of the physicists' rather general notion of a field and are central to the universal property. Of course, this definition of fusion is quite abstract and it is not obvious how to go about actually computing it. One therefore needs a concrete model of this universal gadget to realise the fusion product concretely. We begin to explain this in Sect. 5, working with a model built out of the vector space tensor product of modules, and we explain its shortcomings in terms of convergence issues. By the end of this section, we will have understood Gaberdiel's definition of the fusion product in terms of the universal property of HLZ.

In order to cure the divergences of this model, one can pass to the dual of the vector space tensor product. This is the key to the rigorous formalism developed by HLZ, as we explain in Sect. 6. Here we shall present a succinct exposition of the crucial "$P(z)$-compatibility condition" to build fusion products, although we shall use the complex variable w instead of z to make the relation with the physicists' methods more transparent. We end this section by working out what the HLZ formalism means in the example of the "simplest" possible vertex operator algebras—those associated to commutative associative unital algebras.

Section 7 is devoted to working out the details, using the HLZ formalism, of a specific example of a fusion product. We choose to fuse a certain highest-weight Virasoro module of central charge -2 with itself for historically significant reasons, see [30]. We then introduce, in Sect. 8, some basic features of the NGK fusion algorithm by dualising the algorithm we followed for HLZ. This is also illustrated with the same example, mostly to make the parallel methodology manifest but also to explain the important roles played by Nahm's "special subspaces" (rediscovered by mathematicians as C_1-spaces) and "spurious states". We mention that the NGK algorithm has not only been successfully applied to a multitude of Virasoro fusion products [13, 25, 54, 55, 60, 63, 65], but has also been used to calculate fusion rules for the $N = 1$ superconformal algebras [4, 5], triplet algebras [26, 27, 70, 71], fractional-level Wess–Zumino–Witten models [10, 24, 66] and bosonic ghosts [67].

We shall close this paper in Sect. 9 by providing a quick summary of some of the other approaches to fusion, highlighting their similarities and differences to the NGK and HLZ approaches. Our lack of expertise with these alternative methods means that this section is far less detailed and we ask the reader for forgiveness in case of unintentional omissions. In any case, we hope that the literature we do mention will be of some use to those interested in studying these alternative approaches and will perhaps inspire future comparisons between them and NGK or HLZ.

2 Some Notation and Conventions

Our aim is to explain some of the deep constructions of Huang–Lepowsky–Zhang (HLZ) [41], in an manner accessible to physicists, while providing mathematicians with an opportunity to grasp the ideas of Nahm and Gaberdiel–Kausch (NGK) [25, 61]. The former provides a path to make fusion rigorous and the latter shows how to algorithmically implement the computation and identification of fusion products. There is, of course, significant overlap, but the language used by the authors (being mathematicians and physicists, respectfully) differs markedly. Even before we present precise definitions, which we shall do in due course, we feel that it will be useful to address this divide by providing a short dictionary of terms and notations that shall be used below.

Let V be a vertex operator algebra. By the state-field correspondence, there is a bijection between the space of fields $v(z)$ and the space of states $v \in V$. This state space is, of course, the vertex operator algebra. As a V-module, it is also commonly referred to as the vacuum module. In the mathematical literature, the field $v(z)$ is typically denoted by $Y(v, z)$. We shall use both notations interchangeably. While the axiomatic treatment presented in, for example, [16, 18, 44, 51] often regards z as a formal variable, we shall adhere to the physicists' convention of always working with a complex variable z, unless otherwise indicated.

In any vertex operator algebra, there are two distinguished states: the vacuum, which we shall denote by Ω, and the conformal vector, which we shall denote by ω. The former corresponds to the identity field 1, while the latter yields the energy-momentum (or stress-energy) tensor $T(z)$ whose Fourier modes (see (3) below) are identified with the Virasoro generators $L(n)$, $n \in \mathbb{Z}$. The most common notation, particularly in the physics literature, for the vacuum state is of course $|0\rangle$. However, we shall avoid using "bra-ket notation" entirely, reserving $\langle \cdot, \cdot \rangle$ to denote the usual pairing (whose output is a complex number) between a vector space and its dual.

It is important to note that physicists also use the term "field" to denote objects $\psi(z)$ that "correspond" (in the same sense as that of the state-field correspondence above) to the elements ψ of a given V-module M. More precisely, the fields $\psi(z)$ should be regarded as the chiral, or holomorphic, part of the "bulk fields" of the conformal field theory, the latter being objects $\psi(z, \overline{z})$ that correspond to linear combinations of elements in certain vector space tensor products $M \otimes \overline{M}$ of V-modules.[3] It is also important to note that physicists do not explain what (bulk) fields actually are—fields are fundamental objects in quantum field theory and so need not be explained as long as one can calculate with them. The mathematicians' approach to (the holomorphic part of these) fields will be discussed below in Definition 4.1. We are only interested in such holomorphic parts $\psi(z)$ here.

---

[3]In general, one might instead have tensor products $M \otimes \overline{M}$, where M is a V-module and $\overline{M}$ is a module over another vertex operator algebra, $\overline{V}$ say. In either case, M is responsible for the z-dependence of the field $\psi(z, \overline{z})$ while the $\overline{z}$-dependence comes from $\overline{M}$.

Another major notational difference between the physics and mathematics literature concerns the Fourier coefficients of the fields. In mathematics, the modes are frequently given as follows ($\mathfrak{m}$ stands for mathematics):

$$Y(v, z) = \sum_{n \in \mathbb{Z}} v_n^{(\mathfrak{m})} z^{-n-1}. \tag{1}$$

This has some advantages. First, it applies uniformly to all fields in the vertex operator algebra. Second, the shift by -1 in the exponent of z makes residue calculations particularly easy. Unfortunately, it also has some computational disadvantages, chief among which is that mode indices are not conserved. For example, derivatives and normally ordered products of fields have modes satisfying

$$(\partial v)_n^{(\mathfrak{m})} = -n v_{n-1}^{(\mathfrak{m})} \quad \text{and} \quad :vv':_n^{(\mathfrak{m})} = \sum_{r \leqslant -1} v_r^{(\mathfrak{m})} v_{n-1-r}^{(\mathfrak{m})} + \sum_{r > -1} v_{n-1-r}^{(\mathfrak{m})} v_r^{(\mathfrak{m})}, \tag{2}$$

respectively. This moreover leads to a conflict with the established convention for the Virasoro generators and so mathematicians almost always modify (1) for the energy-momentum tensor:

$$Y(\omega, z) = \sum_{n \in \mathbb{Z}} L(n) z^{-n-2}. \tag{3}$$

The expansion (3) is in fact an example of the convention used universally by physicists. Given a state v of $L(0)$-eigenvalue h_v, called the conformal dimension or conformal weight of v, the corresponding field is expanded as ($\mathfrak{p}$ stands for physics)

$$v(z) = \sum_{n \in \mathbb{Z} - h_v} v_n^{(\mathfrak{p})} z^{-n-h_v}. \tag{4}$$

We therefore have

$$v_n^{(\mathfrak{m})} = v_{n+1-h_v}^{(\mathfrak{p})} \quad \text{and so} \quad v_n^{(\mathfrak{p})} = v_{n-1+h_v}^{(\mathfrak{m})}. \tag{5}$$

The obvious disadvantage is that this only applies to states and fields of definite conformal weight and must be extended by linearity in general. It also means that residue calculations require a lot of h_v factors. The main advantage, aside from (3) being the natural expansion of $T(z)$, is that mode indices reflect the $L(0)$-grading:

$$[L(0), v_n^{(\mathfrak{p})}] = -n\, v_n^{(\mathfrak{p})}. \tag{6}$$

The physicists' analogues of (2) illustrate the consequent conservation of mode indices as well as the omnipresent h_v factors:

$$(\partial v)_n^{(\mathfrak{p})} = -(n + h_v) v_n^{(\mathfrak{p})} \quad \text{and} \quad :vv':_n^{(\mathfrak{p})} = \sum_{r \leqslant -h_v} v_r^{(\mathfrak{p})} v_{n-r}^{(\mathfrak{p})} + \sum_{r > -h_v} v_{n-r}^{(\mathfrak{p})} v_r^{(\mathfrak{p})}. \tag{7}$$

We also mention that the "zero modes" that play such an important role in classifying modules over vertex operator algebras [15, 19, 73], are only modes with index 0 if we employ the physicists' convention. Despite these computational advantages, our study of fusion will require a number of residue computations that turn out to be much cleaner with the mathematics convention. The default convention we use is therefore that of (1). We shall drop the superscript $(\mathfrak{m})$ in what follows for brevity: $v_n \equiv v_n^{(\mathfrak{m})}$.

3 Fusion: The Physicists' Approach

Fusion was originally introduced by physicists studying rational conformal field theory in order to keep track of which primary fields appeared in the operator product expansion of two primary fields. We recall that a primary field is one that corresponds to a (Virasoro) highest-weight vector under the state-field correspondence (extended to include non-vacuum modules). The standard means for computing the fusion of these two primary fields was then to calculate every correlation function of three primary fields in which two were the primaries being fused. If the result was found to be non-zero, then the fusion would include (the conjugate of) the third primary field. This led to *fusion rules* that were expressed as follows:

$$\psi_i \times \psi_j = \sum_k \mathcal{N}_{ij}{}^k \psi_k. \tag{8}$$

Here, the ψ_k are the primary fields, indexed by some (discrete) set, $\times$ is the *fusion product*, and the $\mathcal{N}_{ij}{}^k \in \mathbb{Z}_{\geqslant 0}$ are the *fusion coefficients* or *fusion multiplicities*. The meaning of these coefficients, when not 0 or 1, is somewhat obscure in this framework. We shall now reinterpret fusion rules in a manner that makes this meaning transparent.

In rational theories, conformal invariance allows one to compute correlation functions involving non-primary fields from primary ones. Physicists would therefore speak of the fusion of "conformal families". It is now not difficult to realise that the physicists' notion of a conformal family is morally identical to the mathematician's notion of a highest-weight module. Inevitably, the idea arose that the fusion product should actually be regarded as a product of modules. More precisely, any two modules M and N over a given vertex operator algebra V should admit a fusion product, which we shall denote by $M \boxtimes N$ to avoid confusion with direct products, that is also (naturally) a V-module. In rational conformal field theory, the original fusion product (8) of primary fields is thus upgraded to the following fusion product of irreducible highest-weight V-modules M_k:

$$M_i \boxtimes M_j \cong \bigoplus_k \mathcal{N}_{ij}{}^k M_k. \tag{9}$$

We remark that with this formulation, there is no longer any reason to require that the theory be rational nor that the M_k be irreducible and highest-weight.[4]

As an aside, we mention that mathematicians have since appropriated some of the terminology invented by physicists for fusion, but unfortunately use it in a different, and potentially confusing, way. In particular, the non-negative integers $\mathcal{N}_{ij}{}^{k}$ have come to be known as fusion rules in the mathematical literature. In what follows, we shall eschew this and always follow the nomenclature introduced by physicists. While there is a clear case to be made for respecting original terminology, which is anyway universally used in the physics literature, we also mention that using "fusion rule" for the explicit decomposition (9) of a fusion product into (indecomposable) V-modules accords well with the universally accepted usage of the group-theoretic term "branching rule" for the corresponding decompositions of restrictions of modules.

In any case, once one has decided to view fusion as a product on an appropriate category of V-modules, instead of in terms of correlators and operator product expansions of primary fields, several questions naturally arise:

1. What is the precise definition of fusion?
2. How can one actually compute it?
3. Is the result again a V-module?
4. Which modules can actually be fused?

The last two questions, which are actually about the module category, are perhaps outside the physicists' remit for clearly one must be able to fuse modules appearing in a conformal field theory in order to have a consistent theory. However, the first two were tackled by Gaberdiel in [20, 21] who credits unpublished work of Borcherds and the widely influential work of Moore and Seiberg [59] for inspiration. We review his answers in the rest of this section, referring to the original papers for further details. Our treatment follows [4, Appendix A].

Gaberdiel's algebraic reformulation of fusion begins by considering the following contour integral:

$$\oint_{0,w_1,w_2} \langle \psi_3', v(z)\psi_1(w_1)\psi_2(w_2)\Omega\rangle z^n \frac{\mathrm{d}z}{2\pi\mathrm{i}}. \tag{10}$$

Here, ψ_3' is an arbitrary vector in the space dual to the states of the conformal field theory (see Sect. 4 below for a precise definition), $v(z) = \sum_{m\in\mathbb{Z}} v_m z^{-m-1}$ is an arbitrary field of the vertex operator algebra V (note mathematics mode convention in force!), $\psi_1(w_1)$ and $\psi_2(w_2)$ are arbitrary (chiral) fields corresponding to states in the V-modules M_1 and M_2, respectively, $n \in \mathbb{Z}$ is arbitrary, and the contour integral indicates that the result is the sum of the residues of the integrand at $z = 0$, $z = w_1$ and $z = w_2$. In other words, we consider a simple positively oriented contour that encloses the points 0, w_1 and w_2. When we discuss the mathematical definition of the fields $\psi_i(w_i)$ below, we shall see that they are closely related to certain "intertwining maps"

[4] Note that a non-rational theory may also possess fields that are not generated from primary fields. The original approach to computing fusion from primary correlators will therefore produce incorrect fusion rules in general. Unfortunately, this approach is still widely employed, without comment, in the non-rational physics literature.

evaluated at the states ψ_i. The idea behind considering (10) is that this expression defines a natural action of the mode

$$v_n = \oint_0 v(z) z^n \frac{\mathrm{d}z}{2\pi \mathrm{i}} \tag{11}$$

on the product $\psi_1(w_1)\psi_2(w_2)$ of fields in a correlator and thence on the tensor product states $\psi_1 \otimes \psi_2$ in the vector space tensor product $M_1 \otimes M_2$. Note that the usual radial ordering prescription of conformal field theory requires us to assume that $|z| > |w_1| > |w_2|$ in (10).

We suppose that each field $\psi_i(w_i)$, $i = 1, 2$, is mutually local with respect to $v(z)$, meaning that

$$v(z)\psi_i(w_i) = \psi_i(w_i)v(z), \quad i = 1, 2. \tag{12}$$

In general, this equation would be modified by adding a coefficient μ_i on the right-hand side. For example, we would have $\mu_i = 1$, if $v(z)$ and $\psi_i(w_i)$ are mutually bosonic, and $\mu_i = -1$ if they are mutually fermionic. More complicated mutual localities are of course possible (and easily accommodated in the derivation to follow). For simplicity, we shall assume bosonic statistics ($\mu_i = 1$) throughout. We also suppose that both M_1 and M_2 are untwisted as V-modules, meaning that the operator product expansions of their fields with those of V have trivial monodromy. In particular, we have

$$v(z)\psi_i(w_i) = \sum_{m \in \mathbb{Z}} (v_m \psi_i)(w_i)(z - w_i)^{-m-1}, \quad i = 1, 2, \tag{13}$$

where $(v_m \psi_i)(w_i)$ denotes the field corresponding to the state $v_m \psi_i \in M_i$. Note that requiring M_1 and M_2 to be untwisted excludes, for example, modules from the Ramond sectors of theories with fermions. Gaberdiel's formalism can, of course, be generalised to accommodate twisted modules though it becomes significantly more unwieldy, see [5, 22].

Consider the contribution to (10) corresponding to the residue at w_1. If we substitute the $i = 1$ operator product expansion (13) into this contribution, we (formally) obtain

$$\oint_{w_1} \langle \psi_3', v(z)\psi_1(w_1)\psi_2(w_2)\Omega \rangle z^n \frac{\mathrm{d}z}{2\pi \mathrm{i}}$$
$$= \sum_{m \in \mathbb{Z}} \langle \psi_3', (v_m \psi_1)(w_1)\psi_2(w_2)\Omega \rangle \oint_{w_1} (z - w_1)^{-m-1} z^n \frac{\mathrm{d}z}{2\pi \mathrm{i}} \tag{14a}$$
$$= \sum_{m=0}^{\infty} \binom{n}{m} w_1^{n-m} \langle \psi_3', (v_m \psi_1)(w_1)\psi_2(w_2)\Omega \rangle. \tag{14b}$$

To get the contribution from w_2, we instead substitute the $i = 2$ operator product expansion, after first applying (12). We mention the easily overlooked fact that radial ordering on the left-hand side requires $|z| > |w_1|$, but re-writing the integral using (12) requires $|w_1| > |z|$. The result is

$$\oint_{w_2} \langle \psi_3', v(z)\psi_1(w_1)\psi_2(w_2)\Omega\rangle z^n \frac{\mathrm{d}z}{2\pi \mathrm{i}} = \sum_{m=0}^{\infty} \binom{n}{m} w_2^{n-m} \langle \psi_3', \psi_1(w_1)(v_m\psi_2)(w_2)\Omega\rangle. \tag{15}$$

If $n \geqslant 0$, either substitution shows that the residue at 0 vanishes, hence that (10) is the sum of (14b) and (15). Gaberdiel's conclusion is that the arbitrariness of ψ_3' means that we should interpret this sum as the action of v_n on the tensor product state $\psi_1 \otimes \psi_2 \in M_1 \otimes M_2$ corresponding to the product $\psi_1(w_1)\psi_2(w_2)$ under an extended state-field correspondence:

$$\begin{aligned} \Delta_{w_1,w_2}(v_n)(\psi_1 \otimes \psi_2) &= \sum_{m=0}^{n} \binom{n}{m} \Big[w_1^{n-m} (v_m\psi_1) \otimes \psi_2 + w_2^{n-m} \psi_1 \otimes (v_m\psi_2) \Big] \\ \Rightarrow \quad \Delta_{w_1,w_2}(v_n) &= \sum_{m=0}^{n} \binom{n}{m} \Big[w_1^{n-m} (v_m \otimes \mathbb{1}) + w_2^{n-m} (\mathbb{1} \otimes v_m) \Big] \qquad (n \geqslant 0). \end{aligned} \tag{16}$$

Here, $\mathbb{1}$ denotes the identity operator acting on M_1 or M_2, as appropriate. This action first appeared (with $w_2 = 0$) in work of Moore and Seiberg, see [59, Eq. (2.4)].

Suppose now that $n < 0$. Then, the contribution to (10) from 0 need not vanish. To compute it, we have to use one of the two operator product expansions (13). If we use that with $i = 1$, then the result is the same as in (14a) except that the residue is evaluated at $z = 0$. The sum of the contributions from 0 and w_1 may therefore be expressed as a sum over $m \in \mathbb{Z}$ of terms proportional to

$$\begin{aligned} \oint_{0,w_1} (z - w_1)^{-m-1} z^n \frac{\mathrm{d}z}{2\pi \mathrm{i}} &= -\oint_{\infty} (z - w_1)^{-m-1} z^n \frac{\mathrm{d}z}{2\pi \mathrm{i}} \\ &= \oint_0 (1 - w_1 y)^{-m-1} y^{m-n-1} \frac{\mathrm{d}y}{2\pi \mathrm{i}} \quad (y = z^{-1}), \end{aligned} \tag{17}$$

which vanishes for $m > n$. Adding the $m \leqslant n$ contributions to that from w_2, given in (15), we obtain a formula for the action of the v_n:

$$\Delta^{(1)}_{w_1,w_2}(v_n) = \sum_{m=-\infty}^{n} \binom{-m-1}{-n-1} (-w_1)^{n-m} (v_m \otimes \mathbb{1}) + \sum_{m=0}^{\infty} \binom{n}{m} w_2^{n-m} (\mathbb{1} \otimes v_m) \quad (n < 0). \tag{18}$$

Note that the upper limit on the first sum may be changed to -1 because $\binom{-m-1}{-n-1} = 0$ if $n < m < 0$.

In (18), we have added the label (1) to the action of v_n because, in deriving it, we made a choice to use (13) with $i = 1$ to evaluate the contribution from 0. If we had instead chosen to take $i = 2$, then we would have deduced a seemingly different action:

$$\Delta^{(2)}_{w_1,w_2}(v_n) = \sum_{m=0}^{\infty} \binom{n}{m} w_1^{n-m} (v_m \otimes \mathbb{1}) + \sum_{m=-\infty}^{n} \binom{-m-1}{-n-1} (-w_2)^{n-m} (\mathbb{1} \otimes v_m) \quad (n < 0). \tag{19}$$

Gaberdiel's definition of the fusion product of M_1 and M_2 is then the largest quotient of the vector space tensor product on which these two actions agree. We shall make this manifest shortly after discussing the role of the insertion points w_1 and w_2.

The symbol Δ was chosen by Gaberdiel to indicate these actions because they define coproducts, albeit ones that depend on the two points w_1 and w_2. Indeed, he proved coassociativity[5] in the form

$$(\Delta^{(i)}_{w_1-w,w_2-w} \otimes \mathbb{1}) \circ \Delta^{(i)}_{w,w_3} = (\mathbb{1} \otimes \Delta^{(i)}_{w_2-w,w_3-w}) \circ \Delta^{(i)}_{w_1,w}, \quad (i = 1, 2). \tag{20}$$

He moreover showed that conjugating with translation and dilation operators allows one to replace the points w_1 and w_2 in these coproducts with any two distinct points on the Riemann sphere without changing the equivalence class of the coproduct actions.

Doing so has an immediate practical advantage. While the modes v_m of any field of V must annihilate an arbitrary fixed state, when m is sufficiently large, there is no such requirement for m sufficiently small. It follows that the second infinite sum of (18) and the first infinite sum of (19) are both truncated to finite sums when acting on any state. Unfortunately, no such truncation occurs for the first infinite sum of (18) or the second infinite sum of (19). These coproduct actions therefore take elements of $M_1 \otimes M_2$ to some, as yet uncharacterised, completion of this vector space tensor product. However, we may avoid having to introduce this completion by choosing $w_1 = 0$ in (18) and $w_2 = 0$ in (19):

$$\Delta^{(1)}_{0,w_2}(v_n) = (v_n \otimes \mathbb{1}) + \sum_{m=0}^{\infty} \binom{n}{m} w_2^{n-m} (\mathbb{1} \otimes v_m) \quad (n < 0), \tag{21a}$$

$$\Delta^{(2)}_{w_1,0}(v_n) = \sum_{m=0}^{\infty} \binom{n}{m} w_1^{n-m} (v_m \otimes \mathbb{1}) + (\mathbb{1} \otimes v_n) \quad (n < 0). \tag{21b}$$

These coproduct formulae thus give well defined actions on $M_1 \otimes M_2$. Moreover, they now agree with the corresponding specialisations of the $n \geqslant 0$ formulae of (16). Note that the contribution from 0 to the $n < 0$ formula for $\Delta^{(i)}_{w_1,w_2}$ was derived by inserting an operator product expansion at w_i. We could have therefore insisted from the outset that w_i be required to be close to 0. Setting w_i to 0 at the conclusion is thus very natural.

Example 1 We illustrate a few coproduct formulae in order to appreciate their complexity. For brevity, we use (16), with $w_1 = 1$ and $w_2 = 0$, and (21b), with $w_1 = 1$:

---

[5]To be precise, Gaberdiel showed how to prove it for Virasoro vertex operator algebras in [20, Appendix B]. Coassociativity in general is stated to follow similarly in [21, Sect. 2].

$$\Delta_{1,0}(v_2) = (v_0 \otimes \mathbb{1}) + 2(v_1 \otimes \mathbb{1}) + (v_2 \otimes \mathbb{1}) + (\mathbb{1} \otimes v_2), \tag{22a}$$
$$\Delta_{1,0}(v_1) = (v_0 \otimes \mathbb{1}) + (v_1 \otimes \mathbb{1}) + (\mathbb{1} \otimes v_1), \tag{22b}$$
$$\Delta_{1,0}(v_0) = (v_0 \otimes \mathbb{1}) + (\mathbb{1} \otimes v_0), \tag{22c}$$
$$\Delta_{1,0}^{(2)}(v_{-1}) = (v_0 \otimes \mathbb{1}) - (v_1 \otimes \mathbb{1}) + (v_2 \otimes \mathbb{1}) - \cdots + (\mathbb{1} \otimes v_{-1}), \tag{22d}$$
$$\Delta_{1,0}^{(2)}(v_{-2}) = (v_0 \otimes \mathbb{1}) - 2(v_1 \otimes \mathbb{1}) + 3(v_2 \otimes \mathbb{1}) - \cdots + (\mathbb{1} \otimes v_{-2}). \tag{22e}$$

Note that the coproduct formula for the (mathematics convention!) "zero modes" matches that used for Lie algebra representations. These "zero modes" include the Virasoro mode $L(-1) = \omega_0$. □

We can now make Gaberdiel's definition of the fusion product of M_1 and M_2 precise. Taking $w_1 = 1$ and $w_2 = -1$ in (21), for maximum brevity, the insertion points of the two coproducts are related by a rigid translation. Imposing the equality of the two coproduct actions therefore amounts to setting

$$\Delta_{0,-1}^{(1)}(v_n) = \Delta_{0,-1}^{(2)}(v_n) = \Delta_{1,0}^{(2)}(\mathsf{e}^{L(-1)} v_n \mathsf{e}^{-L(-1)}), \tag{23}$$

for all fields $v(z)$ and all $n < 0$. The fusion product is therefore defined to be the vector space

$$M_1 \boxtimes M_2 = \frac{M_1 \otimes M_2}{\left\langle (\Delta_{0,-1}^{(1)}(v_n) - \Delta_{1,0}^{(2)}(\mathsf{e}^{L(-1)} v_n \mathsf{e}^{-L(-1)}))(M_1 \otimes M_2) \right\rangle}, \tag{24}$$

equipped with the action of V defined by the coproduct formulae (16) and either (18) or (19). Here, the quotient is by the sum of the images for all fields $v(z)$ of V and all $n < 0$. As mentioned above, the point is that the fusion product is constructed to be the largest quotient of the vector space tensor product on which the natural V-action, derived from operator product expansions and locality, is well defined. This foreshadows the idea that the definition should be reinterpreted in terms of a universality property.

Unfortunately, fusion is not all tea and biscuits. While the issue of having to specify a completion of $M_1 \otimes M_2$ was neatly sidestepped by choosing insertion points carefully, a nastier problem now rears its ugly head: the translations required to compare the two coproducts. Specifically, the Lie bracket $[L(-1), v_n] = -n v_{n-1}$ implies that (23) may be expanded into

$$\Delta_{0,-1}^{(1)}(v_n) = \sum_{m=0}^{\infty} (-1)^m \binom{n}{m} \Delta_{1,0}^{(2)}(v_{n-m}). \tag{25}$$

Inserting the coproduct formula on the right-hand side, we find that the sum does not converge for $n \leqslant -1$, not even when acting on $M_1 \otimes M_2$ and taking a completion:

$$\Delta^{(1)}_{0,-1}(v_n) = \sum_{m=0}^{\infty}(-1)^m\binom{n}{m}(v_0 \otimes \mathbb{1}) + \sum_{m=0}^{\infty}(-1)^m\binom{n}{m}(n-m)(v_1 \otimes \mathbb{1})$$
$$+ \sum_{m=0}^{\infty}(-1)^m\binom{n}{m}\binom{n-m}{2}(v_2 \otimes \mathbb{1}) + \cdots + \sum_{m=0}^{\infty}(-1)^m\binom{n}{m}(\mathbb{1} \otimes v_{n-m}). \tag{26}$$

Gaberdiel was certainly aware that issues like this arose whenever one combined coproducts and translations, see [20, Eq. (2.14)] and the subsequent discussion. However, he did not offer any solutions. Nahm, in his seminal paper on the definition and computation of fusion [61], noted that such convergence issues may be resolved by either working with dual modules or by redefining the fusion product as a projective limit of finite-dimensional truncations of $M_1 \otimes M_2$, see Sect. 8.[6] At around the same time, Huang and Lepowsky [35, 36] were likewise formulating their definition of fusion, which we shall review shortly, using dual spaces.

However, Nahm's truncation idea turned out to be much more interesting to physicists as it was subsequently generalised by Gaberdiel and Kausch [25] and used to form the basis of a practical algorithm to explicitly construct (truncations of) fusion products. We shall discuss this algorithm, now known as the Nahm–Gaberdiel–Kausch fusion algorithm, in Sect. 8. First, however, we shall compare Gaberdiel's definition of fusion with the dual definition of Huang and Lepowsky.

4 The Mathematician's Approach I: A Universal Definition

There are many essentially equivalent definitions of vertex operator algebras, which the reader may find in references such as [16, 18, 44]. The main axiom defining modules for a given vertex operator algebra V is what is called the Jacobi identity in [51]. We will be interested in V-modules M which are graded by the zero mode $L(0)$ of the Virasoro field. More precisely, we require that M be spanned, and thus graded, by its *generalised* $L(0)$-eigenspaces—in particular, we allow non-diagonalisable actions of $L(0)$. We denote the generalised eigenspace of M corresponding to the eigenvalue $h \in \mathbb{C}$ by $M_{[h]}$; this eigenspace decomposition is called the *conformal grading*. We shall insist, with a caveat to be discussed below, that each $M_{[h]}$ is finite-dimensional. Then, restricting $L(0)$ to $M_{[h]}$ results in finite-rank Jordan blocks.

One is also required to truncate the conformal grading from below, in some sense, because the action of a field on a state of a module should be expressed as a Laurent series with poles, but no essential singularities. This translates into the following requirement: For every $v \in V$ and $\psi \in M$, we have $v_n\psi = 0$ for all sufficiently large n. However, one frequently finds stronger conditions being assumed in the mathematical literature, for example that for any given $h \in \mathbb{C}$, the spaces $M_{[h-n]}$ are zero for all

[6] This inverse limit approach to fusion has also reappeared in the work of Miyamoto [57] and Tsuchiya and Wood [70].

sufficiently large integers n. This stronger condition is met when V is C_2-cofinite (see [1]), but can be inappropriate in general. In particular, the "staggered/logarithmic" modules of the admissible-level affine [2, 24, 66] and bosonic ghost [68] vertex operator algebras all fail to meet this condition.

It may also be highly beneficial, or indeed necessary, to introduce additional gradings by other zero modes. The most frequent (and natural) occurrence of this arises when V includes a Heisenberg vertex operator subalgebra whose zero modes act semisimply on an appropriate class of V-modules. Such additional gradings are used, even when V is rational, to refine characters so that they may be used to distinguish inequivalent irreducible V-modules. When V is not C_2-cofinite, they may be required for characters to even be defined because the generalised $L(0)$-eigenspaces may be infinite-dimensional. Insisting on finite-dimensional homogeneous spaces with respect to the conformal and additional gradings then saves the notion of character while also preserving the finite-rank property for Jordan blocks. The theorems of HLZ [41] are designed to handle such additional gradings. However, for the sake of simplicity, we will not emphasise this level of generality in what follows.

To proceed with the rigorous definitions, we first say what we mean by the *restricted dual* M' of a graded V-module M. As a vector space, the definition is straightforward: one merely takes M' to be the direct sum of the duals of the generalised $L(0)$-eigenspaces. In symbols,

$$M = \bigoplus_{h \in \mathbb{C}} M_{[h]} \quad \Rightarrow \quad M' = \bigoplus_{h \in \mathbb{C}} M^*_{[h]}. \tag{27}$$

Here, * denotes the ordinary vector space dual.[7] This generalises in the obvious way when there are additional grading in force. In Sect. 6, we shall refine this notion by endowing the restricted dual M' with the structure of a V-module.

We shall also need to introduce the notion of a suitable completion of a module. Generically, the field $Y(v, z)$ (written in mathematical notation), with z being a non-zero complex parameter and $v \in V$, acts on $\psi \in M$ to give an infinite sum of elements in M. To accommodate for this (and other analogous infinite sums), we will often work with a completion $\overline{M}$ of a module, in which the direct sum of the generalised $L(0)$-eigenspaces of M is replaced by the direct product.

We now come to the definition of an intertwining map, the objects that form the backbone of the tensor product theory of HLZ. In mathematics, tensor products are often defined abstractly using "universal properties" before proving that the (sometimes) obvious construction of the product satisfies these properties. For example, this is how tensor products of vector spaces are defined in many mathematical textbooks. The main advantages of this universal approach include capturing the uniqueness properties of the tensor product construction as well as its relation to other algebraic and categorical gadgets. In the setup of HLZ [41], the fusion product introduced by physicists is recast as a universal tensor product object with respect

---

[7]Because we assume that the homogeneous subspaces $M_{[h]}$ are all finite-dimensional, we may safely ignore all questions regarding the topological nature of these duals.

to intertwining maps, as we now explain. A closely related notion, namely that of *intertwining operators*, was introduced much earlier in [17] (see [56] for *logarithmic* intertwining operators). These are formal variable cousins of the intertwining maps that we consider here.

Definition 4.1 Fix $w \in \mathbb{C}^\times$. Given V-modules M_1, M_2 and M_3, a *$P(w)$-intertwining map of type* $\begin{pmatrix} M_3 \\ M_1\, M_2 \end{pmatrix}$ is a bilinear map $I\colon M_1 \otimes M_2 \to \overline{M_3}$ that satisfies the following properties:

1. For any $\psi_1 \in M_1$ and $\psi_2 \in M_2$, $\pi_h(I[\psi_1 \otimes \psi_2]) = 0$ for all $h \ll 0$, where π_h denotes the projection onto the generalised eigenspace $(M_3)_{[h]}$ of $L(0)$-eigenvalue h.[8]
2. For any $\psi_1 \in M_1$, $\psi_2 \in M_2$ and $\psi_3' \in M_3'$, the series defined by

$$\begin{gathered}\Big\langle \psi_3', Y_3(v,z) I[\psi_1 \otimes \psi_2] \Big\rangle, \quad \Big\langle \psi_3', I[Y_1(v,z-w)\psi_1 \otimes \psi_2] \Big\rangle \\ \text{and} \quad \Big\langle \psi_3', I[\psi_1 \otimes Y_2(v,z)\psi_2] \Big\rangle \end{gathered} \tag{28}$$

   are absolutely convergent in the regions $|z| > |w| > 0$, $|w| > |z-w| > 0$ and $|w| > |z| > 0$, respectively.[9] (The subscript under each Y indicates the module being acted upon.)
3. Given any $f(t) \in R_{P(w)} = \mathbb{C}[t, t^{-1}, (t-w)^{-1}]$, the field of rational functions whose poles lie in some subset of $\{0, w, \infty\}$, we have the *Cauchy–Jacobi identity*:

$$\begin{aligned}\oint_{0,w} f(z) \Big\langle \psi_3', Y_3(v,z) I[\psi_1 \otimes \psi_2] \Big\rangle \frac{\mathrm{d}z}{2\pi \mathrm{i}} &= \oint_{w} f(z) \Big\langle \psi_3', I[Y_1(v,z-w)\psi_1 \otimes \psi_2] \Big\rangle \frac{\mathrm{d}z}{2\pi \mathrm{i}} \\ &\quad + \oint_{0} f(z) \Big\langle \psi_3', I[\psi_1 \otimes Y_2(v,z)\psi_2] \Big\rangle \frac{\mathrm{d}z}{2\pi \mathrm{i}}. \end{aligned} \tag{29}$$

   Here, as in Sect. 3, the subscript on each integrals indicate the points in $\{0, w, \infty\}$ that must be enclosed by the simple positively oriented contours.

The nomenclature "$P(w)$" may look strange, but it emphasises the fact that we are working on the Riemann sphere with three punctures at 0, w and ∞. The first two are designated as being "incoming", or positively oriented, while the ∞ puncture is

---

[8] This requirement will clearly need refining when it is necessary (or desirable) to include additional gradings on the V-modules.

[9] In the physics literature, it is customary to take z and w to be the first and second insertion points, respectively, in an operator product expansion. In line with this convention, quantities like (28) naturally lead us to speak of $P(w)$-intertwining maps as opposed to the $P(z)$-intertwining maps that are ubiquitous in the mathematics literature.

"outgoing", or negatively oriented. We also have a preferred choice of local coordinates around the punctures: about 0, we take $z \mapsto z$; about w, we take $z \mapsto z - w$; and about ∞, we take $z \mapsto z^{-1}$. The space $R_{P(w)}$ introduced above is precisely the field of rational functions on this punctured sphere. Below, we shall expand these functions as Laurent series centred at the punctures, implicitly using the provided local coordinates. We mention that punctured spheres are ubiquitous in mathematical approaches to fusion (and conformal field theory in general), see for example Huang's book [32] and the seminal work of Kazhdan and Lusztig [46–49].

Let us also mention that the sewing of such spheres—outgoing punctures to incoming punctures, respecting the local coordinates—plays a central role in building the rest of the tensor category structure, most importantly the associativity morphisms. We shall not discuss this further since our focus is only on the tensor (fusion) product itself. Interested readers may consult [32, 35] for further details.

We now connect the notion of intertwining maps to the objects that were introduced in Sect. 3. The entity $\psi_1(w)\psi_2$, in the notation of that section, is essentially a $P(w)$-intertwining map, evaluated on $\psi_1 \otimes \psi_2$. This raises two immediate questions:

1. As per the definition, an intertwining map *comes equipped* with a fixed choice of w, but in a field such as $\psi_1(w)$, we are free to let w vary. How do we reconcile this?
2. In the physics literature, the action of the field $\psi_1(w)$, with $\psi_1 \in M_1$, on the state $\psi_2 \in M_2$ may be determined from the corresponding operator product expansion, by applying the (generalised) state-field correspondence. The result of this application is therefore in (the completion of) the fusion product of M_1 and M_2. How then does the fusion product relate to the target spaces M_3 of the corresponding intertwining maps?

For the first question, it is possible to pass from a $P(w)$-intertwining map (in which w is fixed!) to a formal variable intertwining operator. This is interesting in itself, and is explained in [41]. Once this is done, any other non-zero complex number may be substituted in place of the formal variable. This must be done with great care—it is necessary to choose branches because the series expansion of the intertwining operator will usually involve non-integer powers and logarithms. The second question is likewise very interesting. It leads us to a definition of the fusion product in terms of a universal tensor product module, $M_1 \boxtimes M_2$, which we shall define below.

First, however, note that an intertwining map is defined on the whole of the module $M_1 \otimes M_2$, and not just on their "top spaces" or primary vectors (or any other "special" subspaces). Nevertheless, a fact that will prove crucial later is that an intertwining map is often completely determined by its definition on certain subspaces and/or quotients. This is quite delicate however. Given a map on appropriate subspaces and/or quotients, it may require a great deal of effort to prove that it extends to an intertwining map on the entire module, if at all.

Example 2 Consider the rank-1 Heisenberg vertex operator algebra, known to physicists as the free boson on a one-dimensional non-compact spacetime. (All of this generalises naturally to higher ranks and dimensions.) Its irreducible highest-weight modules, known as Fock spaces, are parametrised by complex numbers: for each $\lambda \in \mathbb{C}$, we have a Fock space F_λ, generated by a highest-weight vector f_λ. Fix $w \in \mathbb{C}^\times$. It can be proved easily that there are no $P(w)$-intertwining maps of type $\begin{pmatrix} F_\nu \\ F_\lambda\, F_\mu \end{pmatrix}$ unless $\nu = \lambda + \mu$. (In the language of physics, this is conservation of momentum.) We leave it as an exercise for the reader to demonstrate, as in Sect. 7, to prove that there is at most one $P(w)$-intertwining map (up to scalar multiples) of the type $\begin{pmatrix} F_{\lambda+\mu} \\ F_\lambda\, F_\mu \end{pmatrix}$. There indeed exists a non-zero $P(w)$-intertwining map. Up to normalisation, it acts as

$$I_{\lambda,\mu}[f_\lambda \otimes f_\mu] = w^{\lambda\mu} f_{\lambda+\mu} + \cdots \in \overline{F_{\lambda+\mu}}, \tag{30}$$

where the ellipses indicate terms involving factors $w^{\lambda\mu+n}$, with n a positive integer. Physicists will indeed recognise this in terms of the operator product expansion of the primary fields corresponding to f_λ (at w) and f_μ (at 0).[10] As mentioned above, one still needs to prove that this definition extends to all of $F_\lambda \otimes F_\mu$ (the details may be found in [11]). Note that it is important to make use of a specific branch of the logarithm to make sense of terms like $w^{\lambda\mu}$. However, in this case, different choices only lead to scalar multiples of $I_{\lambda,\mu}$. □

We now give a precise definition of the tensor product by a universal property, as promised above. Let $\mathscr{C}$ be a category of modules for a vertex operator algebra V. We shall define the $P(w)$-tensor product of two modules $M_1, M_2 \in \mathscr{C}$ as a universal object, denoted by $M_1 \boxtimes_{P(w)} M_2 \in \mathscr{C}$, with respect to the $P(w)$-intertwining maps $I\colon M_1 \otimes M_2 \to \overline{M_3}$, for all M_3 in $\mathscr{C}$. Comparing with the physics approach of Sect. 3, it is clear that this universal approach depends upon the choice of category $\mathscr{C}$. In the next two sections, we will separate the categorical and non-categorical constraints. By focusing only on the latter, we shall produce a rigorous constructive definition of the fusion product of two modules in the formalism of HLZ that can be compared directly with computations performed by physicists.

Definition 4.2 For each $M_1, M_2 \in \mathscr{C}$, the pair $(M_1 \boxtimes_{P(w)} M_2, \boxtimes_{P(w)})$, where $M_1 \boxtimes_{P(w)} M_2 \in \mathscr{C}$ and $\boxtimes_{P(w)}$ is a $P(w)$-intertwining map of type $\begin{pmatrix} M_1 \boxtimes_{P(w)} M_2 \\ M_1\, M_2 \end{pmatrix}$, is called the *$P(w)$-tensor product* of M_1 and M_2 if for any $M \in \mathscr{C}$ and $P(w)$-

[10] We remark that it is these primary fields (and only these primary fields) that are called *vertex operators* in the physics literature. In the setting of (non-compact) free bosons, they are therefore not fields of the Heisenberg vertex operator algebra. The term *vertex operator algebra* itself presumably arose in the mathematical literature because early work concentrated on examples related to lattices (compactified free bosons) in which certain vertex operators are promoted to fields of an extended vertex operator algebra.

intertwining map I of type $\binom{M}{M_1\,M_2}$, there exists a unique $\mathscr{C}$-morphism $\eta\colon M_1 \boxtimes_{P(w)} M_2 \to M$ such that

$$(\overline{\eta} \circ \boxtimes_{P(w)})[\psi_1 \otimes \psi_2] = I[\psi_1 \otimes \psi_2], \tag{31}$$

for all $\psi_1 \in M_1$ and $\psi_2 \in M_2$. Here, $\overline{\eta}$ denotes the extension of η to a map between the completions of $M_1 \boxtimes_{P(w)} M_2$ and M (necessary because the image of I is typically in the completion of the target module). In terms of a diagram, the following should commute:

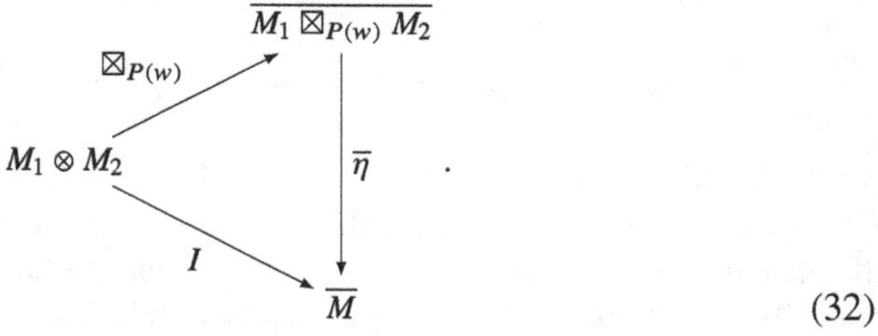

(32)

There is of course the very natural question of whether this definition of $P(w)$-tensor products actually depends on the choice of $w \in \mathbb{C}^\times$. As expected, HLZ answer this in the negative [41, Rem. 4.22]. By abuse of notation, we shall denote $\boxtimes_{P(w)}(\psi_1 \otimes \psi_2)$ by $\psi_1 \boxtimes_{P(w)} \psi_2$, for $\psi_1 \in M_1$ and $\psi_2 \in M_2$, keeping in mind that this element is in the completion $\overline{M_1 \boxtimes_{P(w)} M_2}$ and may not be in the module itself. This subtlety, among others, necessitates the need for analytic arguments throughout [41].

Example 3 It is shown in [41] that $V \boxtimes_{P(w)} M = M$ for all V-modules M, with the universal intertwining map $\boxtimes_{P(w)}$ of type $\binom{M}{V\,M}$ being given by $v \boxtimes_{P(w)} \psi = Y(v, w)\psi \in \overline{M}$. On the other hand, we also have $M \boxtimes_{P(w)} V = M$, with $\psi \boxtimes_{P(w)} v = \mathrm{e}^{wL(-1)} Y(v, -w)\psi$. The vacuum module V is thus a unit for the $P(w)$-tensor product, in accordance with expectations for the fusion product. □

Example 4 Returning to the rank-1 Heisenberg vertex operator algebra of Example 2, it can be proved [7] that if $\mathscr{C}$ is the semisimple category whose objects are finite direct sums of Fock spaces, then one may take $F_\lambda \boxtimes_{P(w)} F_\mu = F_{\lambda+\mu}$, with universal intertwining map $\boxtimes_{P(w)} = I_{\lambda,\mu}$ given by (30). This also agrees with the well-known ($\mathscr{C}$-independent) fusion rules known to physicists. □

5 The Mathematician's Approach II: A Model for Tensor Products

Universal definitions are all well and good, for some purposes (mostly abstract ones). But sometimes, one needs an alternative definition that comes with an honest construction. Mathematicians often refer to such constructions as *models* for the universal definition. The basic philosophy behind building models satisfying the universal properties is to first take the "biggest" candidate possible and then to cut it down by imposing relations arising out of the constraints given by the "test" conditions.

With a view towards the vertex-algebraic picture to be presented later, let us review a basic example of a model: the explicit construction of the tensor product $M_1 \otimes_A M_2$ of A-modules M_1 and M_2, where A is a commutative associative unital algebra over $\mathbb{C}$ (say), as a quotient of the vector space tensor product $M_1 \otimes M_2 \equiv M_1 \otimes_{\mathbb{C}} M_2$. The universal definition of $\otimes_A$ says that if we are given an A-bilinear map $B \colon M_1 \otimes M_2 \to M$, where M is an arbitrary A-module, then there exists a unique A-linear map $f : M_1 \otimes_A M_2 \to M$ such that

$$B(m_1 \otimes m_2) = (f \circ \otimes_A)(m_1, m_2) = f(m_1 \otimes_A m_2), \tag{33}$$

for all $m_1 \in M_1$ and $m_2 \in M_2$. In other words, the following diagram must commute:

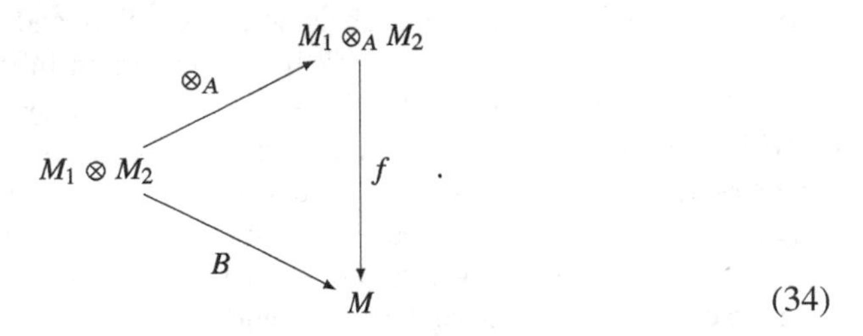

(34)

To construct $M_1 \otimes_A M_2$, we first recall that the plain old vector space tensor product $M_1 \otimes M_2$ is naturally an A-module under the action $a \cdot (m_1 \otimes m_2) = m_1 \otimes (a \cdot m_2)$, $a \in A$.[11] But, the bilinear map B is constrained by one more property, namely that $B(a \cdot m_1 \otimes m_2) = B(m_1 \otimes a \cdot m_2)$. We therefore quotient the A-module $M_1 \otimes M_2$ by the ideal corresponding to imposing $(a \cdot m_1) \otimes m_2 = m_1 \otimes (a \cdot m_2)$. Defining $\Delta^{(1)}(a) = a \otimes \mathbb{1}$ and $\Delta^{(2)}(a) = \mathbb{1} \otimes a$, we therefore have the following (hypothetical) model for the universal definition:

$$M_1 \otimes_A M_2 = \frac{M_1 \otimes M_2}{\langle (\Delta^{(1)}(a) - \Delta^{(2)}(a))(M_1 \otimes M_2) : a \in A \rangle}. \tag{35}$$

One can of course verify that this definition does satisfy the universal property (34) and so is indeed a model of the A-tensor product.

In all the mathematics and physics literature, this essential idea is behind the explicit constructions of tensor product modules. In particular, Gaberdiel's original definition (24) of fusion is now clearly identified as an attempt to construct a model for the fusion product of two V-modules M_1 and M_2. However, as vertex operator algebras are highly non-classical objects, one quickly runs into obstacles to making this rigorous. We also discussed these briefly in Sect. 3. Let us recall them again:

[11] Here, we choose to act on m_2, rather than m_1, in order to keep in line with the vertex-algebraic generalisation to follow.

1. If one tries to work out certain relations analogous to $(a \cdot m_1) \otimes m_2 = m_1 \otimes (a \cdot m_2)$ in the vertex-algebraic setting, one finds that they do not converge, even when completions are taken into account, because some coefficients are found to be formally infinite, see (26).
2. Unlike the commutative ring case, where $a \cdot (m_1 \otimes_A m_2) = m_1 \otimes_A (a \cdot m_2)$ trivially satisfies the axioms for the algebra action on a module, it takes much more effort to prove the analogous theorem in the vertex-algebraic setting. This theorem would answer question 3 in the material before (10) and we shall say a bit more about this towards the end of Sect. 6.

In the rest of this section, we shall derive the relations and V-action that (morally) should define a model for the universal $P(w)$-tensor product, ignoring these obstacles to making the construction rigorous. The aim is to draw parallels with Gaberdiel's work, as reviewed in Sect. 3, before describing the rigorous formalism developed to this end by HLZ in Sect. 6.

For now, it will be highly beneficial to detach the v and the n in the (mathematicians') notation for modes $v_n = v_n^{(\mathrm{m})}$ by writing them in the form $v \otimes t^n$, where t is some auxiliary formal variable. This is not merely a syntactic vinegar—it opens up wider possibilities. In particular, we are now working in $V \otimes \mathbb{C}[t, t^{-1}]$ and we have room to accommodate other regular functions on the 3-punctured sphere, for instance $v \otimes (t-w)^n$, $n \in \mathbb{Z}$, by enlarging further to $V \otimes \mathbb{C}[t, t^{-1}, (t-w)^{-1}]$ or $V \otimes \mathbb{C}(t)$ (rational functions in t) or even $V \otimes \mathbb{C}((t))$ (Laurent series in t). It will often be convenient to ignore the $\otimes$ sign and write simply $v f(t)$ instead of $v \otimes f(t)$.

Once again, choose $w \in \mathbb{C}$. It is convenient to define a translation map

$$T_w \colon \mathbb{C}(t) \to \mathbb{C}(t), \quad \text{by} \quad f(t) \mapsto f(t+w), \tag{36}$$

and two expansion maps

$$\iota_+ \colon \mathbb{C}(t) \hookrightarrow \mathbb{C}((t)) \quad \text{and} \quad \iota_- \colon \mathbb{C}(t) \hookrightarrow \mathbb{C}((t^{-1})) \tag{37}$$

that expand a given rational function in t as a power series around $t = 0$ and $t = \infty$, respectively.

We turn to constructing a model for the fusion product $M_1 \boxtimes M_2$, now identified as the $P(w)$-tensor product $M_1 \boxtimes_{P(w)} M_2$ defined above. The relations required to cut $M_1 \otimes M_2$ down to the fusion product naturally arise out of the Cauchy–Jacobi identity (29) for the intertwining maps, since these are essentially the only constraints we have. Let us therefore analyse this identity closely.

As in (29), let $v \in V$, $\psi_1 \in M_1$, $\psi_2 \in M_2$, $\psi_3' \in M_3'$ and let I be a $P(w)$-intertwining map of type $\binom{M_3}{M_1\, M_2}$. Let $f(z) = z^a (z-w)^b$, where a and b are arbitrary integers. The term on the left-hand side of (29) may be expanded formally as

$$\begin{aligned}
&\oint_{0,w} f(z)\left\langle \psi_3', Y_3(v,z) I[\psi_1 \otimes \psi_2]\right\rangle \frac{\mathrm{d}z}{2\pi\mathrm{i}} \\
&\qquad = -\oint_{\infty} z^a (z-w)^b \left\langle \psi_3', Y_3(v,z) I[\psi_1 \otimes \psi_2]\right\rangle \frac{\mathrm{d}z}{2\pi\mathrm{i}} \\
&\qquad = \oint_0 \left(y^{-a-2} (y^{-1} - w)^b \left\langle \psi_3', Y_3(v, y^{-1}) I[\psi_1 \otimes \psi_2]\right\rangle\right) \frac{\mathrm{d}y}{2\pi\mathrm{i}} \\
&\qquad = \oint_0 \sum_{m=0}^{\infty} (-1)^m \binom{b}{m} w^m y^{m-a-b-2} \sum_{n\in\mathbb{Z}} \left\langle \psi_3', v_n y^{n+1} I[\psi_1 \otimes \psi_2]\right\rangle \frac{\mathrm{d}y}{2\pi\mathrm{i}} \\
&\qquad = \sum_{m=0}^{\infty} (-1)^m \binom{b}{m} w^m \left\langle \psi_3', v_{a+b-m} I[\psi_1 \otimes \psi_2]\right\rangle \\
&\qquad = \left\langle \psi_3' \big(v\,\iota_-(t^a (t-w)^b)\big) \cdot I[\psi_1 \otimes \psi_2]\right\rangle.
\end{aligned} \tag{38a}$$

The terms on the right-hand side of (29) can similarly be expanded formally as

$$\begin{aligned}
&\oint_w f(z)\left\langle \psi_3', I[Y_1(v, z-w)\psi_1 \otimes \psi_2]\right\rangle \frac{\mathrm{d}z}{2\pi\mathrm{i}} \\
&\qquad = \left\langle \psi_3', I\Big[(v\,\iota_+ T_w(t^a(t-w)^b) \cdot \psi_1) \otimes \psi_2 \Big]\right\rangle
\end{aligned} \tag{38b}$$

and
$$\begin{aligned}
&\oint_0 f(z)\left\langle \psi_3', I[\psi_1 \otimes Y_2(v,z)\psi_2]\right\rangle \frac{\mathrm{d}z}{2\pi\mathrm{i}} \\
&\qquad = \left\langle \psi_3', I\Big[\psi_1 \otimes (v\,\iota_+(t^a(t-w)^b) \cdot \psi_2) \Big]\right\rangle.
\end{aligned} \tag{38c}$$

The Cauchy–Jacobi identity (29) may therefore be written in the form

$$\begin{aligned}
\left\langle \psi_3', v\,\iota_-(f(t)) \cdot I[\psi_1 \otimes \psi_2]\right\rangle &= \left\langle \psi_3', I\big[(v\,\iota_+ T_w(f(t)) \cdot \psi_1) \otimes \psi_2 \big]\right\rangle \\
&\quad + \left\langle \psi_3', I[\psi_1 \otimes (v\,\iota_+(f(t)) \cdot \psi_2)]\right\rangle \quad (f(t) \in R_{P(w)}).
\end{aligned} \tag{39}$$

Recall that we are trying to satisfy the universal property described in (32). If we were to define an "action", denoted by Δ_w, of $V \otimes R_{P(w)}$ on $M_1 \otimes M_2$ by

$$\Delta_w(v\, f(t)) \cdot (\psi_1 \otimes \psi_2) = (v\,\iota_+ T_w\,(f(t)) \cdot \psi_1) \otimes \psi_2 + \psi_1 \otimes (v\,\iota_+\,(f(t)) \cdot \psi_2)\,, \tag{40}$$

then we could write (39) succinctly as

$$\left\langle \psi_3', v\,\iota_-(f(t)) \cdot I[\psi_1 \otimes \psi_2]\right\rangle = \left\langle \psi_3', I[\Delta_w(v\, f(t)) \cdot (\psi_1 \otimes \psi_2)]\right\rangle. \tag{41}$$

In particular, I intertwines the action of $v_n = v\,t^n$ on M_3 and that of $\Delta_w(v_n)$ on $M_1 \otimes M_2$. The case for elements such as $v\, f(t)$, with $f(t) = t^a(t-w)^b$ and $b < 0$, is more subtle and hence more interesting. While (41) continues to hold, we may also expand the term $\iota_-\,(f(t))$ on the left-hand side to get

$$\begin{aligned}
&\left\langle \psi_3', I[\Delta_w(v\, f(t)) \cdot (\psi_1 \otimes \psi_2)] \right\rangle \\
&\qquad = \left\langle \psi_3', v\, \iota_-(t^a (t-w)^b) \cdot I[\psi_1 \otimes \psi_2] \right\rangle \\
&\qquad = \sum_{m=0}^{\infty} \binom{b}{m} (-w)^m \left\langle \psi_3', v\, t^{a+b-m} \cdot I[\psi_1 \otimes \psi_2] \right\rangle \\
&\qquad = \sum_{m=0}^{\infty} \binom{b}{m} (-w)^m \left\langle \psi_3', I[\Delta_w(v\, t^{a+b-m}) \cdot (\psi_1 \otimes \psi_2)] \right\rangle \\
&\qquad = \left\langle \psi_3', I[\Delta_w(v\, \iota_-(f(t))) \cdot (\psi_1 \otimes \psi_2)] \right\rangle,
\end{aligned} \tag{42}$$

assuming that everything converges. If so, we can summarise this simply as saying that the subspace generated by

$$\Delta_w(v\, f(t)) \cdot (\psi_1 \otimes \psi_2) - \Delta_w(v\, \iota_-(f(t))) \cdot (\psi_1 \otimes \psi_2) \tag{43}$$

is in the kernel of any $P(w)$-intertwining map I.

This has a most natural interpretation: If we act on elements $\psi_1 \otimes \psi_2$ in the $P(w)$-tensor product with $v\, f(t)$ or with its expansion about $t = \infty$, then the result must be the same:

$$\Delta_w(v\, f(t)) = \Delta_w(v\, \iota_-(f(t))) \quad \text{on} \quad M_1 \boxtimes_{P(w)} M_2. \tag{44}$$

The natural guess for a model of the universal $P(w)$-tensor product of M_1 and M_2 is therefore

$$M_1 \boxtimes_{P(w)} M_2 = \frac{M_1 \otimes M_2}{\left\langle \Big(\Delta_w(v\, f(t)) - \Delta_w(v\, \iota_-(f(t))) \Big) (M_1 \otimes M_2) \right\rangle}, \tag{45}$$

where the quotient is by the vector space spanned obtained by taking all $v \in V$ and $f(t) \in R_{P(w)}$. Moreover, there is only one natural way to turn this space into a V-module, namely by letting v_n act as $\Delta_w(v\, t^n)$. Notice that this quotient requires us to have extended the action of the modes of V on $M_1 \otimes M_2$ to an action of the "global mode algebra" $V \otimes R_{P(w)}$. This globalisation avoids the need to explicitly translate the action of the modes between different local coordinates, as in (23).

Comparing the definition (45) with Gaberdiel's definition (24) for the fusion product $M_1 \boxtimes M_2$, we find striking similarities as well as one major difference. The latter is the fact that (24) involves two distinct coproduct actions, deduced from locality, while here there is only one. To explain this, recall that (45) was derived from the Cauchy–Jacobi identity (29), the intertwining map analogue of the Jacobi identity of vertex algebras, and that the Jacobi identity naturally subsumes both locality and the operator product expansion. It is therefore natural to expect that we can recover from this formalism the identities that were derived using locality in Sect. 3. Indeed, we shall confirm these expectations shortly.

First, however, let us provide some concrete formulae that capture the Δ_w-action and quotient relations of (45). This will show that this definition of $M_1 \boxtimes_{P(w)} M_2$ essentially coincides with Gaberdiel's definition of the fusion product $M_1 \boxtimes M_2$. In the next section, we shall provide dual versions of these formulae. From the action (40) with $f(t) = t^n$, we obtain

$$\begin{aligned} v_n(\psi_1 \otimes \psi_2) &= \Delta_w(v\,t^n) \cdot (\psi_1 \otimes \psi_2) \\ &= (v\,\iota_+ T_w(t^n) \cdot \psi_1) \otimes \psi_2 + \psi_1 \otimes (v\,\iota_+(t^n) \cdot \psi_2) \\ &= (v\,\iota_+((t+w)^n) \cdot \psi_1) \otimes \psi_2 + \psi_1 \otimes (v\,t^n \cdot \psi_2) \\ &= \sum_{m=0}^{\infty} \binom{n}{m} w^{n-m} (v_m \psi_1) \otimes \psi_2 + \psi_1 \otimes (v_n \psi_2). \end{aligned} \tag{46}$$

This is identical to Gaberdiel's action, given in (16) and (21b), with $w_1 = w$ and $w_2 = 0$.[12] The actions of V on $M_1 \boxtimes M_2$ and $M_1 \boxtimes_{P(w)} M_2$ therefore coincide. As before, the sum acting on ψ_1 is actually finite due to the definition of a module over a vertex operator algebra.

If we explicitly compute the action (40) with $f(t) = (t-w)^n$, we instead arrive at Gaberdiel's other action, given in (16) and (21a), with $w_1 = 0$ and $w_2 = -w$:

$$\Delta_w(v\,(t-w)^n) \cdot (\psi_1 \otimes \psi_2) = (v_n \psi_1) \otimes \psi_2 + \sum_{m=0}^{\infty} \binom{n}{m} (-w)^{n-m} \psi_1 \otimes (v_m \psi_2). \tag{47}$$

Again, this is very reasonable as t^n and $(t-w)^n$ are related by a rigid translation by w while Gaberdiel's two coproducts are also identified up to a rigid translation by w, see (23) (which assumes that $w = 1$). The icing on the cake is the fact that the relation

$$\Delta_w(v\,(t-w)^n) = \Delta_w(v\,\iota_-((t-w)^n)) = \sum_{m=0}^{\infty} \binom{n}{m} (-w)^m \Delta_w(v_{n-m}), \tag{48}$$

imposed by the definition (45) of $M_1 \boxtimes_{P(w)} M_2$, is now seen to reduce to Gaberdiel's translation identity (25) (once w is set to 1). The latter is of course equivalent to the relations that are imposed by his definition (24) of $M_1 \boxtimes M_2$.

It should now be clear that the (formal) manipulations of this section amount to a second derivation of Gaberdiel's definition (24) of the fusion product, here called the $P(w)$-tensor product. Unfortunately, this means that the fruits of this labour suffer from exactly the same problems as before:

---

[12] We recall that the action of a $P(w)$-intertwining map on $\psi_1 \otimes \psi_2$ is (a projection of) $\psi_1(w)\psi_2$, in physics notation, explaining this specialisation of insertion points.

1. The most urgent problem is that

$$\Delta_w(v\,\iota_-(t^a(t-w)^b)) = \sum_{m=0}^{\infty}\binom{b}{m}(-w)^m\,\Delta_w(v_{a+b-m}) \tag{49}$$

involves an infinite sum of coproducts if b is negative. As noted in (26), substituting in the coproduct actions on the right-hand side gives hopelessly divergent results.
2. A second problem is that we have not restricted the targets W_3 of our intertwining maps to lie in the category $\mathscr{C}$ that is provided to us. This means that we may have fewer intertwining operators when a category is specified, thereby implying that the tensor product so-defined may be smaller than it might otherwise be. In particular, the model (45) for $M_1 \boxtimes_{P(w)} M_2$ need not lie in $\mathscr{C}$. As mentioned above, we shall ignore such categorical considerations in this note in order to focus on the algorithmic aspects.
3. More fundamentally, we have not yet addressed the question of whether this model is actually a V-module. If so, then it still remains to construct a universal intertwining map $\boxtimes_{P(w)}$ in order to complete the identification with the universal definition of Sect. 4.

Nevertheless, we now have the advantage of having rephrased Gaberdiel's definition of the fusion product in a language that is mathematically more precise: that of intertwining maps. We shall exploit this in the following section when we turn to the rigorous "double dual approach" of HLZ [41]. This is perhaps the cleanest way to get around the first difficulty mentioned above and thus facilitate addressing the remaining problems. We mention that Nahm [61] was also well aware of this utility of dual spaces.

6 The Huang–Lepowsky–Zhang Approach: Double Duals

In (27), we defined the restricted dual M' of a graded V-module M (V being a vertex operator algebra as usual) as a vector space, promising that we would in time equip M' with the structure of a V-module. That time has now come and the way to equip M' is through an involutive antiautomorphism ^{opp} on the modes v_n, $n \in \mathbb{Z}$, of V that acts as a kind of adjoint for the canonical pairing of M' and M:

$$\langle v_n\psi', \psi\rangle = \left\langle \psi', v_n^{\mathsf{opp}}\psi\right\rangle \quad (\psi \in M,\, \psi' \in M'). \tag{50}$$

This turns M' into a V-module.

It is shown in [17] that a natural choice for this involution, for completely general vertex operator algebras, is given by the following formula. Using mathematics notation for modes and assuming that v has conformal weight h_v, the formula is

$$(v_n^{(\mathfrak{m})})^{\mathsf{opp}} = (-1)^{h_v} \sum_{j \geqslant 0} \frac{1}{j!} (L(1)^j v)^{(\mathfrak{m})}_{-j+2h_v-2-n}. \tag{51a}$$

The formula looks much nicer with physicists' notation:

$$(v_n^{(\mathfrak{p})})^{\mathsf{opp}} = (-1)^{h_v} \sum_{j \geqslant 0} \frac{1}{j!} (L(1)^j v)^{(\mathfrak{p})}_{-n} = (-1)^{h_v} (\mathrm{e}^{L(1)} v)^{(\mathfrak{p})}_{-n}. \tag{51b}$$

We now illustrate this formula with examples in order to clarify this definition for physicists.

Example 5 Suppose that v is *quasiprimary*, meaning in particular that it is annihilated by $L(1)$. Then,

$$(v_n^{(\mathfrak{p})})^{\mathsf{opp}} = (-1)^{h_v} v^{(\mathfrak{p})}_{-n}. \tag{52}$$

If we extend ^{opp} to the field $v(z) = \sum_n v_n^{(\mathfrak{p})} z^{-n-h_v}$, then we have

$$v(z)^{\mathsf{opp}} = (-1)^{h_v} z^{-2h_v} v(z^{-1}). \tag{53}$$

Apart from the sign, which we shall discuss in the following examples, this adjoint for the quasiprimary field $v(z)$ will be very familiar to physicists. □

Example 6 Interesting special cases of (52) include the following:

- The identity operator $\mathbb{1} = \Omega_0^{(\mathfrak{p})}$ is self-adjoint, meaning that $\mathbb{1}^{\mathsf{opp}} = \mathbb{1}$, as expected.
- The Virasoro modes $L(n) = \omega_n^{(\mathfrak{p})}$ satisfy $L(n)^{\mathsf{opp}} = L(-n)$, as expected. In particular, $L(0)$ is self-adjoint.
- The modes of a weight-1 quasiprimary field, for example an affine current of the $\mathsf{SU}(2)_k$ Wess–Zumino–Witten model, satisfy $J_n^{(\mathfrak{p})} = -J^{(\mathfrak{p})}_{-n}$, $J = E, H, F$. In particular, the Cartan zero mode $H_0^{(\mathfrak{p})}$ is *not* self-adjoint and the adjoint of $E_n^{(\mathfrak{p})}$ is *not* $F^{(\mathfrak{p})}_{-n}$, contrary (perhaps) to expectations.

This possibly unexpected behaviour for affine modes is easily explained by noting that one can always twist the definition of ^{opp} by composing with an automorphism of V. In the case where $V = \mathsf{SU}(2)_k$, we can twist by the finite Weyl reflection of $\mathfrak{sl}_2$, which acts as $E \mapsto -F$, $H \mapsto -H$ and $F \mapsto -E$, to recover the expected adjoint.[13] □

Example 7 The case where V is the Heisenberg (free boson) vertex operator algebra and $v = a^{(\mathfrak{p})}_{-1}\Omega$ is the generator likewise gives $(a_0^{(\mathfrak{p})})^{\mathsf{opp}} = -a_0^{(\mathfrak{p})}$, which can also be "fixed" by twisting by the automorphism $a \mapsto -a$ of $\mathfrak{gl}_1$. However, we can appreciate the utility of this sign, and thus of the definition (51), by considering (as in the Coulomb gas formalism) the modified conformal vector

[13] In terms of the classification of real simple Lie algebras using involutions, the definition (51) corresponds to the split real form, while the adjoint familiar to physicists is associated with the compact real form.

$$\omega_\lambda = \frac{1}{2}(a_{-1}^{(\mathfrak{p})})^2\Omega + \frac{1}{2}\lambda a_{-2}^{(\mathfrak{p})}\Omega \quad (\lambda \in \mathbb{R}) \tag{54}$$

and its associated Virasoro zero modes $L_\lambda(0)$. For $\lambda = 0$, a straightforward computation shows that $(a_n^{(\mathfrak{p})})^{\mathsf{opp}} = -a_{-n}^{(\mathfrak{p})}$ and $(a_n^{(\mathfrak{p})})^{\mathsf{opp}} = +a_{-n}^{(\mathfrak{p})}$ both give $L_0(n)^{\mathsf{opp}} = L_0(-n)$, as required. However, if $\lambda \neq 0$, then v is no longer quasiprimary and (51) instead gives

$$(a_n^{(\mathfrak{p})})^{\mathsf{opp}} = \lambda\delta_{n,0}\mathbb{1} - a_{-n}^{(\mathfrak{p})}. \tag{55}$$

Another straightforward calculation now shows that the general Ansatz $(a_n^{(\mathfrak{p})})^{\mathsf{opp}} = \alpha\delta_{n,0}\mathbb{1} + \beta a_{-n}^{(\mathfrak{p})}$ is only consistent with $L_\lambda(n)^{\mathsf{opp}} = L_\lambda(-n)$ if $\alpha = \lambda$ and $\beta = -1$, as in (55). □

We now wish to enlarge this ^{opp} adjoint to be defined on the bigger space $V \otimes R_{P(w)}$. In fact, we will go even further and define it on $V \otimes \mathbb{C}[[t, t^{-1}]]$. Inserting $v_n = v\,t^n$ into (51a), so switching back to mathematician's conventions, we obtain

$$(v\,t^n)^{\mathsf{opp}} = \sum_{m=0}^{\infty} \frac{t^{-j}L(1)^j}{j!}(-t^2)^{h_v} v\,t^{-2-n} = \mathsf{e}^{t^{-1}L(1)}(-t^2)^{L(0)} v\,t^{-2-n}. \tag{56}$$

We shall therefore define

$$v^{\mathsf{opp}} = \mathsf{e}^{t^{-1}L(1)}(-t^2)^{L(0)} v\,t^{-2} \tag{57}$$

and then extend opp linearly to $V \otimes \mathbb{C}[[t, t^{-1}]]$ by

$$(v\,f(t))^{\mathsf{opp}} = v^{\mathsf{opp}} f(t^{-1}). \tag{58}$$

It is not very hard to prove that $((v\,f(t))^{\mathsf{opp}})^{\mathsf{opp}} = v\,f(t)$, hence that ^{opp} is still involutive.

Example 8 Before moving on, let us rewrite the affine and Virasoro examples discussed above in this new notation. For affine modes $J_n = J\,t^n$, we have

$$J^{\mathsf{opp}} = e^{t^{-1}L(1)}(-t^2)^{L(0)} J t^{-2} = -J, \tag{59}$$

hence $J_n^{\mathsf{opp}} = (J\,t^n)^{\mathsf{opp}} = J^{\mathsf{opp}} t^{-n} = -J\,t^{-n} = -J_{-n}$, as before. For Virasoro modes $L(n) = \omega\,t^{n+1}$, we have instead

$$\omega^{\mathsf{opp}} = e^{t^{-1}L(1)}(-t^2)^{L(0)} \omega t^{-2} = \omega\,t^2 \tag{60}$$

and so $L(n)^{\mathsf{opp}} = (\omega\,t^{n+1})^{\mathsf{opp}} = \omega^{\mathsf{opp}} t^{-n-1} = \omega\,t^{-n+1} = L(-n)$, also as before. □

We are now ready to transfer the Δ_w-action from the previous section to dual modules. Consider the full dual $(M_1 \otimes M_2)^*$. While this space is undoubtedly too big, we recall that the notion dual to taking a quotient of $M_1 \otimes M_2$ is identifying an appropriate subspace of $(M_1 \otimes M_2)^*$. This subspace shall be identified after we have dualised the action (40) of $V \otimes R_{P(w)}$, remembering to replace t by t^{-1}, and have thus obtained an action of $V \otimes \mathbb{C}[t^{-1}, t, (t^{-1} - w)^{-1}]$ on $(M_1 \otimes M_2)^*$. Note that because $(t^{-1} - w)^{-1} = -w^{-1} t (t - w^{-1})^{-1}$, we may regard this as an action of $V \otimes R_{P(w^{-1})}$.

Let ψ^* be a generic element of $(M_1 \otimes M_2)^*$. For $v \in V$ and $f(t) \in R_{P(w^{-1})}$, we define

$$\begin{aligned}&\langle v\, f(t) \cdot \psi^*, \psi_1 \otimes \psi_2\rangle = \langle \psi^*, \Delta_w((v\, f(t))^{\mathsf{opp}}) \cdot (\psi_1 \otimes \psi_2)\rangle \\ &= \left\langle \psi^*, \left(v^{\mathsf{opp}}\, \iota_+ T_w(f(t^{-1})) \cdot \psi_1\right) \otimes \psi_2\right\rangle + \left\langle \psi^*, \psi_1 \otimes \left(v^{\mathsf{opp}}\, \iota_+(f(t^{-1})) \cdot \psi_2\right)\right\rangle,\end{aligned} \tag{61}$$

for all $\psi_1 \in M_1$ and $\psi_2 \in M_2$. Again, we pause to give examples.

Example 9 In the case of affine modes, this dual action of $V \otimes R_{P(w^{-1})}$ specialises to

$$\begin{aligned}\langle J_n \cdot \psi^*, \psi_1 \otimes \psi_2\rangle &= \langle J\, t^n \cdot \psi^*, \psi_1 \otimes \psi_2\rangle \\ &= \left\langle \psi^*, (-J\, \iota_+ T_w(t^{-n}) \cdot \psi_1) \otimes \psi_2\right\rangle + \left\langle \psi^*, \psi_1 \otimes (-J\, \iota_+(t^{-n}) \cdot \psi_2)\right\rangle \\ &= -\left\langle \psi^*, (J\, \iota_+((t+w)^{-n}) \cdot \psi_1) \otimes \psi_2\right\rangle - \left\langle \psi^*, \psi_1 \otimes (J\, t^{-n} \cdot \psi_2)\right\rangle \\ &= -\sum_{m=0}^{\infty} \binom{-n}{m} w^{-n-m} \langle \psi^*, (J_m \psi_1) \otimes \psi_2\rangle - \langle \psi^*, \psi_1 \otimes (J_{-n}\psi_2)\rangle.\end{aligned} \tag{62}$$

The Virasoro version is

$$\begin{aligned}&\langle L(n) \cdot \psi^*, \psi_1 \otimes \psi_2\rangle = \left\langle \omega\, t^{n+1} \cdot \psi^*, \psi_1 \otimes \psi_2\right\rangle \\ &= \sum_{m=0}^{\infty} \binom{-n+1}{m} w^{-n+1-m} \langle \psi^*, (L(m-1)\psi_1) \otimes \psi_2\rangle + \langle \psi^*, \psi_1 \otimes (L(-n)\psi_2)\rangle.\end{aligned} \tag{63}$$

A few special cases are worth noting:

$$\langle L(1) \cdot \psi^*, \psi_1 \otimes \psi_2\rangle = \langle \psi^*, (L(-1)\psi_1) \otimes \psi_2\rangle + \langle \psi^*, \psi_1 \otimes (L(-1)\psi_2)\rangle, \tag{64a}$$

$$\langle L(0) \cdot \psi^*, \psi_1 \otimes \psi_2 \rangle = w \langle \psi^*, (L(-1)\psi_1) \otimes \psi_2 \rangle + \langle \psi^*, (L(0)\psi_1) \otimes \psi_2 \rangle + \langle \psi^*, \psi_1 \otimes (L(0)\psi_2) \rangle, \tag{64b}$$

$$\langle L(-1) \cdot \psi^*, \psi_1 \otimes \psi_2 \rangle = w^2 \langle \psi^*, (L(-1)\psi_1) \otimes \psi_2 \rangle + 2w \langle \psi^*, (L(0)\psi_1) \otimes \psi_2 \rangle + \langle \psi^*, (L(1)\psi_1) \otimes \psi_2 \rangle + \langle \psi^*, \psi_1 \otimes (L(1)\psi_2) \rangle. \tag{64c}$$

Recall now from (45) that we should impose the consistency relation

$$\Delta_w(v\, f(t)) \cdot (\psi_1 \otimes \psi_2) = \Delta_w(v\, \iota_-(f(t))) \cdot (\psi_1 \otimes \psi_2) \quad (f(t) \in R_{P(w)}). \tag{65}$$

More precisely, we should impose its dual version, namely that we should restrict to the subspace of those $\psi^* \in (M_1 \otimes M_2)^*$ for which

$$\langle \psi^*, \Delta_w(v\, f(t)) \cdot (\psi_1 \otimes \psi_2) \rangle = \langle \psi^*, \Delta_w(v\, \iota_-(f(t))) \cdot (\psi_1 \otimes \psi_2) \rangle, \tag{66}$$

for all $v\, f(t) \in R_{P(w)}$, $\psi_1 \in M_1$ and $\psi_2 \in M_2$. Equivalently, the desired subspace consists of those ψ^* for which

$$v^{\mathsf{opp}}\, f(t^{-1}) \cdot \psi^* = v^{\mathsf{opp}}\, \iota_+(f(t^{-1})) \cdot \psi^*, \tag{67}$$

for all $v\, f(t) \in R_{P(w)}$, because expanding about $t = \infty$ (ι_-) is dual to expanding about $t = 0$ (ι_+). Since ^{opp} is involutive, this consistency condition may be written in the form

$$v\, f(t) \cdot \psi^* = v\, \iota_+(f(t)) \cdot \psi^* \quad (v\, f(t) \in V \otimes R_{P(w^{-1})}). \tag{68a}$$

In other words, the dual action should be compatible with any infinite sums that are obtained by expanding $f(t)$ using ι_+. This is one of the *$P(w)$-compatibility conditions* of HLZ, see [41, Eq. (5.141)]. The other condition is in fact nothing more than requiring that

$$v_n \cdot \psi^* = v\, t^n \cdot \psi^* = 0 \quad (v \in V \text{ and } n \gg 0). \tag{68b}$$

This is, in general, required for the subspace of $(M_1 \otimes M_2)^*$ satisfying (68a) to stand any chance of being a V-module. Moreover, (68b) also guarantees that any infinite sum in (68a) is actually finite. The condition (68a) is thus an equality in $(M_1 \otimes M_2)^*$, rather than in its completion.

The functionals that we are really after here are those that act as $\langle \psi^*, \psi_1 \otimes \psi_2 \rangle = \langle \psi_3', I[\psi_1 \otimes \psi_2] \rangle$, where I is any intertwining operator of type $\begin{pmatrix} M_3 \\ M_1\, M_2 \end{pmatrix}$ (M_3 is some test module) and ψ_3' is an arbitrary element of the restricted dual M_3'. These functionals are capable of acting non-trivially on large subspaces of $M_1 \otimes M_2$. Therefore, the full dual $(M_1 \otimes M_2)^*$ first provides us with enough room to work with such functionals (without having to deal with completions!) and second, the convergence issues that plagued our previous fusion product definitions, (24) and (45), are neatly sidestepped by the natural "lower truncation" requirement (68b).

Note that (68a) is dual to (44) which, as we saw in (48), is equivalent to Gaberdiel's "coproduct equality" (23). In other words, the cure for the divergences that rendered Gaberdiel's definition of fusion meaningless is just (68b). Note also that, as defined, one must check the $P(w)$-compatibility condition (68a) for all $v\, f(t) \in V \otimes R_{P(w^{-1})}$. We can however do better! Zhang essentially proved in [72] that it is enough to check this condition for $v \in S$, where S is a set of strong generators of V.

The HLZ definition for the $P(w)$-tensor product of the V-modules M_1 and M_2 can now be stated as follows:

- Determine the subspace of $(M_1 \otimes M_2)^*$ consisting of elements ψ^* that satisfy both $P(w)$-compatibility conditions (68).
- In this subspace, consider the subspace of "finite-energy" vectors that are spanned by the generalised $L(0)$-eigenvectors.
- Define the $P(w)$-tensor product $M_1 \boxtimes_{P(w)} M_2$ to be the restricted dual of this finite-energy subspace.

This "double dual" approach then provides a candidate model for the universal $P(w)$-tensor product of Definition 4.2. There is one caveat: we have again ignored the category $\mathscr{C}$ completely. In [41], HLZ impose additional conditions beyond (68) so that the resulting $P(w)$-tensor product indeed lies in $\mathscr{C}$. Because our aim is to compare with the physicists' fusion product, which is category-agnostic, we do not discuss these details.

So far, we have discussed how this double dual approach overcomes or avoids the first two problems with the proposed model (45). It therefore remains to prove that the space we have identified above as the $P(w)$-tensor product is actually a V-module. This essentially boils down to questions about the given action of V on $(M_1 \otimes M_2)^*$ and the subspace of compatible functionals. Is this actually an action of V? Is the subspace indeed stable under this action?

As noted in Sect. 3, a proof that the action respects commutation rules and associativity was indicated by Gaberdiel for the case in which V is a universal affine Kac-Moody algebra or the Virasoro algebra. A completely general proof is given by HLZ in [41], lifting commutation rules to Borcherds identities (also known as generalised commutation relations), with no restrictions on the vertex operator algebra or its module category. There, it is also proved that the space of functionals ψ^* satisfying the compatibility conditions (68) is indeed stable under the given action of $V \otimes R_{P(w^{-1})}$. These proofs actually use formal variables and formal delta functions. It would be interesting to prove them completely in the complex-analytic setting.

We round out this section by going back to the example, discussed at the start of Sect. 5, of tensor products for modules M_1 and M_2 over a commutative associative unital algebra A. We may regard A as a vertex operator algebra with $Y(a, z)b = a \cdot b$ and $\omega = 0$. In effect, a is identified with the constant term $a_{-1} = a\,t^{-1}$ of $Y(a, z)$, all other terms being zero, and we have a trivial conformal structure: the central charge is 0 and every vector has conformal weight 0. In addition, any A-module M is naturally a module for this vertex operator algebra via $Y_M(a, z)\psi = a \cdot \psi$.

What does the double dual construction of HLZ[14] give for this class of vertex operator algebras? We first note that $(a\, f(t))^{\mathsf{opp}} = a\, t^{-2} f(t^{-1})$, so in particular we have $a^{\mathsf{opp}} = (a\, t^{-1})^{\mathsf{opp}} = a_{-1}\, t^{-1} = a$. The A-action on the dual module M^* is therefore defined by $\langle a \cdot \psi', \psi \rangle = \langle \psi', a \cdot \psi \rangle$, consistent with expectations. The Δ_w-action on $M_1 \otimes M_2$ is therefore given by

$$\begin{aligned}\Delta_w(a\, t^n) \cdot (\psi_1 \otimes \psi_2) &= (a\, \iota_+ T_w(t^n) \cdot \psi_1) \otimes \psi_2 + \psi_1 \otimes (a\, \iota_+(t^n)) \cdot \psi_2 \\ &= \delta_{n=-1}\, \psi_1 \otimes (a \cdot \psi_2),\end{aligned} \tag{69}$$

since the coefficient of t^{-1} in the expansion of $(t + w)^n$ about $t = 0$ is always 0. It follows that

$$\langle a\, t^n \cdot \psi^*, \psi_1 \otimes \psi_2 \rangle = \langle \psi^*, \Delta_w(a\, t^{-n-2}) \cdot (\psi_1 \otimes \psi_2) \rangle = \delta_{n=-1} \langle \psi^*, \psi_1 \otimes (a \cdot \psi_2) \rangle. \tag{70}$$

A similar, but more involved computation, gives

$$\begin{aligned}\langle a\, t^j (t^{-1} - w)^k \cdot \psi^*, \psi_1 \otimes \psi_2 \rangle &= \binom{-j-2}{-k-1} w^{k-j-1} \langle \psi^*, (a \cdot \psi_1) \otimes \psi_2 \rangle \\ &\quad + \binom{k}{j+1} (-w)^{k-j-1} \langle \psi^*, \psi_1 \otimes (a \cdot \psi_2) \rangle,\end{aligned} \tag{71}$$

where we note the commonly employed convention that $\binom{n}{r} = 0$ if r is a negative integer.

However, the $P(w)$-compatibility condition (68a) lets us take a different route by first using ι_+ to expand $t^j (t^{-1} - w)^k$ on the left-hand side and then using (70):

$$\begin{aligned}\langle a\, t^j (t^{-1} - w)^k \cdot \psi^*, \psi_1 \otimes \psi_2 \rangle &= \sum_{m=0}^{\infty} \binom{k}{m} (-w)^m \langle a\, t^{j-k+m} \cdot \psi^*, \psi_1 \otimes \psi_2 \rangle \\ &= \binom{k}{k-j-1} (-w)^{k-j-1} \langle \psi^*, \psi_1 \otimes (a \cdot \psi_2) \rangle.\end{aligned} \tag{72}$$

[14]The alert reader will notice that this double dual construction is overkill here because the fields $Y(a, z)$ are independent of z. Nevertheless, we feel it helps to unpack this abstract machinery in the simplest case and see that it works. We shall consider a less straightforward example in the next section.

Therefore, we must have

$$\binom{-j-2}{-k-1}(-1)^{k-j-1}\langle\psi^*, (a\cdot\psi_1)\otimes\psi_2\rangle = \left[\binom{k}{k-j-1}-\binom{k}{j+1}\right]\langle\psi^*, \psi_1\otimes(a\cdot\psi_2)\rangle. \tag{73}$$

Note first that the left-hand side is 0 if $k \geqslant 0$, while the binomial coefficients on the right-hand side are either equal (if $0 \leqslant j+1 \leqslant k$) or both 0 (for all other j). We may therefore restrict to $k < 0$ and consider the following four possibilities for j:

1. $j \leqslant k-1$ and $j \geqslant -1$ is impossible, so at most one of the binomial coefficients on the right-hand side may be non-zero.
2. $j > k-1$ and $j < -1$ is possible, but then $-k-1 > -j-2 \geqslant 0$ and both sides are zero.
3. If $j \leqslant k-1$ and $j < -1$, then the second binomial coefficient on the right-hand side is zero. Since $-j-2 \geqslant -k-1 \geqslant 0$, we have

$$\binom{-j-2}{-k-1}(-1)^{k-j-1} = \binom{-j-2}{k-j-1}(-1)^{k-j-1} = \binom{k}{k-j-1} \tag{74}$$

   and so (73) reduces to

$$\langle\psi^*, (a\cdot\psi_1)\otimes\psi_2\rangle = \langle\psi^*, \psi_1\otimes(a\cdot\psi_2)\rangle. \tag{75}$$

4. A similar calculation for $j > k-1$ and $j \geqslant -1$ likewise reduces to this same equation.

It is quite enlightening to redo this analysis using generating functions (thereby considering all values of j and k simultaneously) and the properties of formal delta functions, as in [40]. Either way, we conclude that the $P(w)$-compatibility conditions for commutative associative unital algebras pick out precisely the ψ^* that implement the constraint that reduces $M_1 \otimes M_2$ to $M_1 \otimes_A M_2$. The double dual model of the HLZ $P(w)$-tensor product is, for these vertex operator algebras, just the usual tensor product over A. There are other quirks about this example, mainly arising from the fact that the central charge is zero, which are discussed in detail in [40].

7 An Explicit Example of a Fusion Calculation: Virasoro at $c = -2$

Let V be the universal vertex operator algebra associated with the Virasoro algebra at central charge $c = -2$. At this central charge, the universal Virasoro vertex operator algebra is actually simple. Let M be the irreducible highest-weight Virasoro module of central charge -2 whose highest-weight vector γ has conformal weight $-\frac{1}{8}$. It follows [19] that M is a V-module. Let ψ_1 and ψ_2 be arbitrary elements of M and, finally, let ψ^* be an arbitrary element of $(M \otimes M)^*$.

Our aim is, obviously, to calculate the fusion product $M \boxtimes M$, from here on identified with the $P(w)$-tensor product $M \boxtimes_{P(w)} M$ (with $w \in \mathbb{C}^\times$ arbitrary). We shall do so using the formalism of HLZ, identifying $M \boxtimes M$ as a Virasoro module. For this, we need to determine the possibilities for the ψ^* that satisfy HLZ's $P(w)$-compatibility conditions (68). In fact, we will only derive some restrictions on ψ^* that follow as consequences of being $P(w)$-compatible. It is in general a hard task to prove that such ψ^* indeed exist.

Experts will note that this example has historical significance, being the very first calculation performed [25] using the computational method that is now known as the Nahm–Gaberdiel–Kausch algorithm. The motivation behind this first calculation was, of course, the expectation that the fusion product would exhibit a non-diagonalisable action of $L(0)$, a fact that had been previously been established by Gurarie [30] using correlation functions. We shall discuss the calculation of [25], and this algorithm, in Sect. 8. In what follows, one may take $w = 1$ for convenience, but this is not at all necessary. We will also freely use the notations that were introduced in previous sections.

Consider $v\, f(t) \in V \otimes R_{P(w^{-1})}$. By a result of Zhang [72], it is enough for us to take $\xi = \omega\, f(t)$, since ω strongly generates V. By way of preparation, we compute the actions

$$\begin{aligned}
&\langle L(n) \cdot \psi^*, \psi_1 \otimes \psi_2 \rangle = \left\langle \psi^*, \Delta_w((\omega\, t^{n+1})^{\mathrm{opp}}) \cdot (\psi_1 \otimes \psi_2) \right\rangle \\
&= \left\langle \psi^*, \Delta_w(\omega\, t^{-n+1}) \cdot (\psi_1 \otimes \psi_2) \right\rangle \\
&= \left\langle \psi^*, \left(\omega\, \iota_+((t+w)^{-n+1}) \cdot \psi_1\right) \otimes \psi_2 \right\rangle + \left\langle \psi^*, \psi_1 \otimes \left(\omega\, \iota_+(t^{-n+1}) \cdot \psi_2\right) \right\rangle \\
&= \sum_{m=0}^{\infty} \binom{-n+1}{m} w^{-n-m+1} \langle \psi^*, (L(m-1)\psi_1) \otimes \psi_2 \rangle + \langle \psi^*, \psi_1 \otimes (L(-n)\psi_2) \rangle
\end{aligned} \tag{76}$$

and (in the same way)

$$\begin{aligned}
\left\langle \omega\, t(t^{-1} - w)^n \cdot \psi^*, \psi_1 \otimes \psi_2 \right\rangle &= \left\langle \psi^*, \left((wL(n-1) + L(n))\psi_1 \right) \otimes \psi_2 \right\rangle \\
&\quad + \sum_{m=0}^{\infty} \binom{n}{m} (-w)^{n-m} \langle \psi^*, \psi_1 \otimes (L(m)\psi_2) \rangle .
\end{aligned} \tag{77}$$

To begin with, let us assume that $\psi^* \in (M \otimes M)^*$ satisfies both the $P(w)$-compatibility conditions (68) and, in addition, $L(n) \cdot \psi^* = 0$ for all $n \geqslant 1$.

1. For $n \geqslant 1$, (76) gives

$$0 = \langle L(n) \cdot \psi^*, \psi_1 \otimes \psi_2 \rangle$$
$$= \sum_{m=0}^{\infty} \binom{-n+1}{m} w^{-n-m+1} \langle \psi^*, (L(m-1)\psi_1) \otimes \psi_2 \rangle + \langle \psi^*, \psi_1 \otimes (L(-n)\psi_2) \rangle. \quad (78)$$

This implies that the values $\langle \psi^*, M \otimes \psi_2 \rangle$ are determined by the values $\langle \psi^*, M \otimes \gamma \rangle$, because the action of any negative mode $L(-n)$ in the second tensor factor may be traded for an action on the first tensor factor. (We recall that γ is the highest weight vector of M.)

2. Now we investigate $\omega\, t(t^{-1} - w)^n \cdot \psi^*$ for $n \leqslant -1$. On the one hand, the $P(w)$-compatibility condition (68a) requires that

$$\omega\, t(t^{-1} - w)^n \cdot \psi^* = \omega\, \iota_+(t(t^{-1} - w)^n) \cdot \psi^*$$
$$= \sum_{m=0}^{\infty} \binom{n}{m} (-w)^m L(m-n) \cdot \psi^* = 0, \quad (79)$$

because $m - n$ is strictly positive. On the other hand, combining this with (77) now gives

$$w\Big\langle \psi^*, (L(n-1)\psi_1) \otimes \gamma \Big\rangle + \Big\langle \psi^*, (L(n)\psi_1) \otimes \gamma \Big\rangle$$
$$+ (-w)^n \Big\langle \psi^*, \psi_1 \otimes (L(0)\gamma) \Big\rangle = 0. \quad (80)$$

Since we have assumed that $n \leqslant -1$, the action of $L(-2), L(-3), \ldots$ on ψ_1 may be traded for that of $L(-1), L(-2), \ldots$ respectively. It therefore follows that $\langle \psi^*, M \otimes \gamma \rangle$ is determined by the numbers $\langle \psi^*, (L(-1)^k \gamma) \otimes \gamma \rangle$, for $k \in \mathbb{Z}_{\geqslant 0}$. However, the highest-weight vector γ satisfies

$$L(-1)^2 \gamma = \frac{1}{2} L(-2) \gamma \quad (81)$$

by virtue of M being irreducible [3]. Thus, $\langle \psi^*, M \otimes \gamma \rangle$ is actually determined by the numbers $\langle \psi^*, \gamma \otimes \gamma \rangle$ and $\langle \psi^*, (L(-1)\gamma) \otimes \gamma \rangle$. Consequently, the space of $P(w)$-compatible ψ^* satisfying $L(n) \cdot \psi^* = 0$, for all $n \geqslant 1$, is at most 2-dimensional.

3. We now investigate the action of $L(0)$ on such ψ^*. As $L(n)L(0) \cdot \psi^* = 0$, for all $n \geqslant 1$, and the $P(w)$-compatible elements form a V-submodule [41], we can apply all the discussion above with $L(0) \cdot \psi^*$ in place of ψ^*. In particular, (64b) gives

$$\begin{aligned}\langle L(0)\cdot\psi^*, \gamma\otimes\gamma\rangle &= \langle\psi^*, (wL(-1)\gamma)\otimes\gamma + (L(0)\gamma)\otimes\gamma + \gamma\otimes(L(0)\gamma)\rangle \\ &= -\frac{1}{4}\langle\psi^*, \gamma\otimes\gamma\rangle + w\langle\psi^*, (L(-1)\gamma)\otimes\gamma\rangle \end{aligned} \tag{82a}$$

and, using also (81), then (80) with $n = -1$ and $\psi_1 = \gamma$,

$$\begin{aligned}&\langle L(0)\cdot\psi^*, (L(-1)\gamma)\otimes\gamma\rangle \\ &\quad= \langle\psi^*, (wL(-1)^2\gamma)\otimes\gamma + (L(0)L(-1)\gamma)\otimes\gamma) + (L(-1)\gamma)\otimes(L(0)\gamma)\rangle \\ &\quad= \frac{1}{2}w\langle\psi^*, (L(-2)\gamma)\otimes\gamma\rangle + \frac{3}{4}\langle\psi^*, (L(-1)\gamma)\otimes\gamma\rangle \\ &\quad= \frac{1}{4}\langle\psi^*, (L(-1)\gamma)\otimes\gamma\rangle + \frac{1}{2}w^{-1}\langle\psi^*, \gamma\otimes(L(0)\gamma)\rangle \\ &\quad= -\frac{1}{16}w^{-1}\langle\psi^*, \gamma\otimes\gamma\rangle + \frac{1}{4}\langle\psi^*, (L(-1)\gamma)\otimes\gamma\rangle. \end{aligned} \tag{82b}$$

4. In effect, we have shown that

$$\begin{bmatrix}\langle L(0)\cdot\psi^*, \gamma\otimes\gamma\rangle \\ \langle L(0)\cdot\psi^*, (L(-1)\gamma)\otimes\gamma\rangle\end{bmatrix} = \begin{bmatrix}-1/4 & w \\ -w^{-1}/16 & 1/4\end{bmatrix}\begin{bmatrix}\langle\psi^*, \gamma\otimes\gamma\rangle \\ \langle\psi^*, (L(-1)\gamma)\otimes\gamma\rangle\end{bmatrix}. \tag{83}$$

The 2×2 matrix representing this action of $L(0)$ has zero trace and zero determinant, hence consists of a rank 2 Jordan block with eigenvalue 0. However, recall that we have not shown that there are two linearly independent $P(w)$-compatible ψ^* that are annihilated by the $L(n)$ with $n \geqslant 1$, just that there are at most 2. Consequently, this space of ψ^* could actually be 1- or even 0-dimensional.
In principle, we should test these ψ^* for $P(w)$-compatibility using $\omega\, f(t)$, for completely general $f(t) \in R_{P(w^{-1})}$. However, it is very difficult to show that no further constraints arise. Instead, one can resort to information obtained through indirect means. In particular, Gurarie showed in [30] that the correlation function involving four copies of the field $\gamma(z) = Y_{\boxtimes}(\gamma, z)$ (at four different insertion points) possesses logarithmic singularities. As this cannot happen if $L(0)$ acts diagonalisably on the fusion product $M \boxtimes M$, this strongly suggests that there are no further constraints. We shall assume the truth of this statement, hence that we have correctly identified a rank 2 Jordan block in the $L(0)$-action. It is useful to indicate this conclusion pictorially as follows:

$$\bullet \xleftarrow{L(0)} \bullet\,. \tag{84}$$

We emphasise that this picture indicates just a part of the space F of of all functionals on $M \otimes M$ that satisfy the $P(w)$-compatibility conditions (68). Equation (83) demonstrates that this part describes finite-energy vectors in F.

5. We next consider $L(-1)\cdot\psi^*$, assuming that ψ^* is $P(w)$-compatible and that $L(n)\cdot\psi^*=0$ for all $n\geqslant 0$. In other words, we wish to compute the action of $L(-1)$ on the ψ^* that corresponds to the $L(0)$-eigenvector in the Jordan block identified above. However, the $P(w)$-compatible elements form a V-module [41] and

$$L(n)L(-1)\cdot\psi^* = L(-1)L(n)\cdot\psi^* + (n+1)L(n-1)\cdot\psi^* = 0, \tag{85}$$

for all $n\geqslant 1$. The above discussion therefore applies *mutatis mutandis* with $L(-1)\cdot\psi^*$ in place of ψ^*. In particular, the conclusion that such ψ^* are generalised $L(0)$-eigenvectors of eigenvalue 0 must apply to $L(-1)\cdot\psi^*$. As this is impossible (the $L(0)$-eigenvalue is clearly 1), the only way out is to have $\langle L(-1)\psi^*, M\otimes M\rangle$ are determined by the numbers $L(-1)\cdot\psi^*=0$.
This conclusion does not apply if we relax the condition that $L(0)\cdot\psi^*=0$, that is if we consider ψ^* to be one of the generalised eigenvectors of $L(0)$, because $L(1)L(-1)\cdot\psi^* = 2L(0)\cdot\psi^* \neq 0$. To identify if $L(-1)\cdot\psi^*$ is zero or not, we would need to identify a basis of the subspace of $P(w)$-compatible elements that are annihilated by the $L(n)$ with $n\geqslant 2$. This can be done, though we shall not do so here, and the result is that $L(-1)\cdot\psi^*$ is not zero when $L(0)\cdot\psi^*\neq 0$. We add this information to our pictorial representation of F thusly:

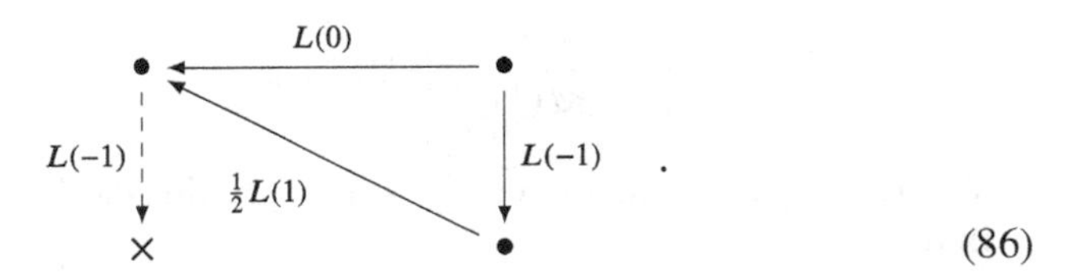

(86)

Here, the $\times$ indicates that the target state is 0.
6. Using the well-known structure theory for Virasoro highest-weight modules, we conclude that the submodule of F generated by the $L(0)$-eigenvector of eigenvalue 0 is irreducible and is therefore isomorphic to $M(0)/M(1)$, where $M(h)$ denotes the $c=-2$ Virasoro Verma module of conformal weight h. In particular, it is isomorphic to the vacuum module V of the vertex operator algebra. The generalised $L(0)$-eigenvectors do not generate highest-weight modules of course. However, each becomes a genuine eigenvector in the quotient module F/V and therefore generates a highest-weight module H of conformal weight 0 in this quotient. Unfortunately, the information we have does not allow us to identify H—all we can say is that it is not isomorphic to V.
To identify H, we must delve deeper into the structure of the fusion product. Considering the Virasoro action on the subspace of $P(w)$-compatible ψ^* that are annihilated by the $L(n)$ with $n\geqslant 4$, some tedious computation verifies that there is a second vanishing relation akin to the relation $L(-1)\cdot\psi^*=0$ established above. The image of this relation in F/V is

$$\big(L(-1)^2 - 2L(-2)\big)L(-1) \cdot [\psi^*] = 0, \tag{87}$$

where $[\psi^*]$ is the image of any generalised $L(0)$-eigenvector of conformal weight 0. This vanishing, along with the structure theory of Virasoro highest-weight modules now identifies H as the quotient $M(0)/M(3)$.
It is therefore natural to conjecture that the finite-energy submodule $\underline{F}$ of F is characterised by the following non-split short exact sequence of V-modules:

$$0 \longrightarrow \frac{M(0)}{M(1)} \longrightarrow \underline{F} \longrightarrow \frac{M(0)}{M(3)} \longrightarrow 0. \tag{88}$$

Taking restricted duals gives the following non-split short exact sequence for the fusion product:

$$0 \longrightarrow \left(\frac{M(0)}{M(3)}\right)' \longrightarrow M \boxtimes M \longrightarrow \frac{M(0)}{M(1)} \longrightarrow 0. \tag{89}$$

Here, we have noted that $M(0)/M(1) \cong V$ is irreducible, hence self-dual. This can, however, be reorganised so as to arrive at the same non-split short exact sequence as for $\underline{F}$:

$$0 \longrightarrow \frac{M(0)}{M(1)} \longrightarrow M \boxtimes M \longrightarrow \frac{M(0)}{M(3)} \longrightarrow 0. \tag{90}$$

We indicate this reorganisation pictorially by taking the dual (reversing the arrows) in (86):

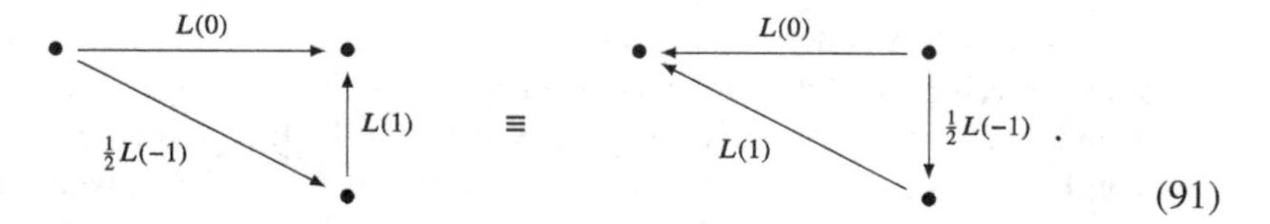

(91)

If this conjecture is true, then the sequence (90), along with the non-diagonalisable $L(0)$-action, makes $M \boxtimes M$ a *staggered module* [69] and the structure theory of such modules [50] shows that this non-split sequence completely determines the (self-dual) fusion product $M \boxtimes M$ up to isomorphism (the corresponding extension group is $\mathbb{C}$). To verify this, one needs to rule out the possibility that the fusion product is in fact larger than this staggered module. This can be done with the information currently available, but requires much deeper knowledge of extension groups (in a generalisation of category $\mathscr{O}$ that allows for non-diagonalisable $L(0)$-actions).

With regard to this calculation, there are several points that merit further explanation.

First, note that at every stage, we are only giving *necessary constraints* that are consequences of the $P(w)$-compatibility conditions. This means that at every stage, it may be possible that we have fewer possibilities for the ψ^* than one might naïvely expect. Typically, such additional constraints arise from relations satisfied in the modules being fused, for instance the vanishing of singular vectors, and their interpretation using $P(w)$-compatibility. In the physics literature, such extraneous ψ^* are (dual versions of) the *spurious states* of Nahm [61], see Sect. 8.

Second, recall that HLZ identify a subspace of the (enormous) full dual $(M_1 \otimes M_2)^*$ as the restricted dual of the actual fusion product $M_1 \boxtimes M_2$. It seems like dualising this inclusion leads to the conclusion that the fusion product may be realised as a quotient of $M_1 \otimes M_2$, vindicating the approaches discussed in Sects. 3 and 5. However, we know that these approaches led to unacceptable divergences. The crucial fact to pinpoint here is that "dualising this inclusion" is the source of our confusion—the full dual $(M_1 \otimes M_2)^*$ is not graded by generalised $L(0)$-eigenspaces, so we cannot take its restricted dual. The best that we could conclude then is that the fusion product may be realised as the finite-energy vectors in the double dual $(M_1 \otimes M_2)^{**}$ (which is even more enormous than the full dual).

Lastly, in order to build fusion products algorithmically, for example by considering subspaces of $P(w)$-compatible ψ^* that are annihilated by the $L(n)$ with $n \geqslant d$, for some d, we are naturally led to incorporate filtrations of the mode algebra into the procedure. The obvious filtration is given by conformal weights, but we may have to get more creative in general. We shall have a little more to say on this in the following section.

8 The Fusion Algorithm of Nahm–Gaberdiel–Kausch

We have reviewed Gaberdiel's original approach to defining fusion and shown that it may be interpreted mathematically in terms of $P(w)$-intertwining maps. The resulting "construction" of the fusion product was shown to be formally meaningless due to convergence issues, but we have seen how this can be made rigorous using dual spaces *á la* HLZ. We have even seen how to perform non-trivial calculations in the HLZ formalism that allow one to identify fusion products, in favourable circumstances. Now, we would like to complete the circle and discuss how physicists perform these calculations. The method that we shall explain here was originally proposed by Nahm [61] and was subsequently generalised and implemented by Gaberdiel and Kausch [25]. It is therefore known as the Nahm–Gaberdiel–Kausch fusion algorithm, or NGK algorithm for short.

We shall discuss this algorithm shortly with the aid of the same example (Virasoro at $c = -2$) that was treated in the previous section. Here, we shall be fairly brief because there are already several detailed discussions addressing these practicalities in the physics literature, see for example [4, 9, 13, 25]. The point is really to make manifest that the NGK algorithm is essentially the dual of the computational method that we have outlined above for the HLZ formalism.

An elephantine question now enters the room: How did NGK develop a practical algorithm to construct fusion products when they knew that their definition of fusion suffered from divergences? The answer is of course that physicists have a long history of dealing with divergences, especially when field theory is involved, and so their reaction to this seemingly intransigent block was decidedly meh. Indeed, Nahm pointed out [61] that these divergences could be fixed by working in the dual space. More interestingly, however, he chose to ignore this in favour of a practical approach that avoided duals and instead worked with quotients.

To explain his idea, recall that our HLZ computation above began by considering the *subspace* of $P(w)$-compatible ψ^* in the dual of the tensor product space that were annihilated by the (dual) action of the $L(n)$ with $n \geqslant 1$. Dualising this now leads us to consider the *quotient* of the tensor product space in which we impose annihilation by the action of the $L(-n) = L(n)^{\mathrm{opp}}$ with $n \geqslant 1$. On this quotient, we also need to impose the dual of $P(w)$-compatibility. We have already noted above that the dual of (68a) is Gaberdiel's equality of coproducts (23), itself derived from locality, which is formally divergent on the full tensor product. Let us check how this condition fares on Nahm's quotient. Substituting (16) or (18), with $w_1 = 0$ and $w_2 = -1$, into (25) gives (for $v = \omega$)

$$(L(n) \otimes \mathbb{1}) + \sum_{m=0}^{\infty} \binom{n+1}{m} (-1)^{n+1-m} (\mathbb{1} \otimes L(m-1)) = \sum_{m=0}^{\infty} (-1)^m \binom{n+1}{m} \Delta_{1,0}^{(2)}(L(n-m)). \tag{92}$$

The infinite sum on the left-hand side is harmless when acting on any equivalence class $[\psi_1 \otimes \psi_2]$ because $v_m \psi_2 = 0$ for $m \gg 0$, by definition of a V-module. The novel feature is thus that the infinite sum on the right-hand side is now also rendered finite because we are working in a quotient in which negative modes act as 0. Nahm's quotient has thus cured the divergences in much the same way that (68b) did for HLZ. The divergences are similarly cured for the other v in this Virasoro vertex operator algebra thanks to the magic of normal ordering.

Quotienting by the action of the $L(-n)$ with $n \geqslant 1$ is perfectly fine for identifying fusion products in rational conformal field theory. However, we saw in the previous section that there are fusion products for which this is not going to be true. Gaberdiel and Kausch recognised this in the course of studying such examples (which arise in so-called *logarithmic* conformal field theory) and therefore generalised Nahm's cure to incorporate quotients by the action of the $L(-n)$ with $n > d$, for some "depth" d.[15]

We digress briefly to note that for a general vertex operator algebra V, the depth-0 quotient of any irreducible V-module (whose conformal weights are bounded below) coincides with the image of said module under the Zhu functor of [73]. On the quotient, one can only compute the action of those modes of V that commute

[15] In fact, they took this a step further and discussed quotients by actions of fairly arbitrary subalgebras of the mode algebra of V. Appropriate filtrations by such subalgebras then lead to a consistent framework in which one can evaluate the action of any given mode. The need for quite exotic filtrations is best exemplified by referring to the rather difficult computations that arise when studying fusion products for modules over non-rational affine vertex operator algebras, see [24, 66].

with $L(0)$ (zero modes with physics conventions). This action of course agrees with that of the Zhu algebra. As regards depth-d quotients, aficionados of higher-level Zhu algebras should feel right at home. However, it is possible to define actions of certain non-zero modes that map between quotients of different depths [25]. This does not seem to have been incorporated into higher Zhu theory yet (though perhaps it should).

A natural question to ask now is whether one can honestly define fusion products in terms of these filtered quotients on which the divergence malady has been eradicated. In favourable cases, such as that of the previous section where one has the highest-weight theory of the Virasoro algebra at one's disposal, we only need analyse a finite number of these quotients in order to completely identify the fusion product. However, an abstract definition should apply more generally. Such a definition might go as follows:

- Assemble the quotients into a projective system and take the projective limit.
- Define the fusion product to be the submodule of finite-energy vectors in this projective limit.

This definition is hinted at in Nahm's original paper [61] and is proposed concretely in a paper of Tsuchiya and Wood [70] (though this has not been developed further to the best of our knowledge), see also similar work of Miyamoto [57]. Unfortunately, none of these papers seem to prove that this definition agrees with that of HLZ, even in favourable cases. We are very tempted to conjecture that it does.

One obvious difficulty with this approach is in determining whether the result is independent of the choice of (suitable) filtration and its quotients. For C_2-cofinite vertex operator algebras, one might expect this to be the case with the filtration by conformal weight being perhaps sufficient to completely identify the results. However, applying the NGK algorithm to non-rational affine vertex operator algebras [24, 66] suggests strongly that this is a much more subtle question in general—the presence of sectors that are twisted by spectral flow automorphisms means that filtering by conformal weight definitely does not suffice. In fact, it can happen that all conformal weight filtration quotients lead to zero.

At the end of the day however, what is clear is that we may dualise the methodology we detailed in Sect. 7 to compute fusion products in the HLZ formalism. We shall explain in the rest of this section that this dual formalism is essentially the NGK algorithm. The conclusion is that fusion computations performed using NGK will necessarily agree with those performed using HLZ, despite the fact that the NGK formalism currently has no rigorous definition for the fusion product. Of course, the NGK algorithm has the relative advantage of dispensing with the abstraction of the dual space formalism.

As Feynman advised, we shall now shut up and calculate, again considering the $c = -2$ Virasoro fusion product of M with itself, as discussed in the previous section. We begin by investigating the depth-0 quotient of $M \otimes M$, imposing relations such as Gaberdiel's coproduct equality (92) (which we recall is dual to $P(w)$-compatibility) in order to cut it down to the depth-0 quotient of $M \boxtimes M$. If we take $n \leqslant -1$, then the right-hand side acts as 0 on the depth-0 quotient and we have

$$(L(n) \otimes \mathbb{1}) = \sum_{m=0}^{\infty} \binom{n+1}{m} (-1)^{n-m} (\mathbb{1} \otimes L(m-1)). \tag{93}$$

For $n \leqslant -2$, this lets us swap the action of $L(n)$ on the first tensor factor for a linear combination of terms in which $L(-1)$, $L(0)$, $L(1)$ and so on act on the second tensor factor. By swapping all such actions onto the second factor, we are left with only $L(-1)$-modes acting on the first. However, two of these may be replaced by an $L(-2)$, by (81), which may then be swapped for an action on the second factor. It follows that (the images of) $\gamma \otimes M$ and $(L(-1)\gamma) \otimes M$ together form a spanning set for the zero-depth quotient.

We can do even better by using (21b) for the action of the negative Virasoro modes, again because this action is required to vanish. This formula simplifies, for $n \leqslant -1$, to

$$(\mathbb{1} \otimes L(n)) = -\sum_{m=0}^{\infty} \binom{n+1}{m} (L(m-1) \otimes \mathbb{1}), \tag{94}$$

which allows us to swap the action of a negative mode of the second factor for $L(-1)$, $L(0)$, $L(1)$ and so on acting on the first. With some careful bookkeeping, it is easy to see that by bouncing the actions from one factor to the other and back again, we get the spanning set

$$\{\gamma \otimes \gamma, (L(-1)\gamma) \otimes \gamma\} \tag{95}$$

for the zero-depth quotient of $M \boxtimes M$.

As in Sect. 7, it is not easy to tell if this spanning set is a basis or not, but we can rephrase our conclusion in a more general (and hopefully enlightening) manner. Let us call the span of γ and $L(-1)\gamma$ the *special subspace* M^{ss} of M. We shall denote the depth-0 quotient of an $L(0)$-graded module N by $N^{(0)}$. The conclusion is then that the depth-0 quotient of $M \boxtimes M$ may be expressed as a quotient of the tensor product of the special subspace of M and its depth-0 quotient:

$$M^{\text{ss}} \otimes M^{(0)} \twoheadrightarrow (M \boxtimes M)^{(0)}. \tag{96}$$

As these are just vector spaces, it would also be reasonable to consider the depth-0 quotient of the fusion product as a subspace of the tensor product. Nahm's first deep insight into fusion was to realise that this is a quite general fact. For an arbitrary vertex operator algebra V and an arbitrary $L(0)$-graded V-module M, let $C_1(M)$ denote the image in M of the action of the negative modes v_n, for all $v \in V$ and $n \geqslant 1$ (using mathematics conventions!). The *special subspace* M^{ss} is then defined [61] to be the vector space quotient $M/C_1(M)$. Nahm then argues that

$$M^{\text{ss}} \otimes N^{(0)} \twoheadrightarrow (M \boxtimes N)^{(0)}, \tag{97}$$

for V-modules M and N.

In the example at hand, we can determine the action of $L(0)$ on the candidate space for $(M \boxtimes M)^{(0)}$ by applying (16) and bouncing the actions back and forth between the tensor factors until we again have a linear combination of the elements of (95). For example, (16), (81), (93) and (94) give

$$\begin{aligned}
&\Delta_{1,0}(L(0))(L(-1)\gamma \otimes \gamma) \\
&\qquad = (L(-1)^2\gamma \otimes \gamma) + (L(0)L(-1)\gamma \otimes \gamma) + (L(-1)\gamma \otimes L(0)\gamma) \\
&\qquad = \frac{1}{2}(L(-2)\gamma \otimes \gamma) + \frac{3}{4}(L(-1)\gamma \otimes \gamma) \\
&\qquad = \frac{1}{2}\Big[(\gamma \otimes L(-1)\gamma) + (\gamma \otimes L(0)\gamma)\Big] + \frac{3}{4}(L(-1)\gamma \otimes \gamma) \\
&\qquad = -\frac{1}{16}(\gamma \otimes \gamma) + \frac{1}{4}(\gamma \otimes L(-1)\gamma),
\end{aligned} \tag{98}$$

which of course agrees with the result obtained in (82) using the HLZ double dual formalism. Here, as there, we find a non-diagonalisable action of $L(0)$ which, when combined with Gurarie's observation about logarithmic singularities in correlation functions, shows that (95) is a basis for $(M \boxtimes M)^{(0)}$.

This game of bouncing actions between tensor factors is not confined to the depth-0 world. Gaberdiel and Kausch showed in [25] that (97) generalises readily to depth-d, at least for the Virasoro case with filtration by conformal weight:

$$M^{\text{ss}} \otimes N^{(d)} \twoheadrightarrow (M \boxtimes N)^{(d)}. \tag{99}$$

This has since been verified for several other vertex operator algebras [4, 22, 24, 26, 67]. As in Sect. 7, we shall look quickly at what happens to our example when $d = 1$. First, note that $\{\gamma, L(-1)\gamma\}$ is a basis for both M^{ss} and $M^{(1)}$, hence that $(M \boxtimes M)^{(1)}$ is a quotient of a four-dimensional space. We saw in Sect. 7 that this fusion quotient was actually only three-dimensional. There, this was established using some straightforward abstract reasoning. Here, we want to take the time to see how it also follows from a relation satisfied by the Virasoro action on M.

Recall that to identify the special subspace of M, we used the singular vector relation (81). This relation also holds for the second factor of the tensor product $M \otimes M$, hence may be exploited to deduce additional relations to impose on $M^{\text{ss}} \otimes M^{(1)}$. In particular, $\Delta_{1,0}(L(-1))^2 = 0$ and (81) give

$$\begin{aligned}
0 &= \Delta_{1,0}(L(-1))^2(\gamma \otimes \gamma) \\
&= (L(-1)^2\gamma \otimes \gamma) + 2(L(-1)\gamma \otimes L(-1)\gamma) + (\gamma \otimes L(-1)^2\gamma) \\
&= \frac{1}{2}(L(-2)\gamma \otimes \gamma) + 2(L(-1)\gamma \otimes L(-1)\gamma) + \frac{1}{2}(\gamma \otimes L(-2)\gamma).
\end{aligned} \tag{100}$$

The first term is simplified using (92):

$$(L(-2)\gamma \otimes \gamma) = (\gamma \otimes L(-1)\gamma) + \frac{1}{8}(\gamma \otimes \gamma). \tag{101}$$

The second is fine as is, so we simplify the third by using $\Delta_{1,0}(L(-2)) = 0$:

$$0 = \Delta_{1,0}(L(-2))(\gamma \otimes \gamma) = (L(-1)\gamma \otimes \gamma) + \frac{1}{8}(\gamma \otimes \gamma) + (\gamma \otimes L(-2)\gamma). \tag{102}$$

Equation (100) therefore becomes

$$2(L(-1)\gamma \otimes L(-1)\gamma) - \frac{1}{2}(L(-1)\gamma \otimes \gamma) + \frac{1}{2}(\gamma \otimes L(-1)\gamma) = 0, \tag{103}$$

an additional relation that reduces the dimensionality of $(M \boxtimes M)^{(1)}$ from 4 down to 3. Nahm refers to the left-hand side of (103) as a *spurious state*.

The full identification of the fusion product $M \boxtimes M$ in the NGK formalism now proceeds in a similar fashion to the HLZ computation discussed in Sect. 7. We will not repeat the details here, but instead comment on certain mild differences. As we have seen, the determination of spurious states seemed somewhat easier with the HLZ formalism. This is because the depth-0 part of the HLZ algorithm captured states in the (dual) fusion product that were annihilated by the positive modes, thereby characterising all singular vectors at once. On the other hand, we remarked that some high-powered extension group results were needed to conclude that the HLZ identification did not miss states in the fusion product not generated by highest-weight vectors. By contrast, this conclusion is easy with the NGK formalism because the depth-0 calculation captures all states that are not obtained by acting with a negative mode. It would be very interesting to combine the two formalisms and see if a hybrid algorithm could efficiently take advantage of these observations. We shall not do so here of course, leaving such speculations for future work.

9 A (Very Brief) Summary of Other Approaches to Fusion

If one is working with rational conformal field theories, for which the representation theory of the underlying vertex operator algebra V is semisimple with finitely many simple objects, then the fusion product has the simple form (9). In mathematical language, we can rewrite this in the form

$$M_i \boxtimes_{P(w)} M_j \cong \bigoplus_{k \in S} \dim \left[\mathcal{N}_{P(w)} \binom{M_k}{M_i \, M_j} \right] M_k, \tag{104}$$

where the M_k, with $k \in S$, enumerate the irreducible V-modules up to isomorphism and $\mathcal{N}_{P(w)}$ denotes the space of $P(w)$-intertwining maps of the indicated type.[16] Here, we are assuming that the fusion coefficients appearing in (104) are all finite. Now the problem of finding the fusion product is equivalent to computing the various fusion coefficients (these dimensions). This problem can be solved, in principle, using the technology of Zhu algebras, as was stated by Frenkel and Zhu [19], with the proof given by Li [52, 53]. The formula is as follows:

$$\mathcal{N}_{P(w)} \begin{pmatrix} M_k \\ M_i \; M_j \end{pmatrix} \cong \hom_{A(V)}(A(M_i) \otimes_{A(V)} M_j^{\mathrm{top}}, A(M_k)). \tag{105}$$

Here, $A(V)$ denotes the Zhu algebra of V, $A(M)$ denotes the image of Zhu's functor from V-modules to $A(V)$-bimodules, and M^{top} is the left-$A(V)$-submodule of M spanned by the vectors of minimal conformal weight. The hom-space in this formula corresponds to $A(V)$-bimodule homomorphisms. The proof of this theorem is fairly technical, but uses essentially familiar ideas: showing that the action of an intertwining operator is fully determined by its action on "small enough" spaces, namely $A(M_1) \otimes M_2^{\mathrm{top}}$. The hard part is of course going back: to build an intertwining operator consistently, given only its action on such small spaces.

During the course of his proof of the Frenkel–Zhu bimodule theorem, Haisheng Li in [52, 53] presented another "abstract" construction of the fusion product philosophically similar to the one we have given above. The setting is also non-logarithmic. The fused module $M_1 \boxtimes M_2$ is spanned by "modes" $(\psi_1)_n(\psi_2)$, where ψ_1 and ψ_2 run through the respective modules and n runs over the complex numbers. Here, $(\psi_1)_n$ is the mode corresponding to the "universal intertwining operator". Li thus builds an abstract vector space spanned by all the entities $(\psi_1)_n(\psi_2)$ and cuts it down by imposing relations satisfied by intertwining operators. To the best of our knowledge, the analogous construction has not been carried out in the logarithmic setting. However, a generalisation of the Frenkel–Zhu theorem to the logarithmic world exists [42]. Unsurprisingly, it involves the higher Zhu algebras of [12].

Huang and Lepowsky's first series of papers on the theory of fusion products for rational conformal field theories was, at least in part, inspired by Kazhdan and Lusztig's series of papers [46–49] that provided a tensor structure on the category of ordinary modules over an affine Lie algebra at levels k satisfying $k + \mathsf{h}^\vee \notin \mathbb{Q}_{\geqslant 0}$. For a comparison of the Huang–Lepowsky and Kazhdan–Lusztig approaches for these cases, we refer to [72]. Note that certain aspects of HLZ's opus [41] are decidedly harder than in [46–49]—it requires much more effort to prove in general that the fused object is actually a module for the *vertex operator algebra* as compared to a *Lie algebra*. However, Kazhdan and Lusztig were also able to prove that their categories close under fusion, while Huang and Lepowsky actually build their tensor product theory by *assuming* that fusion closes on a suitable category. Currently, the most

[16]We caution that despite this formula, an explicit construction of the fusion product using $P(w)$-compatibility conditions, among other things, is required in the work of Huang and Lepowsky [36–38] in order to build a braided tensor structure.

general result confirming such closure under fusion is from [33] where it is proved that categories of finite length modules over a C_2-cofinite (also known as *lisse*) vertex operator algebra are closed under fusion (see also [31]). Results pertaining to certain non-C_2-cofinite situations are also available, see, for example, [6].

One can find many other approaches to fusion in the literature, some of which we have already mentioned. Tsuchiya and Wood have announced [70] a rigorous theory of fusion in terms of a projective limit of NGK quotients. However, the proofs have not yet appeared. Miyamoto has constructed [57] a similar theory in which it is asserted that C_1-cofinite modules close under fusion. This bears a strong resemblance to the main result in Nahm's original paper [61] in which C_1-cofiniteness goes by the name of quasirationality (the reader may recall that this term was mentioned in the quote in the introduction).

We also want to mention an approach that has been developed by the statistical physics community, see [62, 64] for example, in which one computes "fusion products" for finite discretisations of the conformal field theory and then takes the continuum scaling limit in order to recover information about the actual fusion products. This approach is currently far from rigorous, but there is a concrete proposal [29] for this discretised fusion based on categories of modules over the Temperley–Lieb algebras and their generalisations. One of the main issues here is the fascinating link between the Temperley–Lieb and Virasoro algebras as dictated by scaling limits [28]. Comparisons between the discretised and conformal fusion results in the logarithmic case [60] indicates that this link exhibits subtle structure that is still poorly understood. Nevertheless, fusion gives us a powerful tools to better understand scaling limits. A rigorous theory of scaling limits would certainly be a jewel in the crown for mathematical physics.

Acknowledgements This paper was made possible by an Endeavour Research Fellowship, ID 6127_2017, awarded to SK by the Australian Government's Department of Education and Training. SK wishes to express sincere gratitude towards the School of Mathematics and Statistics at the University of Melbourne, where this project was undertaken, for their generous hospitality. SK is presently supported by a start-up grant provided by University of Denver. DR's research is supported by the Australian Research Council Discovery Project DP160101520 and the Australian Research Council Centre of Excellence for Mathematical and Statistical Frontiers CE140100049.

It is our privilege to thank our fellow "fusion club" members Arun Ram and Kazuya Kawasetsu for the many hours that we spent together working through the details of the approaches of NGK, HLZ, Kazhdan–Lusztig and Miyamoto. We also thank Thomas Creutzig, Hubert Saleur and Simon Wood for encouraging us to complete this article when time was lacking and deadlines were passing. We similarly thank Dražen Adamović and Paolo Papi for generous amounts of leeway in regard to this last point.

References

1. Abe, T., Buhl, G., Dong, C.: Rationality, regularity, and C_2-cofiniteness. Trans. Am. Math. Soc. **356**(8), 3391–3402 (2004)
2. Adamović, D., Milas, A.: Lattice construction of logarithmic modules for certain vertex algebras. Selecta Math. New Ser. **15**, 535–561 (2009). arXiv:0902.3417 [math.QA]

3. Astashkevich, A.: On the structure of Verma modules over Virasoro and Neveu-Schwarz algebras. Commun. Math. Phys. **186**, 531–562 (1997). arXiv:hep-th/9511032
4. Canagasabey, M., Rasmussen, J., Ridout, D.: Fusion rules for the logarithmic $N = 1$ superconformal minimal models I: the Neveu-Schwarz sector. J. Phys. A **48**, 415402 (2015). arXiv:1504.03155 [hep-th]
5. Canagasabey, M., Ridout, D.: Fusion rules for the logarithmic $N = 1$ superconformal minimal models II: including the Ramond sector. Nucl. Phys. B **905**, 132–187 (2016). arXiv:1512.05837 [hep-th]
6. Creutzig, T., Huang, Y.Z., Yang, J.: Braided tensor categories of admissible modules for affine Lie algebras. Commun. Math. Phys. **362**, 827–854 (2018). arXiv:1709.01865 [math.QA]
7. Creutzig, T., Kanade, S., Linshaw, A., Ridout, D.: Schur-Weyl duality for Heisenberg cosets. Transform. Groups **24**(2), 301–354 (2019). arXiv:1611.00305 [math.QA]
8. Creutzig, T., Kanade, S., McRae, R.: Tensor categories for vertex operator superalgebra extensions. arXiv:1705.05017 [math.QA]
9. Creutzig, T., Ridout, D.: Logarithmic conformal field theory: beyond an introduction. J. Phys. A **46**, 494006 (2013). arXiv:1303.0847 [hep-th]
10. Creutzig, T., Ridout, D.: Relating the archetypes of logarithmic conformal field theory. Nucl. Phys. B **872**, 348–391 (2013). arXiv:1107.2135 [hep-th]
11. Dong, C., Lepowsky, J.: Generalized Vertex Algebras and Relative Vertex Operators. Progress in Mathematics, vol. 112. Birkhäuser Boston Inc, Boston (1993)
12. Dong, C., Li, H., Mason, G.: Vertex operator algebras and associative algebras. J. Algebra **206**, 67–96 (1998). arXiv:q-alg/9612010
13. Eberle, H., Flohr, M.: Virasoro representations and fusion for general augmented minimal models. J. Phys. A **39**, 15245–15286 (2006). arXiv:hep-th/0604097
14. Feigin, B., Fuchs, D.: Cohomology of some nilpotent subalgebras of the Virasoro and Kac-Moody lie algebras. J. Geom. Phys. **5**, 209–235 (1988)
15. Feigin, B., Nakanishi, T., Ooguri, H.: The annihilating ideals of minimal models. Int. J. Mod. Phys. A **7**, 217–238 (1992)
16. Frenkel, E., Ben-Zvi, D.: Vertex Algebras and Algebraic Curves, Mathematical Surveys and Monographs, vol. 88. American Mathematical Society, Providence (2001)
17. Frenkel, I., Huang, Y.Z., Lepowsky, J.: On axiomatic approaches to vertex operator algebras and modules. Mem. Am. Math. Soc.**104**, viii+64 (1993)
18. Frenkel, I., Lepowsky, J., Meurman, A.: Vertex Operator Algebras and the Monster, Pure and Applied Mathematics, vol. 134. Academic Press, Boston (1988)
19. Frenkel, I., Zhu, Y.: Vertex operator algebras associated to representations of affine and Virasoro algebras. Duke Math. J. **66**, 123–168 (1992)
20. Gaberdiel, M.: Fusion in conformal field theory as the tensor product of the symmetry algebra. Int. J. Mod. Phys. A **9**, 4619–4636 (1994). arXiv:hep-th/9307183
21. Gaberdiel, M.: Fusion rules of chiral algebras. Nucl. Phys. B **417**, 130–150 (1994). arXiv:hep-th/9309105
22. Gaberdiel, M.: Fusion of twisted representations. Int. J. Mod. Phys. A **12**, 5183–5208 (1997). arXiv:hep-th/9607036
23. Gaberdiel, M.: An introduction to conformal field theory. Rep. Prog. Phys. **63**, 607–667 (2000). arXiv:hep-th/9910156
24. Gaberdiel, M.: Fusion rules and logarithmic representations of a WZW model at fractional level. Nucl. Phys. B **618**, 407–436 (2001). arXiv:hep-th/0105046
25. Gaberdiel, M., Kausch, H.: Indecomposable fusion products. Nucl. Phys. **B477**, 293–318 (1996). arXiv:hep-th/9604026
26. Gaberdiel, M., Kausch, H.: A rational logarithmic conformal field theory. Phys. Lett. B **386**, 131–137 (1996). arXiv:hep-th/9606050
27. Gaberdiel, M., Runkel, I., Wood, S.: Fusion rules and boundary conditions in the $c = 0$ triplet model. J. Phys. A **42**, 325403 (2009). arXiv:0905.0916 [hep-th]
28. Gainutdinov, A., Jacobsen, J., Read, N., Saleur, H., Vasseur, R.: Logarithmic conformal field theory: a lattice approach. J. Phys. A **46**, 494012 (2013). arXiv:1303.2082 [hep-th]

29. Gainutdinov, A., Vasseur, R.: Lattice fusion rules and logarithmic operator product expansions. Nucl. Phys. B **868**, 223–270 (2013). arXiv:1203.6289 [hep-th]
30. Gurarie, V.: Logarithmic operators in conformal field theory. Nucl. Phys. B **410**, 535–549 (1993). arXiv:hep-th/9303160
31. Huang, Y.Z.: On the applicability of logarithmic tensor category theory. arXiv:1702.00133 [math.QA]
32. Huang, Y.Z.: Two-Dimensional Conformal Geometry and Vertex Operator Algebras. Progress in Mathematics, vol. 148. Birkhäuser, Boston (1997)
33. Huang, Y.Z.: Cofiniteness conditions, projective covers and the logarithmic tensor product theory. J. Pure Appl. Algebra **213**, 458–475 (2009). arXiv:0712.4109 [math.QA]
34. Huang, Y.Z., Jr, A.K., Lepowsky, J.: Braided tensor categories and extensions of vertex operator algebras. Commun. Math. Phys. **337**, 1143–1159 (2015). arXiv:1406.3420 [math.QA]
35. Huang, Y.Z., Lepowsky, J.: Tensor products of modules for a vertex operator algebra and vertex tensor categories. In: Lie Theory and Geometry, Progress in Mathematics, vol. 123, pp. 349–383. Birkhäuser, Boston (1994). arXiv:hep-th/9401119
36. Huang, Y.Z., Lepowsky, J.: A theory of tensor products for module categories for a vertex operator algebra I. Selecta Math. New Ser. **1**(4), 699–756 (1995). arXiv:hep-th/9309076
37. Huang, Y.Z., Lepowsky, J.: A theory of tensor products for module categories for a vertex operator algebra II. Selecta Math. New Ser. **1**(4), 757–786 (1995). arXiv:hep-th/9309159
38. Huang, Y.Z., Lepowsky, J.: A theory of tensor products for module categories for a vertex operator algebra III. J. Pure Appl. Algebra **100**(1–3), 141–171 (1995). arXiv:hep-th/9505018
39. Huang, Y.Z., Lepowsky, J.: Tensor categories and the mathematics of rational and logarithmic conformal field theory. J. Phys. A **46**, 494009 (2013). arXiv:1304.7556 [hep-th]
40. Huang, Y.Z., Lepowsky, J., Li, H., Zhang, L.: On the concepts of intertwining operator and tensor product module in vertex operator algebra theory. J. Pure Appl. Algebra **204**, 507–535 (2006). arXiv:math.QA/0409364
41. Huang, Y.Z., Lepowsky, J., Zhang, L.: Logarithmic tensor product theory I–VIII. arXiv:1012.4193 [math.QA], arXiv:1012.4196 [math.QA], arXiv:1012.4197 [math.QA], arXiv:1012.4198 [math.QA], arXiv:1012.4199 [math.QA], arXiv:1012.4202 [math.QA], arXiv:1110.1929 [math.QA], arXiv:1110.1931 [math.QA]
42. Huang, Y.Z., Yang, J.: Logarithmic intertwining operators and associative algebras. J. Pure Appl. Algebra **216**, 1467–1492 (2012). arXiv:1104.4679 [math.QA]
43. Jr, A.K., Ostrik, V.: On a q-analogue of the McKay correspondence and the ADE classification of $\mathfrak{sl}_2$ conformal field theories. Adv. Math. **171**, 183–227 (2002). arXiv:math.QA/0101219
44. Kac, V.: Vertex Algebras for Beginners. University Lecture Series, vol. 10. American Mathematical Society, Providence (1996)
45. Kazhdan, D., Lusztig, G.: Affine Lie algebras and quantum groups. Int. Math. Res. Not. **1991**, 21–29 (1991)
46. Kazhdan, D., Lusztig, G.: Tensor structures arising from affine Lie algebras I. J. Am. Math. Soc. **6**, 905–947 (1993)
47. Kazhdan, D., Lusztig, G.: Tensor structures arising from affine Lie algebras II. J. Am. Math. Soc. **6**, 949–1011 (1993)
48. Kazhdan, D., Lusztig, G.: Tensor structures arising from affine Lie algebras III. J. Am. Math. Soc. **7**, 335–381 (1994)
49. Kazhdan, D., Lusztig, G.: Tensor structures arising from affine Lie algebras IV. J. Am. Math. Soc. **7**, 383–453 (1994)
50. Kytölä, K., Ridout, D.: On staggered indecomposable Virasoro modules. J. Math. Phys. **50**, 123503 (2009). arXiv:0905.0108 [math-ph]
51. Lepowsky, J., Li, H.: Introduction to Vertex Operator Algebras and their Representations. Progress in Mathematics, vol. 227. Birkhäuser, Boston (2004)
52. Li, H.: Representation theory and tensor product theory for vertex operator algebras. Ph.D. Thesis, Rutgers University (1994). arXiv:hep-th/9406211
53. Li, H.: An analogue of the Hom functor and a generalized nuclear democracy theorem. Duke Math. J. **93**, 73–114 (1998). arXiv:q-alg/9706012

54. Mathieu, P., Ridout, D.: From percolation to logarithmic conformal field theory. Phys. Lett. B **657**, 120–129 (2007). arXiv:0708.0802 [hep-th]
55. Mathieu, P., Ridout, D.: Logarithmic $M(2, p)$ minimal models, their logarithmic couplings, and duality. Nucl. Phys. B **801**, 268–295 (2008). arXiv:0711.3541 [hep-th]
56. Milas, A.: Weak modules and logarithmic intertwining operators for vertex operator algebras. In: Recent developments in infinite-dimensional Lie algebras and conformal field theory, Contemporary Mathematics, vol. 297, pp. 201–225. American Mathematical Society (2002). arXiv:math.QA/0101167
57. Miyamoto, M.: C_1-cofiniteness and fusion products for vertex operator algebras. In: Conformal field theories and tensor categories, Mathematical Lectures from Peking University, pp. 271–279. Springer, Heidelberg (2014). arXiv:1305.3008 [math.QA]
58. Moore, G., Seiberg, N.: Polynomial equations for rational conformal field theories. Phys. Lett. B **212**, 451–460 (1988)
59. Moore, G., Seiberg, N.: Classical and quantum conformal field theory. Commun. Math. Phys. **123**, 177–254 (1989)
60. Morin-Duchesne, A., Rasmussen, J., Ridout, D.: Boundary algebras and Kac modules for logarithmic minimal models. Nucl. Phys. B **899**, 677–769 (2015). arXiv:1503.07584 [hep-th]
61. Nahm, W.: Quasirational fusion products. Int. J. Mod. Phys. B **8**, 3693–3702 (1994). arXiv:hep-th/9402039
62. Pearce, P., Rasmussen, J., Zuber, J.B.: Logarithmic minimal models. J. Stat. Mech. **0611**, P11017 (2006). arXiv:0607232 [hep-th]
63. Rasmussen, J.: Classification of Kac representations in the logarithmic minimal models $LM(1, p)$. Nucl. Phys. B **853**, 404–435 (2011). arXiv:1012.5190 [hep-th]
64. Read, N., Saleur, H.: Associative-algebraic approach to logarithmic conformal field theories. Nucl. Phys. B **777**, 316–351 (2007). arXiv:hep-th/0701117
65. Ridout, D.: On the percolation BCFT and the crossing probability of Watts. Nucl. Phys. B **810**, 503–526 (2009). arXiv:0808.3530 [hep-th]
66. Ridout, D.: Fusion in fractional level $\widehat{\mathfrak{sl}}(2)$-theories with $k = -\frac{1}{2}$. Nucl. Phys. B **848**, 216–250 (2011). arXiv:1012.2905 [hep-th]
67. Ridout, D., Wood, S.: Bosonic ghosts at $c = 2$ as a logarithmic CFT. Lett. Math. Phys. **105**, 279–307 (2015). arXiv:1408.4185 [hep-th]
68. Ridout, D., Wood, S.: The Verlinde formula in logarithmic CFT. J. Phys. Conf. Ser. **597**, 012065 (2015). arXiv:1409.0670 [hep-th]
69. Rohsiepe, F.: On reducible but indecomposable representations of the Virasoro algebra. arXiv:hep-th/9611160
70. Tsuchiya, A., Wood, S.: The tensor structure on the representation category of the $\mathcal{W}_p$ triplet algebra. J. Phys. A **46**, 445203 (2013). arXiv:1201.0419 [hep-th]
71. Wood, S.: Fusion rules of the $W(p, q)$ triplet models. J. Phys. A **43**, 045212 (2010). arXiv:0907.4421 [hep-th]
72. Zhang, L.: Vertex tensor category structure on a category of Kazhdan-Lusztig. New York J. Math. **14**, 261–284 (2008). arXiv:math.QA/0701260
73. Zhu, Y.: Modular invariance of characters of vertex operator algebras. J. Am. Math. Soc. **9**, 237–302 (1996)

Permutation Orbifolds of Rank Three Fermionic Vertex Superalgebras

Antun Milas, Michael Penn and Josh Wauchope

Abstract We describe the structure of the permutation orbifold of the rank three free fermion vertex superalgebra (of central charge $\frac{3}{2}$) and of the rank three symplectic fermion vertex superalgebra (of central charge -6).

Keywords Fermions · Vertex algebras · $\mathcal{W}$-algebras

1 Introduction

Invariant subalgebras of free fields vertex algebras and superalgebras are rich sources of interesting simple vertex algebras. There is already a substantial body of work on this subject, especially from the perspective of $\mathcal{W}$-algebras. These approaches are primarily based on application of classical invariant theory (as in [7, 15, 17]). Interesting $\mathcal{W}$-algebras that arise from finite orbifolds show up in the classification of $c = 1$ rational vertex algebras (cf. [9]). Similarly, new examples of C_2-cofinite vertex algebras come from for the triplet vertex algebra [1] and its ADE orbifolds [2, 3, 18]. For some related work in this direction see [5, 11].

When it comes to permutation orbifolds (fixed under the *full* symmetric group S_n) very little is known except for $n = 2$. Recently, the first two authors, jointly with Shao, have investigated the structure of the permutation orbifold of the rank three Heisenberg algebra under the full symmetric group, denoted by $\mathcal{H}(3)^{S_3}$ [19]. They proved that this is a $\mathcal{W}$-algebra of type $(1, 2, 3, 4, 5, 6^2)$. In this note we establish similar results for fermionic vertex superalgebras.

Let us recall here

$$\mathcal{F} = \Lambda(\phi(-1/2), \phi(-3/2), \ldots),$$

A. Milas (✉) · J. Wauchope
Department of Mathematics and Statistics, SUNY-Albany, 1400 Washington Avenue, Albany, NY, USA
e-mail: amilas@albany.edu

M. Penn
Randolph College, 2500 Rivermont Avenue, Lynchburg, VA, USA
e-mail: mpenn@randolphcollege.edui

D. Adamović and P. Papi (eds.), *Affine, Vertex and W-algebras*,
Springer INdAM Series 37, https://doi.org/10.1007/978-3-030-32906-8_8

the rank one free fermion vertex superalgebra generated by an odd field $\phi(z)$, with super-brackets

$$[\phi(n), \phi(m)]_+ = \delta_{n+m,0}.$$

We consider the S_3-orbifold of three copies of $\mathcal{F}$:

$$(\mathcal{F} \otimes \mathcal{F} \otimes \mathcal{F})^{S_3},$$

with S_3 permuting the tensor factors (with signs). Our first main result, Theorem 1, pertains to the inner structure of this vertex superalgebra. We prove it is of type $(\frac{1}{2}, 2, 4, \frac{9}{2})$. We also obtain a related result for $(\mathcal{F} \otimes \mathcal{F} \otimes \mathcal{F})^{\mathbb{Z}_3}$ under a 3-cycle permutation of tensor factors. We also give a solely bosonic description of $\mathcal{F}(3)^{S_3}$ as $\mathcal{F} \otimes V_L^+, L = \mathbb{Z}\alpha, \langle \alpha, \alpha \rangle = 9$, with respect to involution $\alpha \to -\alpha$. Another description of this orbifold comes from a coset construction of $\mathfrak{so}(9)$, see Theorem 4. We obtain yet another realization from a certain $\mathcal{W}$-algebra obtained by Drinfeld–Sokolov reduction from $\mathfrak{osp}(1|8)$, see Theorem 5.

In the second part of the paper, we consider $\mathbb{Z}$-graded symplectic fermion vertex operator superalgebras. Recall the rank one symplectic fermion vertex superalgebra

$$SF = \Lambda(e(-1), e(-2), \ldots, f(-1), f(-2), \ldots)$$

generated by odd fields $e(z)$ and $f(z)$ subject to bracket relations

$$[e(i), f(j)]_+ = i\langle e, f \rangle \delta_{i+j,0}$$

where $\langle \, , \, \rangle$ is skew-symmetric. Again we consider the invariant vertex superalgebra

$$(\mathcal{S}F \otimes \mathcal{S}F \otimes \mathcal{S}F)^{S_3}.$$

Our second main result is about the structure of this algebra. We prove in Theorem 6 that this orbifold is of type $(1^2, 2, 3^3, 4^3, 5^5, 6^4)$, meaning that we have a minimal strongly generated set of this type.

Throughout the paper, for brevity, we let $\mathcal{F}(n) := \underbrace{\mathcal{F} \otimes \cdots \otimes \mathcal{F}}_{n-times}$ and $\mathcal{S}\mathcal{F}(n) = \underbrace{\mathcal{S}\mathcal{F} \otimes \cdots \otimes \mathcal{S}\mathcal{F}}_{n-times}$. We also use ϕ_i, $1 \leq i \leq n$, e_i, f_i to denote ith component fermion inside the tensor product of ordinary and symplectic fermions, respectively.

Several computations in this paper are performed by the OPE package [22], for Mathematica.

2 Warm-Up: S_2-Orbifold of $\mathcal{F}(2)$

This orbifold is well-known (see, for instance, [14]). If we set

$$\psi := \frac{1}{\sqrt{2}}(\phi_1 - \phi_2),$$

we see that the generator of S_2 acts via $\psi \mapsto -\psi$. As

$$\frac{1}{\sqrt{2}}(\phi_1 + \phi_2),$$

is fixed under the involution we immediately get

$$\mathcal{F}(2)^{S_2} = \mathcal{F}(1) \otimes \mathcal{F}(1)^{\mathbb{Z}_2} \cong \mathcal{F}(1) \otimes L_{Vir}\left(\frac{1}{2}, 0\right),$$

where $L_{Vir}(\frac{1}{2}, 0)$ is the simple Virasoro VOA of central charge $\frac{1}{2}$. In particular, this orbifold is of type $(\frac{1}{2}, 2)$. The standard proof in [14] uses unitarity and highest weight theory to conclude that the second tensor factor is generated by ω. As above argument cannot be used to other permutation groups, we offer another proof which uses classical invariant theory.

An application of the first fundamental theorem of invariant theory for $\mathcal{O}(1) \cong \mathbb{Z}_2 \cong S_2$ gives us an initial generating set of

$$\{\omega(a, b) | 0 \leq a < b\},$$

where

$$\omega(a, b) := \psi_{-\frac{1}{2}-a}\psi_{-\frac{1}{2}-b}\mathbb{1}.$$

The derivation operator allows us to reduce this generating set to

$$\{\omega(0, 2m + 1) | m \geq 0\}.$$

Now if we set

$$\Omega(a, b) := Y(\omega(a, b), z),$$

we have the following relation

$$:\Omega(0, 2m + 1)\Omega(0, 1): = -\frac{2m + 5}{4m + 6}\Omega(0, 2m + 3) + \frac{2m + 3}{2m + 2}\Omega(1, 2m + 2). \quad (1)$$

Now we replace

$$\Omega(1, 2m + 2) = \frac{\partial}{\partial z}\Omega(0, 2m + 2) - \Omega(0, 2m + 3)$$

which allows us to rewrite (1) as

$$\Omega(0, 2m+3) = \frac{2m+3}{(m+2)(6m+7)}\Big((2m+3)\frac{\partial}{\partial z}\Omega(0, 2m+2) \\ - (2m+2)^{\circ}_{\circ}\Omega(0, 2m+1)\Omega(0, 1)^{\circ}_{\circ}\Big). \quad (2)$$

Equivalently we have the following equation involving the states

$$\omega(0, 2m+3) = \frac{1}{(m+2)(6m+7)}\left((2m+3)\omega(0, 2m+2)_{-2}\mathbb{1} - \omega(0, 2m+1)_{-1}\omega(0, 1)\right). \quad (3)$$

From which it inductively follows that we only need the generator $\omega(0, 1)$, as argued earlier.

3 Structure of the S_3-Orbifold of $\mathcal{F}(3)$

We consider a tensor product of three free fermions $\mathcal{F}(3)$. As before we denote by ϕ_i the ith component fermion so that $\mathcal{F}(3)$ is isomorphic to $\Lambda(\phi_1(-1/2), \phi_2(-1/2), \phi_3(-1/2), \ldots)$. The symmetric groups S_3 now acts via permuting the indices of ϕ_i, $1 \le i \le 3$. In fact, we can view $S_3 \subset \mathcal{O}(3)$, with orthogonal group $\mathcal{O}(3)$ acting in the usual way on $\mathrm{Span}\{\phi_1(-n-1/2), \phi_2(-n-1/2), \phi_3(-n-1/2)\}$.

Lemma 1 *The vertex algebra $\mathcal{F}(3)^{S_3}$ is generated by*

$$\begin{aligned} \omega_1(a) &:= \sum_{i=1}^{3} \phi_i(-a-1/2) \text{ for } a \ge 0 \\ \omega_2(a,b) &:= \sum_{i=1}^{3} \phi_i(-a-1/2)\phi_i(-b-1/2) \text{ for } a > b \ge 0 \\ \omega_3(a,b,c) &:= \sum_{i=1}^{3} \phi_i(-a-1/2)\phi_i(-b-1/2)\phi_i(-c-1/2) \text{ for } a > b > c \ge 0. \end{aligned} \quad (4)$$

Proof We begin with the following change of basis

$$\begin{aligned} \psi_0 &= \frac{1}{\sqrt{3}}(\phi_1 + \phi_2 + \phi_3) \\ \psi_1 &= \frac{1}{\sqrt{3}}(\phi_1 + \eta^2\phi_2 + \eta\phi_3) \\ \psi_2 &= \frac{1}{\sqrt{3}}(\phi_1 + \eta\phi_2 + \eta^2\phi_3), \end{aligned} \quad (5)$$

where η is a primitive third root of unity. In this new basis the vector ψ_0 is clearly fixed and generates a copy of the rank 1 free fermion algebra, $\mathcal{F}(1)$. Moreover the generators of $S_3 \cong D_3$ (viewed as Dihedral group acting on $\mathbb{R}^2$) act as follows

$$\begin{aligned} \tau_{23}\psi_1 &= \psi_2, & \tau_{23}\psi_2 &= \psi_1 \\ \sigma_{123}\psi_1 &= \eta\psi_1, & \sigma_{123}\psi_2 &= \eta^2\psi_2. \end{aligned} \tag{6}$$

It is clear that we have an initial decomposition of

$$\mathcal{F}(3)^{S_3} \cong \mathcal{F}(1) \otimes \mathcal{F}(2)^{D_3}. \tag{7}$$

As such, we will describe an initial set of generators for the orbifold $\mathcal{F}(2)^{D_3}$ and show that together with ψ_0 these may be used to construct (4). Further, these diagonalized generators will be used in our reduction calculations below.

The associated graded algebra of $\mathcal{F}(2)$ is the exterior algebra

$$\mathfrak{F}(2) = \bigwedge(x_1(m_1), x_2(m_2) | m_i \geq 0)$$

where we have the linear isomorphism

$$\pi : \mathcal{F}(2) \to \mathfrak{F}(2) \tag{8}$$

given by

$$\begin{aligned} \psi_1(-m_1 - 1/2) \cdots \psi_1(-m_k - 1/2)\psi_2(-n_1 - 1/2) \cdots \psi_2(-n_\ell - 1/2)\mathbb{1} \\ \mapsto x_1(m_1) \cdots x_1(m_k)x_2(n_1) \cdots x_2(n_\ell) \end{aligned}, \tag{9}$$

for $m_i, n_j \in \mathbb{Z}_{\geq 0}$. Now we recall (see [21] for the even case) that given a finite group acting on $\mathfrak{F}(2)$, the invariant subalgebra $\mathfrak{F}(2)^G$ is generated by the set of orbit sums of monomials. That is, by the elements

$$\mathbf{o}(m) = \sum_{g \in G} g \cdot m. \tag{10}$$

Given an arbitrary monomial $m = x_1(m_1) \cdots x_1(m_k)x_2(n_1) \cdots x_2(n_\ell) \in \mathfrak{F}(2)$, we see that

$$\begin{aligned} \mathbf{o}(m) =& (1 + \eta^{k-\ell} + \eta^{2(k-\ell)}) \\ & (x_1(m_1) \cdots x_1(m_k)x_2(n_1) \cdots x_2(n_\ell) + x_2(m_1) \cdots x_2(m_k)x_1(n_1) \cdots x_1(n_\ell)), \end{aligned} \tag{11}$$

which is nonzero if and only if $k - \ell \equiv 0 \pmod 3$. As such, we see that $\mathfrak{F}(2)^{D_3}$ is generated by

$$\begin{aligned} q_{k,3\ell}(\mathbf{r}, \mathbf{s}, \mathbf{t}) =& (x_1(r_1)x_2(s_1)) \cdots (x_1(r_k)x_2(s_k))x_1(t_1) \cdots x_1(t_{3\ell}) \\ & + (x_2(r_1)x_1(s_1)) \cdots (x_2(r_k)x_1(s_k))x_2(t_1) \cdots x_2(t_{3\ell}). \end{aligned} \tag{12}$$

If $\ell = 0$ then $q_{k,0}(\mathbf{r}, \mathbf{s}, \mathbf{0}) \in \mathfrak{F}(2)^{D_n}$ for all n and is thus in $\mathfrak{F}(2)^{\mathcal{O}(2)}$ (note that $\cup_{n\geq 2} D_n$ is dense in $\mathcal{O}(2)$).

An odd analogue to the first fundamental theorem of invariant theory for $\mathcal{O}(2)$, [16, 24], implies that these terms are in the subalgebra generated by the quadratic binomials

$$q_2(m, n) = x_1(m)x_2(n) + x_2(m)x_1(n). \tag{13}$$

So we have accounted for all of the generators from (12) of the form $q_{k,0}(\mathbf{r}, \mathbf{s}, \mathbf{0})$. Now we move onto those of the form $q_{k,3\ell}(\mathbf{r}, \mathbf{s}, \mathbf{t})$ for $\ell \neq 0$ the first of which is

$$q_3(m_1, m_2, m_3) = x_1(m_1)x_1(m_2)x_1(m_3) + x_2(m_1)x_2(m_2)x_2(m_3). \tag{14}$$

We will inductively show that all other invariants of the form (12) are in the subalgebra generated by the binomials $q_2(m_1, m_2)$ and $q_3(n_1, n_2, n_3)$ for $m_i, n_j \in \mathbb{Z}_{\geq 0}$. Suppose that we have an orbit sum $q_{k,3\ell}(\mathbf{r}, \mathbf{s}, \mathbf{t})$ with $k, \ell \neq 0$, which can be written as

$$(x_1(r_1)x_2(s_1))x_1(t_1)x_1(t_2)x_1(t_3)X_1 + (x_2(r_1)x_1(s_1))x_2(t_1)x_2(t_2)x_2(t_3)X_2, \tag{15}$$

where X_i are monomials appropriate to complete the term. Now denote

$$q_3^X(a, b, c) = x_1(a)x_1(b)x_1(c)X_1 + x_2(a)x_2(b)x_2(c)X_2, \tag{16}$$

where is invariant of degree two less than our orbit sum. Now observe that

$$\begin{aligned} q_{k,3\ell}(\mathbf{r}, \mathbf{s}, \mathbf{t}) =& \frac{1}{2}q_2(r_1, s_1)q_3^X(t_1, t_2, t_3) + \frac{1}{2}q_2(r_1, t_3)q_3^X(s_1, t_1, t_2) \\ &+ \frac{1}{2}q_2(s_1, t_3)q_3^X(r_1, t_2, t_3). \end{aligned} \tag{17}$$

Now suppose that we have an orbit sum $q_{k,3\ell}(\mathbf{r}, \mathbf{s}, \mathbf{t})$ with $\ell \geq 2$, which can be written (up to a sign) as

$$x_1(t_1)\cdots x_1(t_6)X_1 + x_2(t_1)\cdots x_2(t_6)X_2, \tag{18}$$

where X_i are monomials appropriate to complete the term. Now denote

$$\begin{aligned} q_2^X(a, b) &= x_1(a)x_2(b)X_1 + x_2(a)x_1(b)X_2 \\ q_3^X(a, b, c) &= x_1(a)x_1(b)x_1(c)X_1 + x_2(a)x_2(b)x_2(c)X_2 \end{aligned} \tag{19}$$

which are invariants of degree 3 and 2 less than our orbit sum, respectively. Now observe that

$$
\begin{aligned}
q_{k,3\ell}(\mathbf{r}, \mathbf{s}, \mathbf{t}) = & \frac{1}{2} q_2(t_1, t_2) q_2(t_3, t_6) q_2^X(t_4, t_5) - \frac{1}{2} q_2(t_1, t_4) q_2(t_2, t_6) q)2^X(t_3, t_5) \\
& - \frac{1}{2} q_2(t_1, t_6) q_2(t_2, t_3) q_2^X(t_4, t_5) - q_2(t_1, t_6) q_2(t_2, t_4) q_2^X(t_3, t_5) \\
& - q_2(t_1, t_6) q_2(t_2, t_5) q_2^X(t_3, t_4) - \frac{1}{4} q_2(t_2, t_4) q_2(t_3, t_6) q_2^X(t_1, t_5) \\
& - \frac{3}{4} q_2(t_3, t_6) q_2(t_2, t_5) q_2^X(t_1, t_4) + \frac{1}{4} q_2(t_4, t_5) q_2(t_2, t_3) q_2^X(t_1, t_6) \\
& - \frac{1}{4} q_2(t_4, t_5) q_2(t_2, t_3) q_2^X(t_1, t_6) - \frac{1}{4} q_2(t_4, t_5), q_2(t_2, t_6) q_2^X(t_1, t_3) \\
& - \frac{3}{4} q_2(t_4, t_5) q_2(t_3, t_6) q_2^X(t_1, t_2) + \frac{1}{4} q_2(t_4, t_6) q_2(t_3, t_5) q_2^X(t_1, t_2) \\
& + \frac{1}{2} q_2(t_5, t_6) q_2(t_2, t_4) q_2^X(t_1, t_3) + \frac{3}{4} q_2(t_5, t_6) q_2(t_3, t_4) q_2^X(t_1, t_2) \\
& - q_3(t_1, t_2, t_4) q_3^X(t_3, t_5, t_6).
\end{aligned}
\tag{20}
$$

Now, given an arbitrary orbit sum $q_{k,3\ell}(\mathbf{r}, \mathbf{s}, \mathbf{t})$, we can inductively use (17) to reduce it to an algebraic combination of $q_2(m_1, m_2)$ and orbit sums $q_{0,3\ell}(\mathbf{r}', \mathbf{s}', \mathbf{t}')$ and then (20) can be used to write $q_{0,3\ell}(\mathbf{r}', \mathbf{s}', \mathbf{t}')$ in terms of $q_2(m_1, m_2)$ and $q_3(n_1, n_2, n_3)$ implying that these are the only required generators.

The preimages of these terms in $\mathcal{F}(2)$, along with the linear generator, are

$$
\begin{aligned}
\omega_1^0(a) &= \psi_0(-a - 1/2) \\
\omega_2^0(a, b) &= \psi_1(-a - 1/2)\psi_2(-b - 1/2)\mathbb{1} + \psi_2(-a - 1/2)\psi_1(-b - 1/2)\mathbb{1} \\
\omega_3^0(a, b, c) &= \sum_{i=1}^{2} \psi_i(-a - 1/2)\psi_i(-b - 1/2)\psi_i(-c - 1/2)\mathbb{1}.
\end{aligned}
\tag{21}
$$

Now observe that we have the following translation between (4) and (21)

$$
\begin{aligned}
\omega_1(a) =& \sqrt{3}\omega_1^0(a) \\
\omega_2(a, b) =& \omega_2(a, b) + \omega_1^0(a)_{-1}\omega_1^0(b) \\
\omega_3(a, b, c) =& \frac{1}{\sqrt{3}}(\omega_3^0(a, b, c) + \omega_1^0(a)_{-1}\omega_2^0(b, c) - \omega_1^0(b)\omega_2^0(a, c) \\
& + \omega_1^0(c)\omega_2^0(a, b) + \omega_1^0(a)_{-1}\omega_1^0(b)_{-1}\omega_1^0(c)).
\end{aligned}
\tag{22}
$$

□

We now work to minimize this generating set by way of quantum corrections to the odd analogues of the classical relations found in [19]. The results are summarized in the following Lemma.

Lemma 2 *The generators described in (4) may be replaced with the set*

$$\omega_1(0), \omega_2(0,1), \omega_2(0,3), \omega_3(0,1,2). \tag{23}$$

Proof For $a \geq 0$ the reduction of the linear generators to the single vector $\omega_1(0)$ is trivial by way of the translation operator. Moving on to the quadratic generators, using methods from [8] we can initial reduce these to the set $\omega_2(0, 2m+1)$ for $m \geq 0$.

To reduce the quadratic generators further down to $\omega_2(0,1), \omega_2(0,3)$ we can simply use a result of Linshaw [16]. Here we give a direct proof for completeness.

Further, if we set $\Omega_2(a,b) = Y(\omega_2^0(a,b), z)$ we have

$$\begin{aligned}\Omega_2(0,5) = &-\frac{2}{15} {}^\circ_\circ \Omega_2(0,1)\Omega_2(0,1)\Omega_2(0,1){}^\circ_\circ - \frac{14}{15} {}^\circ_\circ \Omega_2(0,1)\Omega_2(0,3){}^\circ_\circ \\ &+ \frac{3}{5} {}^\circ_\circ \partial^2 \Omega_2(0,1)\Omega_2(0,1){}^\circ_\circ + \frac{37}{15}\partial^2\Omega_2(0,3) - \frac{53}{30}\partial^4\Omega_2(0,1).\end{aligned} \tag{24}$$

Also, we have for all $m \geq 2$

$$\Omega_2(0, m+1) = \frac{1}{2m^2+4m} \left(3\Omega_2(0,1)_0 - \Omega_2(0,3)_1\right) \Omega_2(0,m), \tag{25}$$

allowing us to lift the relation (24) to a higher weight and establish the quadratic portion of (23). In the process we use the fact that the modes of quadratic operators form a Lie algebra.

For cubic operators we need a different approach as cubic operators do not close a Lie algebra. We first consider cubic generators beginning with the observation that all of the cubic vectors of weight at most $\frac{13}{2}$ can be written as a linear combination of $\omega_3(0,1,2)$ and $\omega_3(0,1,4)$ with the translation operator applied as necessary and more generally we may only consider vectors $\omega_3(0,a,b)$ with $0 < a < b$. Now we set $\Omega_3(a,b,c) = Y(\omega_3^0(a,b,c), z)$.

Our main tool for reducing the cubic generating set will be the relation among the generators of the associated graded algebra given by

$$\begin{aligned} D_5^C(a_1,a_2,a_3,a_4,a_5) := &\; q_2(a_1,a_2)q_3(a_3,a_4,a_5) + q_2(a_1,a_5)q_3(a_2,a_3,a_4) \\ &- q_2(a_2,a_5)q_3(a_1,a_3,a_4) - q_2(a_3,a_4)q_3(a_1,a_2,a_5) \\ &- q_2(a_3,a_5)q_3(a_1,a_2,a_4) + q_2(a_4,a_5)q_3(a_1,a_2,\alpha_3) \\ &= 0, \end{aligned} \tag{26}$$

which when lifted to fields corresponding to elements in the orbifold yields the expression

$$\begin{aligned} D_5(a_1, a_2, a_3, a_4, a_5) := & {}^{\circ}_{\circ}\Omega_2(a_1, a_2)\Omega_3(a_3, a_4, a_5){}^{\circ}_{\circ} + {}^{\circ}_{\circ}\Omega_2(a_1, a_5), \Omega_3(a_2, a_3, a_4){}^{\circ}_{\circ} \\ & - {}^{\circ}_{\circ}\Omega_2(a_2, a_5)\Omega_3(a_1, a_3, a_4){}^{\circ}_{\circ} - {}^{\circ}_{\circ}\Omega_2(a_3, a_4)\Omega_3(a_1, a_2, a_5){}^{\circ}_{\circ} \\ & - {}^{\circ}_{\circ}\Omega_2(a_3, a_5)\Omega_3(a_1, a_2, a_4){}^{\circ}_{\circ} + {}^{\circ}_{\circ}\Omega_2(a_4, a_5)\Omega_3(a_1, a_2, \alpha_3){}^{\circ}_{\circ}, \end{aligned} \tag{27}$$

that because of (26) we may write

$$D_5(a_1, a_2, a_3, a_4, a_5) = \sum_{\substack{b_1, b_2, b_3 \geq 0 \\ b_1 + b_2 + b_2 = a_1 + \cdots + a_5 + 2}} \lambda_{b_1, b_2, b_3} \Omega_3(b_1, b_2, b_3). \tag{28}$$

It will also be helpful to order the fields $\Omega_3(a, b, c)$ lexicographically by their entries.

Before embarking on our argument involving the expressions $D_5(a_1, a_2, a_3, a_4, a_5)$, we make an initial observation that for $0 \leq a < b < c$ we may write

$$\Omega_3(a, b, c) = \partial\Omega_3(a-1, b, c) - \Omega_3(a-1, b+1, c) - \Omega_3(a-1, b, c+1) \tag{29}$$

which applied iteratively, allows us to initially reduce our generating set to fields of the form $\Omega_3(0, b, c)$ for $1 < b < c$.

We may use $D_5(1, 0, 2, 0, 1)$ to form the lowest weight decoupling relation for the initial generating set (4)

$$\Omega_3(0, 1, 4) = -\frac{12}{17}{}^{\circ}_{\circ}\Omega_2(0, 1)\Omega_3(0, 1, 2){}^{\circ}_{\circ} + \frac{10}{17}\partial^2\Omega_3(0, 1, 2). \tag{30}$$

Expanding the expression $D_5(1, 0, a, 0, b)$, for $1 \leq a \leq b+2$ gives

$$\begin{aligned} D_5(1, 0, a, 0, b) = & \left(-\frac{(-1)^a}{a+2} - \frac{1}{a+2} - \frac{(-1)^a}{a+1} - \frac{1}{a+1}\right)\Omega_3(0, a+2, b) \\ & - \left(\frac{(-1)^a}{a+1} + \frac{(-1)^b}{b+1} + \frac{(-1)^b}{b+1} - 4\right)\Omega_3(0, a+1, b+1) + \cdots \end{aligned} \tag{31}$$

where the missing terms have second entry less than $a+1$. In the case that a is even this expression can be used to solve for $\Omega_3(0, a+2, b)$ in terms of fields lower in the order. If a is odd, the coefficient of $\Omega(0, a+2, b)$ in (31) is zero and thus we may use this equation to write $\Omega_3(0, a+1, b+1)$ in terms of fields lower in the ordering. All that remains is to show that for $a \geq 5$ $\Omega_3(0, 1, a)$ may be eliminated from the generating set. We use a similar argument to the one found in [19], considering the expressions $D_5(1, 0, a-2, 0, 1)$, $D_5(1, 0, a-3, 0, 2)$, $D_5(2, 0, a-3, 0, 1)$, $D_5(2, 0, a-4, 0, 2)$, and $D_5(1, 0, a-4, 0, 2)$ all of which may be written as linear combinations of $\Omega_3(0, k, a+1-k)$ for $1 \leq k \leq 5$ and derivatives of fields of lower weight. We form the matrix A whose (i, j) entry is the coefficient of $\Omega_3(0, i, a+1-i)$ from the jth expression listed above. We have

$$\det A = -\frac{(a+2)\left(1065a^5+3550a^4-47884a^3+108085a^2-77316a+9180\right)}{40(a-3)(a-2)^2(a-1)^2a} \tag{32}$$

if a is even and

$$\det A = -\frac{1065a^5+8620a^4-66306a^3+132611a^2-112314a+42120}{40(a-3)(a-2)(a-1)^2a} \tag{33}$$

if a is odd. In each of these cases, this allows us to write $\Omega_3(0,k,a+1-k)$ for $1 \leq k \leq 5 \leq a$ in terms of fields of lower weight. This along with (31) allows us to inductively eliminate all cubic generators not described in (23).

This leads us to the following result.

Theorem 1 *The orbifold $\mathcal{F}(3)^{S_3}$ is strongly generated by a minimal generating set of vectors of weight $\frac{1}{2}$, 2, 4 and $\frac{9}{2}$. As a vertex algebra it is isomorphic to the tensor product of $\mathcal{F}(1)$ and a $\mathcal{W}$-(super)algebra of type $(2, 4, \frac{9}{2})$.*

In a future work, we will examine the algebra W described in the Theorem 1, which is of central charge 1 and minimally strongly generated by the primary vectors

$$\begin{aligned} \omega &= -\frac{1}{2}\omega_2^0(0,1) \\ j &= \omega_2^0(0,3) - 2\omega_2^0(0,1)_{-3}\mathbb{1} + \frac{7}{3}\omega_2^0(0,1)_{-1}\omega_2^0(0,1) \\ c &= \omega_3^0(0,1,2). \end{aligned} \tag{34}$$

We may also calculate the character of $\mathcal{F}(3)^{S_3}$.

Theorem 2 *We have*

$$\chi_{\mathcal{F}(3)^{S_3}}(q) = \frac{q^{-\frac{1}{16}}}{6}\left(\prod_{n\geq 1}(1+q^{n-\frac{1}{2}})^3 + 3\prod_{n\geq 1}(1-q^{2n-1})(1+q^{n-\frac{1}{2}}) + 2\prod_{n\geq 1}(1+q^{3n-\frac{3}{2}})\right). \tag{35}$$

Proof Proof is analogous to the proof of the character formula of $\mathcal{H}(3)^{S_3}(q)$ in [19] so we omit it here.

Using methods similar to those above it follows that the orbifold $\mathcal{F}(3)^{\mathbb{Z}_3}$ has a minimal strong generating set given by

$$\begin{aligned}\widehat{\omega_1}(0) &= \psi_0\left(-\frac{1}{2}\right)\\ \widehat{\omega_2}(0,0) &= \psi_1\left(-\frac{1}{2}\right)\psi_2\left(-\frac{1}{2}\right)\\ \widehat{\omega_3^1}(0,1,2) &= \psi_1\left(-\frac{5}{2}\right)\psi_1\left(-\frac{3}{2}\right)\psi_1\left(-\frac{1}{2}\right)\\ \widehat{\omega_3^2}(0,1,2) &= \psi_2\left(-\frac{5}{2}\right)\psi_2\left(-\frac{3}{2}\right)\psi_2\left(-\frac{1}{2}\right)\end{aligned} \tag{36}$$

and thus $\mathcal{F}(3)^{\mathbb{Z}_3} \cong \mathcal{F}(1) \otimes W$ where W is of type $1, \frac{9}{2}, \frac{9}{2}$ and

$$\chi_{\mathcal{F}(3)^{\mathbb{Z}_3}}(q) = \frac{q^{-\frac{1}{16}}}{3}\left(\prod_{n\geq 1}(1+q^{n-\frac{1}{2}})^3 + 2\prod_{n\geq 1}(1+q^{3(n-\frac{1}{2})})\right) \tag{37}$$

3.1 Bosonic Description of $\mathcal{F}(3)^{S_3}$

The $\mathbb{Z}_3$-orbifold of $\mathcal{F}(3)$ can be now used to give another description of $\mathcal{F}(3)^{D_3} = \mathcal{F} \otimes \mathcal{F}(2)^{D_3}$.

Theorem 3 *We have*

$$\mathcal{F}(3)^{D_3} \cong \mathcal{F} \otimes V_{3\mathbb{Z}}^+$$

where $V_{3\mathbb{Z}}^+$ is the fixed point subalgebra of the lattice vertex superalgebra $V_{3\mathbb{Z}}$ under the involution $\alpha \to -\alpha$, where α is generator of $3\mathbb{Z}$.

Proof We already established that two generating vectors of weight $\frac{9}{2}$ are primary. It is also easy to see that they are of charge ± 3 highest weight vectors for the generator $h := \psi_1\left(-\frac{1}{2}\right)\psi_2\left(-\frac{1}{2}\right)$. Denote by $M(1)$ the Heisenberg subalgebra generated by h. We have conformal embedding $M(1) \subset \mathcal{F}(2)^{\mathbb{Z}_3}$ which decomposes as a direct sum of irreducible $M(1)$-modules

$$\mathcal{F}(2)^{\mathbb{Z}_3} \cong \oplus_{\lambda\in S} M(1,\lambda),$$

where $\lambda \in \mathbb{Z}$ and $\pm 3 \in S$ for some set S. Since $\mathcal{F}(2)^{\mathbb{Z}_3}$ is generated by h and two highest weight vectors of charge ± 3 and the charge is additive we conclude that $\mathcal{F}(2)^{\mathbb{Z}_3}$ is contained inside $\oplus_{m\in\mathbb{Z}} M(1,3m) \cong V_{3\mathbb{Z}}$, which itself has $\frac{1}{2}\mathbb{Z}$-graded vertex superalgebra structure. The rest follows from the simplicity of $\mathcal{F}(2)^{\mathbb{Z}_3}$ and a uniqueness property of lattice vertex algebras as in [10, Sect. 5], that is

$$\mathcal{F}(2)^{\mathbb{Z}_3} \cong V_{3\mathbb{Z}}$$

as vertex algebras. Finally, we observe that there is an automorphism in D_3 which acts as $h \to -h$, so we have the claim.

Remark 1 Clearly we can give another proof for strong generation for $\mathcal{F}(2)^{D_3}$ using lattice vertex algebra structure.

Comparing two characters of the bosonic and fermionic side for the $\mathbb{Z}_3$-orbifold gives the following q-series identity.

Corollary 1 (Boson-Fermion correspondence for characters)

$$\frac{1}{3}\prod_{n\geq 1}(1+q^{n-\frac{1}{2}})^2 + \frac{2}{3}\prod_{n\geq 1}\frac{(1+q^{3(n-\frac{1}{2})})}{(1+q^{n-\frac{1}{2}})} = \frac{\sum_{n\in\mathbb{Z}} q^{\frac{9n^2}{2}}}{(q;q)_\infty}.$$

3.2 $\mathcal{F}(3)^{S_3}$ as a Coset Vertex Superalgebra

Here we give another description of the orbifold algebra using results of Adamovic and Perse [4]. Similar result was also predicted by physicists [23].

Theorem 4 *As vertex superalgebras,*

$$\mathcal{F}(2)^{D_3} \cong SCom(L_{so(9)_2}, L_{so(9)_1} \otimes L_{so(9)_1}),$$

where $SCom(L_{so(9)_2}, L_{so(9)_1} \otimes L_{so(9)_1})$ *denotes a simple current extension of the vertex operator algebra* $Com(L_{so(9)_2}, L_{so(9)_1} \otimes L_{so(9)_1})$ *and* $L_{so(9)_k}$ *is the simple affine Lie algebra of level k and type* $B_4^{(1)}$.

Proof This follows directly from [4, Sect. 4], specifically Corollary 2:

$$Com(L_{so(9)_2}, L_{so(9)_1} \otimes L_{so(9)_1}) \cong V_{6\mathbb{Z}}^+,$$

together with Theorem 3, and decomposition

$$V_{3\mathbb{Z}} = V_{6\mathbb{Z}} \oplus V_{6\mathbb{Z}+3}.$$

3.3 $\mathcal{F}(3)^{S_3}$ Via Drinfeld–Sokolov Reduction

In the physics literature it is often quoted that the Drinfeld–Sokolov reduction of $\widehat{\mathfrak{osp}}(1|2n)$ Lie superalgebra at level k is a $\mathcal{W}$-superalgebra of type $(2, 4, \ldots, 2n, \frac{n+1}{2})$ (cf. [6]) and that it coincide with Fateev–Lukyanov algebra defined via Miura transformation. This $\mathcal{W}$-algebra exists generically and it can be described as the kernel of screenings—for a rigorous proof see [12].

We are interested in the $n = 4$ case. Following [12, 13] it can be easily shown that the central charge of $\mathcal{W}^k(osp(1|2k), f_{reg})$, where k is the level and f_{reg} is a regular nilpotent element [12], is given by

$$c = \frac{-9(55+14k)(65+16k)}{4(9+2k)}.$$

Solving for $c = 1$ gives $k = -\frac{63}{16}$ as one of the solutions (the other solution is $k = -\frac{73}{18}$). For this value of k this vertex algebra is not simple. More precisely,

Theorem 5

$$\mathcal{F}(2)^{D_3} \cong \mathcal{W}_{-\frac{63}{16}}(\mathfrak{osp}(1|8), f_{reg}),$$

where $W_{-\frac{63}{16}}(\mathfrak{osp}(1|8), f_{reg})$, *is the simple quotient of the universal* $\mathcal{W}$*-algebra* $\mathcal{W}^{-\frac{63}{16}}(\mathfrak{osp}(1|8), f_{reg})$.

Proof The proof is straightforward computation. By Theorem 6.4 of [12], the universal $\mathcal{W}$-algebra $\mathcal{W}^{-\frac{63}{16}}(\mathfrak{osp}(1|8), f)$ is strongly generated by fields of weight $\frac{9}{2}$, 2, 4, 6, and 8 which are denoted by G, W_6, W_4, W_2, and W_0 respectively. The field $\frac{1}{2}W_6$ is the conformal element and we will denote it by L. Further, we relabel the weight 4 generator $W_4 = W$. Furthermore a computer calculation shows that

$$\begin{aligned}\widetilde{W}_2 =& W_2 - \frac{7}{120}\partial^2 W + \frac{3017}{77760}\partial^4 L - \frac{31}{90}{:}LW{:} + \frac{5159}{6480}{:}\partial^2 LL{:}\\ &- \frac{497}{810}{:}\partial L\partial L{:} - \frac{49}{180}{:}LLL{:}\end{aligned} \tag{38}$$

and

$$\begin{aligned}\widetilde{W}_0 =& W_0 - \frac{154105}{639824}\partial^2 W_2 + \frac{59346175}{2487635712}\partial^4 W - \frac{1120613725}{537329313792}\partial^6 L\\ &- \frac{877345}{1439604}{:}LW_2{:} + \frac{964075}{8637624}{:}WW{:} + \frac{170975}{12956436}{:}LLW{:}\\ &- \frac{12284125}{34550496}{:}(\partial^2 L)W{:} - \frac{33422375}{310954464}{:}L(\partial^2 W){:} - \frac{10892875}{103651488}{:}LLLL{:}\\ &+ \frac{200351725}{466431696}{:}(\partial^2 L)LL{:} - \frac{141495725}{466431696}{:}(\partial L)(\partial L)L{:} - \frac{571813025}{22388721408}{:}(\partial^4 L)L{:}\\ &+ \frac{83651225}{1865726784}{:}(\partial^3 L)(\partial L){:} - \frac{366545725}{4975271424}{:}(\partial^2 L)(\partial^2 L){:}\end{aligned} \tag{39}$$

are singular and thus generate a proper VOA ideal inside $\mathcal{W}^{-\frac{63}{16}}(\mathfrak{osp}(1|8), f)$.

The rest is comparing OPEs among 2, 4 and $\frac{9}{2}$ generators. Since they agree, modulo the ideal, we have a surjective map from the universal vertex algebra $W^{-\frac{63}{16}}(\mathfrak{osp}(1|8), f)$ to $\mathcal{F}(2)^{D_3}$, sending generators of weight 2, 4 and 9/2 to the corresponding generators of the orbifold and, in light of (38) and (39), $\widetilde{W}_2$ and $\widetilde{W}_0$ are mapped to zero. This map is well-defined. The rest follows from the simplicity of the orbifold algebra.

4 Symmetric Orbifolds of Symplectic Fermions

The rank 3 symplectic fermion, $\mathcal{SF}(3)$, vertex operator algebra is generated by the odd vectors $e_i(-1)$ and $f_i(-1)$ for $1 \leq i \leq 3$ with vertex operators

$$\begin{aligned} Y(e_i(-1), z) &= \sum_{n\in\mathbb{Z}} e_i(n)z^{-n-1} \\ Y(f_i(-1), z) &= \sum_{n\in\mathbb{Z}} f_i(n)z^{-n-1}, \end{aligned} \tag{40}$$

subject to (anti-)commutation relations

$$\begin{aligned} [e_i(m), e_j(n)]_+ &= [f_i(m), f_j(n)]_+ = 0 \\ [e_i(m), f_j(n)]_+ &= m\delta_{i,j}\delta_{m+n,0}. \end{aligned} \tag{41}$$

Some invariant theory argument implies that we have the following initial set of strong generators for the orbifold $\mathcal{SF}(3)$ may be taken to be

$$\begin{aligned} \omega_e^1(a) &= \sum_{i=1}^{3} e_i(-a-1) \\ \omega_f^1(a) &= \sum_{i=1}^{3} f_i(-a-1) \\ \omega_{e,e}^2(a,b) &= \sum_{i=1}^{3} e_i(-a-1)e_i(-b-1) \\ \omega_{e,f}^2(a,b) &= \sum_{i=1}^{3} e_i(-a-1)f_i(-b-1) \\ \omega_{f,f}^2(a,b) &= \sum_{i=1}^{3} f_i(-a-1)f_i(-b-1) \\ \omega_{e,e,e}^3(a,b,c) &= \sum_{i=1}^{3} e_i(-a-1)e_i(-b-1)e_i(-c-1) \\ \omega_{e,e,f}^3(a,b,c) &= \sum_{i=1}^{3} e_i(-a-1)e_i(-b-1)f_i(-c-1) \\ \omega_{e,f,f}^3(a,b,c) &= \sum_{i=1}^{3} e_i(-a-1)f_i(-b-1)f_i(-c-1) \\ \omega_{f,f,f}^3(a,b,c) &= \sum_{i=1}^{3} f_i(-a-1)f_i(-b-1)f_i(-c-1) \end{aligned} \tag{42}$$

for $a, b, c \geq 0$. Observe that the conformal vector is given by $\omega = \frac{1}{2}\omega_{1,2,2}(0,0)$. The following change of basis of the generating set will allow for an efficient reduction in the initial strong generating set (42).

$$\begin{aligned} E_0(-1) &= \frac{1}{\sqrt{3}}(e_1(-1) + e_2(-1) + e_3(-1)) \\ E_1(-1) &= \frac{1}{\sqrt{3}}(e_1(-1) + \eta^2 e_2(-1) + \eta e_3(-1)) \\ E_2(-1) &= \frac{1}{\sqrt{3}}(e_1(-1) + \eta e_2(-1) + \eta^2 e_3(-1)). \end{aligned} \tag{43}$$

and

$$\begin{aligned} F_0(-1) &= \frac{1}{\sqrt{3}}(f_1(-1) + f_2(-1) + f_3(-1)) \\ F_1(-1) &= \frac{1}{\sqrt{3}}(f_1(-1) + \eta^2 f_2(-1) + \eta f_3(-1)) \\ F_2(-1) &= \frac{1}{\sqrt{3}}(f_1(-1) + \eta f_2(-1) + \eta^2 f_3(-1)). \end{aligned} \tag{44}$$

where η is a primitive third root of unity. Using this generating set we have $\sigma \cdot E_0(-1) = E_0(-1)$ and $\sigma \cdot F_0(-1) = F_0(-1)$ for all $\sigma \in S_3$. Examining the action of the generators of S_3 on these new generators of $\mathcal{SF}(3)$, we have

$$\begin{aligned} \tau_{12} E_1(-1) &= E_2(-1) \\ \tau_{12} F_1 &= F_2(-1). \end{aligned} \tag{45}$$

and

$$\begin{aligned} \sigma_{123} E_1(-1) &= \eta E_1(-1) \\ \sigma_{123} E_2(-1) &= \eta^2 E_2(-1) \\ \sigma_{123} F_1(-1) &= \eta F_1(-1) \\ \sigma_{123} F_2(-1) &= \eta^2 F_2(-1). \end{aligned} \tag{46}$$

From this action, we see that an initial generating set for the orbifold may alternatively taken to be

$$\begin{aligned} \Omega_E^1(a) &= E_0(-a-1) \\ \Omega_F^1(a) &= F_0(-a-1) \\ \Omega_{E,E}^2(a,b) &= E_1(-a-1)E_2(-b-1) + E_2(-a-1)E_1(-b-1) \\ \Omega_{E,F}^2(a,b) &= E_1(-a-1)F_2(-b-1) + E_2(-a-1)F_1(-b-1) \\ \Omega_{F,F}^2(a,b) &= F_1(-a-1)F_2(-b-1) + F_2(-a-1)F_1(-b-1) \end{aligned} \tag{47}$$

$$\Omega^3_{E,E,E}(a,b,c) = \sum_{i=1}^{2} E_i(-a-1)E_i(-b-1)E_i(-c-1)$$

$$\Omega^3_{E,E,F}(a,b,c) = \sum_{i=1}^{2} E_i(-a-1)E_i(-b-1)F_i(-c-1)$$

$$\Omega^3_{E,F,F}(a,b,c) = \sum_{i=1}^{2} E_i(-a-1)F_i(-b-1)F_i(-c-1)$$

$$\Omega^3_{F,F,F}(a,b,c) = \sum_{i=1}^{2} F_i(-a-1)F_i(-b-1)F_i(-c-1)$$

for $a, b, c \geq 0$.

Of course the linear portion of the generating set can be immediately reduced to $\Omega_{1,1}(0)$ and $\Omega_{2,1}(0)$. The following lemmas describe the reduction of the quadratic and cubic terms in the generating set.

Lemma 3 *The quadratic elements in the generating set (47) may be replaced with the homogeneous quadratic generators*

$$\Omega^2_{E,E}(0,1),\ \Omega^2_{E,E}(0,3),\ \Omega^2_{F,F}(0,1),\ \Omega^2_{F,F}(0,3) \tag{48}$$

and the heterogeneous quadratic generators

$$\Omega^2_{E,F}(0,0),\ \Omega^2_{E,F}(0,1),\ \Omega^2_{E,F}(0,2),\ \Omega^2_{E,F}(0,3). \tag{49}$$

Proof We begin by reducing the set of homogeneous generators $\Omega^2_{E,E}(a,b)$ and $\Omega^2_{F,F}(a,b)$ for $a, b \geq 0$ by focusing on the generators $\Omega^2_{E,E}(a,b)$, as the others will follow similarly. We have the following decoupling relation

$$\begin{aligned}\Omega^2_{E,E}(0,5) = &-\frac{1}{140}\Omega^2_{E,E}(0,1)_{-1}\Omega^2_{E,F}(0,0)_{-1}\Omega^2_{E,F}(0,0) - \frac{1}{140}\Omega^2_{E,E}(0,1)_{-1}\Omega^2_{E,F}(0,2)\\ &-\frac{1}{140}\Omega^2_{E,E}(0,1)_{-1}\Omega^2_{E,F}(2,0) - \frac{1}{14}\Omega^2_{E,E}(0,3)_{-1}\Omega^2_{E,F}(0,0).\end{aligned}$$

Moving to the heterogeneous quadratic generators, we begin we the following

$$\begin{aligned}\Omega^2_{E,F}(0,4) = &-\frac{1}{36}\Omega^2_{E,F}(0,0)_{-1}\Omega^2_{E,F}(0,0)_{-1}\Omega^2_{E,F}(0,0) - \frac{1}{12}\Omega^2_{E,F}(0,0)_{-1}\Omega^2_{E,F}(0,2)\\ &-\frac{1}{12}\Omega^2_{E,F}(0,0)_{-1}\Omega^2_{E,F}(2,0) + \frac{1}{2}\Omega^2_{E,F}(0,3)_{-2}\mathbb{1} - \frac{5}{18}\Omega^2_{E,F}(0,2)_{-3}\mathbb{1}\\ &+\frac{1}{6}\Omega^2_{E,F}(0,1)_{-4}\mathbb{1} - \frac{1}{6}\Omega^2_{E,F}(0,0)_{-5}\mathbb{1}.\end{aligned}$$

Finally operators similar to those used in the proof of Lemma 2 can be used to construct higher weight decoupling relations and finish the argument.

Lemma 4 *The cubic elements in the generating set (47) may be replaced with the homogeneous cubic generators*

$$\Omega^3_{E,E,E}(0,1,2),\ \Omega^3_{F,F,F}(0,1,2). \tag{50}$$

and heterogeneous cubic generators

$$\Omega^3_{E,E,F}(0,1,0),\ \Omega^3_{E,E,F}(0,2,0),\ \Omega^3_{E,E,F}(1,2,0) \tag{51}$$

and

$$\Omega^3_{E,F,F}(0,1,0),\ \Omega^3_{E,F,F}(0,2,0),\ \Omega^3_{E,F,F}(0,1,2). \tag{52}$$

Proof The reduction of homogeneous cubic generators follows similarly to the $\mathcal{F}(3)$ case as described above. So we move on to the heterogeneous cubic generators. Observe that all weight 6 vectors among the generators $\Omega^3_{E,E,F}(a,b,c)$ are in the list

$$\Omega^3_{E,E,F}(0,3,0),\ \Omega^3_{E,E,F}(1,2,0),\ \Omega^3_{E,E,F}(0,2,1),\ \Omega^3_{E,E,F}(0,1,2). \tag{53}$$

Taking linear combinations of $\Omega^3_{E,E,F}(0,1,0)_{-3}\mathbb{1}$ and $\Omega^3_{E,E,F}(0,2,0)_{-2}\mathbb{1}$ will allow us to eliminate any two of these from our generating set, further the equation

$$\begin{aligned}\Omega^3_{E,E,F}(0,3,0) = &-\frac{1}{5}\Omega^2_{E,F}(0,0)_{-1}\Omega^3_{E,E,F}(0,1,0) - \frac{1}{20}\Omega^3_{E,E,F}(0,1,0)_{-3}\mathbb{1}\\ &-\frac{1}{20}\Omega^3_{E,E,F}(0,2,0)_{-2}\mathbb{1}\end{aligned} \tag{54}$$

allows us to eliminate $\Omega^3_{E,E,F}(0,3,0)$. We keep $\Omega^3_{E,E,F}(1,2,0)$ as our weight 6 cubic generator.

Now observe that all weight 7 vectors among the generators $\Omega^3_{E,E,F}(a,b,c)$ are in the list

$$\Omega^3_{E,E,F}(0,4,0),\ \Omega^3_{E,E,F}(1,3,0),\ \Omega^3_{E,E,F}(0,3,1),\ \Omega^3_{E,E,F}(0,1,3),\ \Omega^3_{E,E,F}(0,2,2). \tag{55}$$

Taking linear combinations of the translation operator applied to vectors on the list (53) allows us to eliminate all but one of these. Further the equation

$$\Omega^2_{E,F}(0,0)_{-1}\Omega^3_{E,E,F}(0,2,0) + \Omega^3_{E,E,F}(0,2,2) + 3\Omega^3_{E,E,F}(0,4,0) = 0 \tag{56}$$

allows us to eliminate the final weight 7 vector. Finally an argument following the outline of Lemma 2 using relations analogous to $D_5(a_1,a_2,a_3,a_4,a_5)$ allows us to eliminate all higher weight generators.

Theorem 6 *The vertex operator algebra $\mathcal{SF}(3)^{S_3}$ is simple of type $(1^2,2,3^3,4^3,5^5,6^4)$. It is isomorphic to $\mathcal{SF}(1)\otimes W$ where W is of type $(2,3^3,4^3,5^5,6^4)$.*

Again we can easily compute the orbifold character as in [19].

Theorem 7 *We have*

$$\chi_{\mathcal{SF}(3)^{S_3}}(q) = \frac{q^{1/4}}{6}\left(\prod_{n\geq 1}(1+q^n)^6 + 3\prod_{n\geq 1}(1+q^n)^2(1-q^{2n})^2 + 2\prod_{n\geq 1}(1+q^{3n})^2\right).$$

Remark 2 Suppose we have $\mathbb{Z}$-graded $\mathcal{W}$ superalgebra freely generated with even and odd generators with integral weights $e_1, \ldots, e_k$ and $o_1, \ldots, o_l$ respectively. Then

$$\chi_{\mathcal{W}}(q) = \left(\prod_{m=1}^{k}\prod_{n\geq e_m}\frac{1}{1-q^n}\right)\left(\prod_{m=1}^{l}\prod_{n\geq o_m}(1+q^n)\right).$$

From here we can conclude that the free character and $\chi_{\mathcal{SF}(3)^{S_3}}(q)$ agree up to $O(q^7)$. Therefore generators listed in Theorem 6 are minimal.

Remark 3 . Using methods similar to those above it follows that the orbifold $\mathcal{SF}(3)^{\mathbb{Z}_3} \cong \mathcal{SF}(1) \otimes W$, where W has strong generating set of even vectors

$$\begin{aligned}
\widehat{\Omega}^2_{E,E}(0,a) &= E_1(-a-1)E_2\\
\widehat{\Omega}^2_{F,F}(0,a) &= F_1(-a-1)F_2\\
\widehat{\Omega}^{2,1}_{E,F}(0,a) &= E_1(-a-1)F_2\\
\widehat{\Omega}^{2,2}_{E,F}(0,a) &= E_2(-a-1)F_1,
\end{aligned}$$

for $a \in \{0, 1\}$, and odd vectors

$$\begin{aligned}
\widehat{\Omega}^{3,1}_{E,E,F} &= E_1(-a-1)E_1(-1)F_1\\
\widehat{\Omega}^{3,2}_{E,E,F} &= E_2(-a-1)E_2(-1)F_2\\
\widehat{\Omega}^{3,1}_{E,F,F} &= E_1(-1)F_1(-a-1)F_1\\
\widehat{\Omega}^{3,2}_{E,F,F} &= E_2(-1)F_2(-a-1)F_2
\end{aligned}$$

for $a \in \{1, 2\}$. So, W is of type $2_e^4, 3_e^4, 4_o^4, 5_o^4$ (here e = even and o = odd is the partiy). The character of $\mathcal{SF}(3)^{\mathbb{Z}_3}$ is given by

$$\chi_{\mathcal{SF}(3)^{\mathbb{Z}_3}}(q) = \frac{q^{1/4}}{3}\left(\prod_{n\geq 1}(1+q^n)^6 + 2\prod_{n\geq 1}(1+q^{3n})^2\right).$$

5 Directions for Future Work

(a) In this paper we obtained several different description of the S_3-orbifold of the free fermion vertex algebra. It would interesting to find an alternative descriptions for the orbifolds of symplectic fermion vertex algebra.
(b) Another interesting direction is to determine the structure of $L_{\mathfrak{sl}_2}(1,0)(3)^{S_3}$—the S_3-orbifold of the simple level one $\mathfrak{sl}_2$ affine vertex operator algebra. We can prove that this is an extension of the vertex operator algebra $V_{\sqrt{6}\mathbb{Z}} \otimes L_{Vir}(\frac{4}{5},0) \otimes W_{2,3}^{+}$, where $V_{\sqrt{6}\mathbb{Z}}$ denotes the lattice VOA of $\sqrt{6}\mathbb{Z}$, $W_{2,3}^{+}$ is the fixed point subalgebra of the simple Zamolodchikov (2, 3)-algebra of central charge $\frac{6}{5}$ under the non-trivial involution. This vertex algebra is non-generic and rational of type $(2, 6, 8, 10)$. More detailed structure will be determined in our future work [20].
(c) Using the methods developed in [19] it is possible to determine the structure of the S_3 permutation orbifold of the Virasoro vertex algebra. Here we expect a $\mathcal{W}$-algebra of type $(2, 4, 5, 6^3, 7, 8^3, 9^3, 10^2)$ for the $\mathbb{Z}_3$ cyclic orbifold, and of type $(2, 4, 6^2, 8^2, 9, 10^2, 11, 12^3, 14)$ for the S_3-orbifold. This is work in progress.
(d) Everything in this paper can be easily extended to $\mathcal{F}(2)^{D_n}$ and $\mathcal{SF}(2)^{D_n}$ where D_n is dihedral and $n \geq 3$.

Acknowledgements A.M. would like to thank T. Creutzig for discussion regarding Sect. 3.2. He would also like to thank D. Adamovic and P. Papi for invitation and hospitality during the conference *Affine, vertex and W-algebras*, INdAM, Rome, December 11–15, 2017. We also thank the referee for useful comments.

References

1. Adamovic, D., Milas, A.: On the triplet vertex algebra $W(p)$. Adv. Math. **217**, 2664–2699 (2008)
2. Adamovic, D., Lin, X., Milas, A.: ADE subalgebras of the triplet vertex algebra W(p): A-series. Commun. Contemp. Math. **15**, 1350028 (2013) 30 pp
3. Adamovic, D., Lin, X., Milas, A.: ADE subalgebras of the triplet vertex algebra; D-series. Int. J. Math. **25**(1), 1450001 (2014) 34 pp
4. Adamovic, D., Perse, O.: On coset vertex algebras with central charge. Math. Commun. **15**(1), 143–157 (2010)
5. Abe, T.: C_2-cofiniteness of 2-cyclic permutation orbifold models. Commun. Math. Phys. **317**(2), 425–445 (2013)
6. Bouwknegt, P., Schoutens, K.: W-symmetry in conformal field theory. Phys. Rep. **223**(4), 183–276 (1993)
7. Creutzig, T., Linshaw, A.: Orbifolds of symplectic fermion algebras. Trans. Am. Math. Soc. **369**, 467–494 (2017)
8. Chandrasekhar, O., Penn, M., Shao, H.: $\mathbb{Z}_2$ orbifolds of free fermion algebras. Commun. Algebr. **46**, 4201–4222 (2018)
9. Dong, C., Jiang, C.: Representations of the vertex operator algebra V_{L2}^{A4}. J. Algebr. **377**, 76–96 (2013)
10. Dong, C., Mason, G.: Rational vertex operator algebras and the effective central charge. IMRN **56**, 2989–3008 (2004)

11. Dong, C., Xu, F., Yu, N.: The 3-permutation orbifold of a lattice vertex operator algebra. J. Pure Appl. Algebr. (2017)
12. Genra, N.: Screening operators for $\mathcal{W}$-algebras. Selecta Math. **23**(3), 2157–2202 (2017)
13. Kac, V., Roan, S.S., Wakimoto, M.: Quantum reduction for affine superalgebras. Commun. Math. Phys. **241**(2), 307–342 (2003)
14. Kac, V.G., Raina, A.K.: Bombay Lectures on Highest Weight Representations. World scientific, Singapore (1987)
15. Linshaw, A.: A Hilbert theorem for vertex algebras. Transform. Groups **15**(2), 427–448 (2010)
16. Linshaw, A.: The structure of the Kac–Wang–Yan algebra. Commun. Math. Phys. **345**(2), 545–585 (2016)
17. Linshaw, A.: Invariant subalgebras of affine vertex algebras. Adv. Math. **234**, 61–84 (2013)
18. Lin, X.: ADE subalgebras of the triplet vertex algebra W(p): E_6, E_7. Int. Math. Res. Not. **2015**(15), 6752–6792 (2014)
19. Milas, A., Penn, M., Shao, H.: Permutation orbfolds of the Heisenberg vertex algebra, $\mathcal{H}(3)$. J. Math. Phys. (To appear). arXiv.1804.01036
20. Milas, A., Penn, M. in preparation
21. Smith, L.: Polynomial Invariants of Finite Groups. Research Notes in Mathematics, vol. 6. A K Peters/CRC Press, Boca Raton (1995)
22. Theilman, K.: Mathematica Package OPE (1991)
23. Watts, G.M.T.: WB algebra representation theory. Nucl. Phys. B **339**, 177 (1990)
24. Weyl, H.: The Classical Groups: Their Invariants and Representations. Princeton University Press, Princeton (1946)

Some Combinatorial Coincidences for Standard Representations of Affine Lie Algebras

Mirko Primc

Abstract In this note we explain, in terms of finite dimensional representations of Lie algebras $\mathfrak{sp}_{2\ell} \subset \mathfrak{sl}_{2\ell}$, a combinatorial coincidence of difference conditions in two constructions of combinatorial bases for standard representations of symplectic affine Lie algebras.

Keywords Affine Lie algebras · Standard representations · Combinatorial bases · Difference conditions for colored partitions

1 Introduction

The first of two famous Rogers–Ramanujan combinatorial identities states that the number of partitions of m, written as

$$m = \sum_{j \geq 1} j f_j,$$

with parts j congruent to ± 1 mod 5, equals the number of partitions of m such that a difference between any two consecutive parts is at least two. Difference two condition can be written in terms of frequencies f_j of parts j as

$$f_j + f_{j+1} \leq 1, \qquad j \geq 1. \tag{1}$$

J. Lepowsky and S. Milne discovered in [14] that the generating function of partitions satisfying congruence condition is, up to a certain "fudge" factor F, the principally specialized character of certain level 3 standard module $L(\Lambda)$ for affine Lie algebra $\widehat{\mathfrak{sl}}_2$. Lepowsky and R. L. Wilson realized that the factor F is a character of the Fock space for the principal Heisenberg subalgebra of $\widehat{\mathfrak{sl}}_2$, and that the generating function

M. Primc (✉)
Department of Mathematics, University of Zagreb, Bijenička 30,
10000 Zagreb, Croatia
e-mail: primc@math.hr

D. Adamović and P. Papi (eds.), *Affine, Vertex and W-algebras*,
Springer INdAM Series 37, https://doi.org/10.1007/978-3-030-32906-8_9

of partitions satisfying difference two condition is the principally specialized character of the vacuum space of $L(\Lambda)$ for the action of principal Heisenberg subalgebra. In a series of papers—see [16] and the references therein—Lepowsky and Wilson gave a Lie-theoretic proof of both Rogers–Ramanujan identities by constructing combinatorial bases of vacuum spaces for the principal Heisenberg subalgebra parametrized by partitions satisfying difference two conditions. Very roughly speaking, in the case of the first Rogers–Ramanujan identity, the vacuum space Ω is spanned by monomial vectors of the form

$$Z(-s)^{f_s} \dots Z(-2)^{f_2} Z(-1)^{f_1} v_\Lambda, \quad s \geq 0, \; f_j \geq 0, \tag{2}$$

where $v_\Lambda \in \Omega$ is a highest weight vector of $L(\Lambda)$ and $Z(j)$ are certain $\mathcal{Z}$-operators. The degree of monomial vector (2) is

$$-m = -\sum_{j=1}^{s} j f_j.$$

Lepowsky and Wilson discovered that $\mathcal{Z}$-operators $Z(j)$ on Ω satisfy certain relations, roughly of the form

$$\begin{aligned} Z(-j)Z(-j) + 2\sum_{i>0} Z(-j-i)Z(-j+i) \approx 0, \\ Z(-j-1)Z(-j) + \sum_{i>0} Z(j-1-i)Z(-j+i) \approx 0, \end{aligned} \tag{3}$$

so we may replace *the leading terms*

$$Z(-j)Z(-j) \text{ and } Z(-j-1)Z(-j) \tag{4}$$

of relations (3) with "higher terms" $Z(-j-i)Z(-j+i)$ and $Z(-j-1-i)$ $Z(-j+i)$, $i > 0$, and reduce the spanning set (2) of Ω to a spanning set

$$Z(-s)^{f_s} \dots Z(-2)^{f_2} Z(-1)^{f_1} v_0, \quad s \geq 0, \; f_j + f_{j+1} \leq 1 \text{ for all } j \geq 1. \tag{5}$$

By invoking the product formula for principally specialized character of Ω and the first Rogers–Ramanujan identity, we see that vectors in the spanning set (5) are in fact a basis of Ω. Lepowsky and Wilson proved directly linear independence of the spanning set (5) and, by reversing the above argument, proved the combinatorial identity.

Lepowsky-Wilson's approach has been applied in many different situations, here we are interested in particular two constructions of combinatorial bases of level $k \geq 1$ standard modules for affine Lie algebras of higher ranks in "homogeneous picture"—both constructions start with a relation of the form

$$\sum_{j_1+\cdots+j_{k+1}=m} x(j_1)\cdots x(j_{k+1}) = 0, \tag{6}$$

(see (11) below) with the leading term

$$x(-j-1)^b x(-j)^a, \tag{7}$$

$a + b = k + 1$ and $(-j-1)b + (-j)a = m$. For $k = 1$ this is analogous to (3) and (4), and again we may replace the leading terms (7) with higher terms in (6) and reduce a Poincaré–Birkhoff–Witt spanning set of monomial vectors in the given representation space to a smaller spanning set. However, this spanning set is not a basis and, in the case we consider, all the relations needed to reduce the Poincaré–Birkhoff–Witt spanning set to a basis are obtained from (6) by the adjoint action of the finite dimensional Lie algebra $\mathfrak{sp}_{2\ell}$ or $\mathfrak{sl}_{2\ell}$, and all the leading terms which we replace by higher terms are obtained by the adjoint action of the same Lie algebra on the leading terms (7). In analogy to the Rogers–Ramanujan case above, we say that the monomial vectors in constructed combinatorial basis *satisfy difference conditions for a given module.*

In this note we explain, in terms of finite dimensional representations of Lie algebras $\mathfrak{sp}_{2\ell} \subset \mathfrak{sl}_{2\ell}$, a combinatorial coincidence of difference conditions in two constructions [3, 23, 28] of combinatorial bases for standard representations of symplectic affine Lie algebra. In particular, Proposition 1 below and Propositions 1 and 2 in [3] provide an alternative proof of Theorem 6.1 in [24].

2 Affine Lie Algebras

We are interested mainly in symplectic affine Lie algebra, but it will be convenient to have a bit more general notation for an affine Lie algebra $\tilde{\mathfrak{g}}$.

2.1 Affine Lie Algebras

Let $\mathfrak{g}$ be a simple complex Lie algebra, $\mathfrak{h}$ a Cartan subalgebra, R the corresponding root system and θ the maximal root with respect to some fixed basis of the root system. For each root α we fix a root vector x_α in $\mathfrak{g}$. Via a symmetric invariant bilinear form $\langle\ ,\ \rangle$ on $\mathfrak{g}$ we identify $\mathfrak{h}$ and $\mathfrak{h}^*$ and we assume that $\langle\theta, \theta\rangle = 2$. Set

$$\hat{\mathfrak{g}} = \coprod_{j \in \mathbb{Z}} \mathfrak{g} \otimes t^j + \mathbb{C}c, \qquad \tilde{\mathfrak{g}} = \hat{\mathfrak{g}} + \mathbb{C}d.$$

Then $\tilde{\mathfrak{g}}$ is the associated untwisted affine Kac–Moody Lie algebra (cf. [12]) with the commutator

$$[x(i), y(j)] = [x, y](i + j) + i\delta_{i+j,0}\langle x, y\rangle c.$$

Here, as usual, $x(i) = x \otimes t^i$ for $x \in \mathfrak{g}$ and $i \in \mathbb{Z}$, c is the canonical central element, and $[d, x(i)] = ix(i)$. We identify $\mathfrak{g}$ and $\mathfrak{g} \otimes 1$. Let B be a basis of $\mathfrak{g}$ consisting of root vectors and elements of $\mathfrak{h}$. Set

$$\hat{B} = \{x(n) \mid x \in B, n \in \mathbb{Z}\}.$$

Then $\hat{B} \cup \{c\}$ is a basis of $\hat{\mathfrak{g}}$. A given linear order $\preceq$ on B we extend to a linear order on $\hat{B}$ by

$$x(n) \prec y(m) \quad \text{iff} \quad n < m \text{ or } n = m \text{ and } x \prec y.$$

2.2 $\mathbb{Z}$-Gradings of $\mathfrak{g}$ by Minuscule Coweights

Let $R^\vee$ be the dual root system of R and fix a minuscule coweight $\omega \in P(R^\vee)$. Set $\Gamma = \{\alpha \in R \mid \omega(\alpha) = 1\}$ and

$$\mathfrak{g}_0 = \mathfrak{h} + \sum_{\omega(\alpha)=0} \mathfrak{g}_\alpha, \qquad \mathfrak{g}_{\pm 1} = \sum_{\alpha \in \pm\Gamma} \mathfrak{g}_\alpha,$$

$$\hat{\mathfrak{g}}_0 = \mathfrak{g}_0 \otimes \mathbb{C}[t, t^{-1}] \oplus \mathbb{C}c, \qquad \hat{\mathfrak{g}}_{\pm 1} = \mathfrak{g}_{\pm 1} \otimes \mathbb{C}[t, t^{-1}].$$

Then on $\mathfrak{g}$ and $\hat{\mathfrak{g}}$ we have $\mathbb{Z}$-gradings

$$\mathfrak{g} = \mathfrak{g}_{-1} + \mathfrak{g}_0 + \mathfrak{g}_1 \quad \text{and} \quad \hat{\mathfrak{g}} = \hat{\mathfrak{g}}_{-1} + \hat{\mathfrak{g}}_0 + \hat{\mathfrak{g}}_1.$$

Let $\mathfrak{h}'_0 \subset \mathfrak{h}$ be a Cartan subalgebra of $\mathfrak{g}_0' = [\mathfrak{g}_0, \mathfrak{g}_0]$. Note that $\mathfrak{g}_0 = \mathfrak{g}_0' + \mathbb{C}\omega$ is a reductive Lie algebra, $\mathfrak{g}_{\pm 1}$ are commutative subalgebras of $\mathfrak{g}$ and $\mathfrak{g}_0$-modules, and that $\hat{\mathfrak{g}}_{\pm 1}$ are commutative subalgebras of $\hat{\mathfrak{g}}$ and $\hat{\mathfrak{g}}_0$-modules.

For a classical $\mathfrak{g}$ all possible minuscule coweights ω are listed on Dynkin diagrams below: the black dot on a diagram denotes the simple root α for which $\omega(\alpha) = 1$ and the rest is the Dynkin diagram of the semisimple Lie algebra $\mathfrak{g}_0'$.

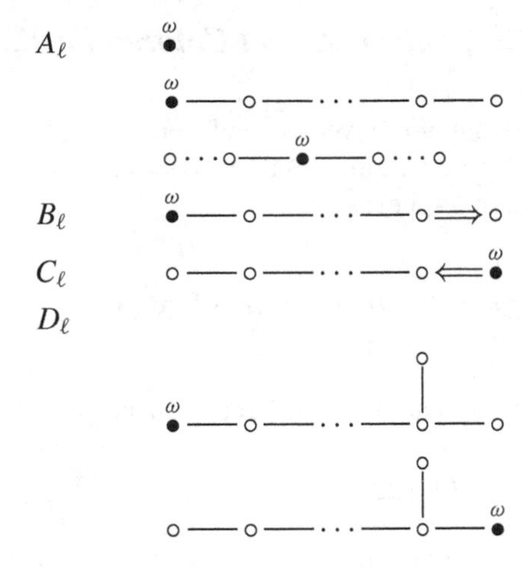

3 Standard Modules and Feigin–Stoyanovsky's Type Subspaces

In this section we describe monomial bases of standard $\tilde{\mathfrak{g}}$-modules $L(k\Lambda_0)$ and monomial bases of Feigin–Stoyanovsky's type subspaces $W(k\Lambda_0)$.

3.1 Standard Modules and Feigin–Stoyanovsky's Type Subspaces

As usual, we denote by $\Lambda_0, \ldots, \Lambda_\ell$ the fundamental weights of $\tilde{\mathfrak{g}}$. For a given integral dominant weight $\Lambda = k_0\Lambda_0 + \cdots + k_\ell\Lambda_\ell$ we denote by $L(\Lambda)$ the standard (i.e. integral highest weight) $\tilde{\mathfrak{g}}$-module with the highest weight Λ, by v_Λ a fixed highest weight vector and by $k = \Lambda(c)$ the level.

For a standard $\tilde{\mathfrak{g}}$-module $L(\Lambda)$ and a given $\mathbb{Z}$-grading on $\hat{\mathfrak{g}}$ as above, we define the corresponding Feigin–Stoyanovsky's type subspace $W(\Lambda)$ as

$$W(\Lambda) = U(\hat{\mathfrak{g}}_1)v_\Lambda \subset L(\Lambda).$$

Note that $U(\hat{\mathfrak{g}}_1) \cong S(\hat{\mathfrak{g}}_1)$ since $\hat{\mathfrak{g}}_1$ is commutative.

3.2 Monomials in Elements of $\hat{B}$ and Colored Partitions

Let B be a basis of $\mathfrak{g}$ consisting of root vectors and elements of $\mathfrak{h}$, with a linear order $\preceq$. We shall write ordered product monomials in $\hat{B} = \{x(n) \mid x \in B, n \in \mathbb{Z}\}$ with factors $x_i(n_i)$ in ascending order, i.e. as

$$\prod_{i=1}^{r} x_i(n_i) = x_1(n_1)x_2(n_2)\dots x_r(n_r), \quad x_1(n_1) \preceq x_2(n_2) \preceq \cdots \preceq x_r(n_r). \tag{8}$$

Sometimes it is convenient to think of the ordered sequence

$$\pi = (x_1(n_1) \preceq x_2(n_2) \preceq \cdots \preceq x_r(n_r)) \tag{9}$$

as a colored partition, i.e., as a "plain partition" $n_1 \leq n_2 \leq \cdots \leq n_r$ colored with "colors" $x_1, x_2, \dots, x_r$. We shall say that r is the length of π and that $n_1 + n_2 + \cdots + n_r$ is the degree of π. Then the monomial in (8) is the corresponding ordered product $u(\pi)$ in the enveloping algebra

$$u(\pi) = x_1(n_1)x_2(n_2)\dots x_r(n_r) \in U(\hat{\mathfrak{g}}). \tag{10}$$

It will is also be convenient to think of the colored partition π as a monomial in a symmetric algebra of $\hat{\mathfrak{g}}$, i.e.

$$\pi = x_1(n_1)x_2(n_2)\dots x_r(n_r) \in S(\hat{\mathfrak{g}}).$$

We denote with $\mathcal{P}$ the set of all colored partitions; it is a monoid with the unit 1—the colored partition with no parts (of length 0 and degree 0). The submonoid $\mathcal{P}_{<0}$ is the set of all colored partitions (9) such that $n_1 \leq n_2 \leq \cdots \leq n_r < 0$.

When ω is a minuscule coweight and B_1 an ordered basis of $\mathfrak{g}_1$, we extend the order to a basis $\hat{B}_1 = \{x_\gamma(n) \mid \gamma \in \Gamma, n \in \mathbb{Z}\}$ of $\hat{\mathfrak{g}}_1$ and, in a similar way as above, we consider colored partitions and ordered monomials in $U(\hat{\mathfrak{g}}_1) \cong S(\hat{\mathfrak{g}}_1))$. We denote the corresponding monoids as $\mathcal{P}^1$ and $\mathcal{P}^1_{<0}$.

3.3 Well Order on Colored Partitions

We extend the order $\preceq$ on $\hat{B}$ to the order on the set of colored partitions (9) first by comparing the lengths—shorter partitions are higher; then by comparing the degrees—partitions with greater degree are higher; then by comparing "plain partitions" $n_1 \leq n_2 \leq \cdots \leq n_r$ in the reverse lexicographical order and, finally, by comparing "the colorings" $x_1, x_2, \dots, x_r$ in the reverse lexicographical order. This order has several good properties:

- On the set of partitions with bounded length and fixed degree $\preceq$ is a well order, so in our arguments we can use induction.
- For $x_1, x_2, \dots, x_r \in \hat{B}$ and any permutation σ two monomials $x_1 x_2 \dots x_r$ and $x_{\sigma(1)} x_{\sigma(2)} \dots x_{\sigma(r)}$ in $U(\hat{\mathfrak{g}})$ differ by a linear combination of $u(\pi)$ with length of π strictly less then r. So in inductive arguments ordered monomials "behave as commutative monomials".
- $\kappa \preceq \lambda$ implies $\kappa\pi \preceq \lambda\pi$ for colored partitions κ, λ, π in $S(\hat{\mathfrak{g}})$. This property enables us to replace the leading terms of relations with higher terms and use induction.

3.4 Relations on $L(\Lambda)$ and $W(\Lambda)$

On level k standard module $L(\Lambda)$ we have vertex operator relations (cf. [15, 18, 19])

$$x_\theta(z)^{k+1} = \sum_{n\in\mathbb{Z}} \Big(\sum_{j_1+\cdots+j_{k+1}=n} x_\theta(j_1) \dots x_\theta(j_{k+1}) \Big) z^{-n-k-1} = 0. \tag{11}$$

With the adjoint action of $\mathfrak{g}$ we get a finite dimensional $\mathfrak{g}$-module

$$U(\mathfrak{g}) \cdot x_\theta(z)^{k+1} \cong L_{\mathfrak{g}}((k+1)\theta).$$

The set $\bar{R}_k$ of all coefficients in $U(\mathfrak{g}) \cdot x_\theta(z)^{k+1}$ we call *relations on* $L(\Lambda)$.

For a minuscule coweight ω we have $\omega(\theta) = 1$. Hence $x_\theta \in \mathfrak{g}_1$ and the vertex operator relation (11) is a relation on the Feigin–Stoyanovsky's type subspace $W(\Lambda)$. Since $\mathfrak{g}_1$ is a $\mathfrak{g}_0$-module, by the adjoint action of $\mathfrak{g}_0$ on the vertex operator relation (11) we get a finite dimensional $\mathfrak{g}_0$-module

$$U(\mathfrak{g}_0) \cdot x_\theta(z)^{k+1} \cong L_{\mathfrak{g}_0'}((k+1)\theta|_{\mathfrak{h}_0'}),$$

where $\mathfrak{h}_0' \subset \mathfrak{h}$ is a Cartan subalgebra of $\mathfrak{g}_0'$. The set $\bar{R}_k^0 \subset \bar{R}_k$ of all coefficients in $U(\mathfrak{g}_0) \cdot x_\theta(z)^{k+1}$ we call *relations on* $W(\Lambda)$.

3.5 The Leading Terms of Relations on $L(\Lambda)$ and $W(\Lambda)$

Every relation $r \in \bar{R}$, $r \neq 0$, can be written as

$$r = c_\rho u(\rho) + \sum_{\pi \succ \rho} c_\pi u(\pi), \qquad c_\rho \neq 0.$$

We say that the colored partition ρ is *the leading term of* r and we write $\rho = \ell t(r)$, and sometimes we shall also say that the ordered monomial $u(\rho)$ is the leading term of the relation $r = 0$. The leading terms of coefficients of (11) are colored partitions of the form

$$x_\theta(-j-1)^b x_\theta(-j)^a \in S(\hat{\mathfrak{g}}), \tag{12}$$

$a + b = k + 1$ and $(-j-1)b + (-j)a = -n$. All leading terms of relations $\bar{R}_k$ and $\bar{R}_k^0$ are obtained as leading terms of finite dimensional spaces

$$\ell t \left(U(\mathfrak{g}) \cdot x_\theta(-j-1)^b x_\theta(-j)^a\right) \quad \text{and} \quad \ell t \left(U(\mathfrak{g}_0) \cdot x_\theta(-j-1)^b x_\theta(-j)^a\right). \tag{13}$$

3.6 Monomial Bases of Feigin–Stoyanovsky's Type Subspaces $W(k\Lambda_0)$

By Poincaré–Birkhoff–Witt theorem the set of monomial vectors

$$u(\pi)v_{k\Lambda_0}, \qquad \pi \in \mathcal{P}^1_{<0}$$

is a spanning set for Feigin–Stoyanovsky's type subspace $W(k\Lambda_0)$. By expressing the leading terms of relations in $\bar{R}_k^0$ with higher monomials, we can reduce this PBW spanning set of $W(k\Lambda_0)$ to a spanning set

$$u(\pi)v_{k\Lambda_0}, \qquad \pi \in \mathcal{P}^1_{<0}, \quad \pi \notin \mathcal{P}^1 \ell t\left(\bar{R}_k^0\right). \tag{14}$$

This set is a basis in many cases with a suitable choice of order $\preceq$ and B: for all positive integer levels k and $\mathfrak{g}$ of the type A_ℓ for all ℓ and all minuscule coweights (see [5, 6, 8, 10, 19, 21, 26, 27]) and for $\mathfrak{g}$ of the type C_ℓ for all ℓ (see [3, 22]); for level $k = 1$ for all classical $\mathfrak{g}$ for all ℓ and all minuscule coweights (see [20]); and for D_4 level $k = 1$ and 2 (see [2]). The monomial bases of Feigin–Stoyanovsky's type subspaces are used in the construction of semi-infinite monomial bases of entire standard modules (cf. [28]).

Different methods are used to prove linear independence of the spanning set (14), but in all proofs the explicit combinatorial description of colored partitions π which satisfy the difference condition $\pi \notin \mathcal{P}^1 \ell t\left(\bar{R}_k^0\right)$ is needed. For A_1 and $k \geq 1$ partitions π have only one color and the difference conditions are the difference two conditions which appear in Rogers–Ramanujan and Gordon combinatorial identities. For A_ℓ and $\omega = \omega_1$ colored partitions π appear as $(k, \ell + 1)$-admissible configurations. For all classical Lie algebras $\mathfrak{g}$ and all minuscule coweights difference conditions for $W(\Lambda_0)$ are given by the energy function of a perfect crystal (cf. [13]) corresponding to the $\mathfrak{g}_0$-module $\mathfrak{g}_1$.

3.7 Monomial Bases of Standard Modules

By Poincaré–Birkhoff–Witt theorem the set of monomial vectors

$$u(\pi)v_{k\Lambda_0}, \qquad \pi \in \mathcal{P}_{<0}$$

is a spanning set for the standard module $L(k\Lambda_0)$. By expressing the leading terms of relations in $\bar{R}_k$ with higher monomials, we can reduce this PBW spanning set of $L(k\Lambda_0)$ to a spanning set

$$u(\pi)v_{k\Lambda_0}, \qquad \pi \in \mathcal{P}_{<0}, \quad \pi \notin \mathcal{P}\ell t\left(\bar{R}_k\right). \tag{15}$$

This set is a basis in several cases with a suitable choice of order $\preceq$ and B: for all positive integer levels k and $\mathfrak{g}$ of the type A_1 (see [9, 11, 17, 18]) and for level $k = 1$ and $\mathfrak{g}$ of the type C_ℓ for all $\ell \geq 2$ (see [23, 25]). In this setting we can think $A_1 \cong C_1$, so (15) is a basis of $L(k\Lambda_0)$ for $\mathfrak{g}$ of the type C_ℓ for $k = 1$ or $\ell = 1$. In [24] we conjecture that (15) is a basis for all $k \geq 1$ and $\ell \geq 1$.

4 Some Combinatorial Coincidences for Two Constructions

In this section we describe the combinatorial coincidence of the leading terms $\ell t\left(\bar{R}_k^0\right)$ for $C_{2\ell}$ and the leading terms $\ell t\left(\bar{R}_k\right)$ for C_ℓ which happens for all $k \geq 1$ and $\ell \geq 1$.

4.1 Combinatorial Coincidence of $\ell t\left(\bar{R}_k^0\right)$ for B_2 and $\ell t\left(\bar{R}_k\right)$ for A_1

Let $\mathfrak{g}$ of the type $B_2 \cong C_2$. Then $\mathfrak{g}_0{}'$ is of the type A_1. We identify the root system with

$$R = \{\pm(\varepsilon_1 - \varepsilon_2), \pm(\varepsilon_1 + \varepsilon_2), \pm\varepsilon_1, \pm\varepsilon_2)\}.$$

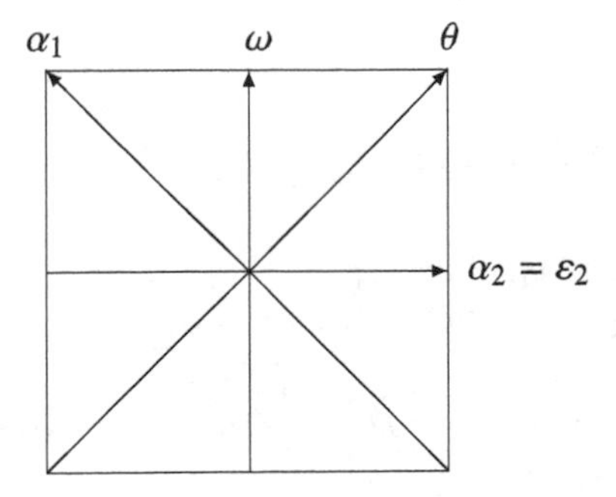

Then $\omega = \omega_1$ and $\Gamma = \{\varepsilon_1 - \varepsilon_2, \varepsilon_1, \varepsilon_1 + \varepsilon_2\}$ with the the corresponding root vectors in $\mathfrak{g}_1$ denoted as $x_{\underline{2}}$, x_0, x_2 with the order $x_{\underline{2}} \prec x_0 \prec x_2$. The consecutive action of $x_{-\varepsilon_2}$ in $\mathfrak{g}_0$ on (12) gives the leading terms

$$\begin{aligned} &x_2(-j-1)^{a_{j+1}} x_0(-j)^{b_j} x_2(-j)^{a_j}, && a_{j+1} + b_j + a_j = k+1, \\ &x_2(-j-1)^{a_{j+1}} x_{\underline{2}}(-j)^{c_j} x_0(-j)^{b_j}, && a_{j+1} + c_j + b_j = k+1, \\ &x_0(-j-1)^{b_{j+1}} x_2(-j-1)^{a_{j+1}} x_{\underline{2}}(-j)^{c_j}, && b_{j+1} + a_{j+1} + c_j = k+1, \\ &x_{\underline{2}}(-j-1)^{c_{j+1}} x_0(-j-1)^{b_{j+1}} x_{\underline{2}}(-j)^{c_j}, && c_{j+1} + b_{j+1} + c_j = k+1, \end{aligned} \tag{16}$$

and the spanning set

$$\prod_{j<0} x_{\underline{2}}(-j)^{c_j} x_0(-j)^{b_j} x_2(-j)^{a_j} v_{k\Lambda_0} \tag{17}$$

of monomial vectors in $W_{B_2^{(1)}}(k\Lambda_0)$ is reduced to a basis (14) if we impose the difference conditions

$$\begin{aligned} c_{j+1} + b_{j+1} + c_j &\leq k, \\ b_{j+1} + a_{j+1} + c_j &\leq k, \\ a_{j+1} + c_j + b_j &\leq k, \\ a_{j+1} + b_j + a_j &\leq k \end{aligned} \tag{18}$$

for all $j \in \mathbb{Z}$ (see [3, 22]). It is clear that the same argument for leading terms applies in the case of the standard module $L_{A_1^{(1)}}(k\Lambda_0)$ (see [9, 18]) if we use the standard ordered basis $x_{-\alpha} \prec h \prec x_\alpha$ of $\mathfrak{sl}_2$ and identify the corresponding bases elements

$$\begin{aligned} &x_{-\alpha} \leftrightarrow x_{\varepsilon_1-\varepsilon_2},\ h \leftrightarrow x_{\varepsilon_1},\ x_\alpha \leftrightarrow x_{\varepsilon_1+\varepsilon_2} \quad \text{and,} \quad \text{for all } j \in \mathbb{Z}, \\ &x_{-\alpha}(j) \leftrightarrow x_{\varepsilon_1-\varepsilon_2}(j),\ h(j) \leftrightarrow x_{\varepsilon_1}(j),\ x_\alpha(j) \leftrightarrow x_{\varepsilon_1+\varepsilon_2}(j). \end{aligned} \tag{19}$$

Therefore, the Feigin–Stoyanovsky's type subspace $W_{B_2^{(1)}}(k\Lambda_0)$ *and the standard module* $L_{A_1^{(1)}}(k\Lambda_0)$ *have the same combinatorial description of the leading terms of relations, and the same combinatorial description of the difference conditions for colored partitions parameterizing monomial bases.*

4.2 Simple Lie Algebra $\mathfrak{g}$ of Type C_ℓ

We fix a simple Lie algebra $\mathfrak{g}$ of the type C_ℓ, $\ell \geq 2$. For a given Cartan subalgebra $\mathfrak{h}$ and the corresponding root system R we can write

$$R = \{\pm(\varepsilon_i \pm \varepsilon_j) \mid i, j = 1, ..., \ell\} \setminus \{0\}.$$

We choose simple roots as in [4]

$$\alpha_1 = \varepsilon_1 - \varepsilon_2, \ \alpha_2 = \varepsilon_2 - \varepsilon_3, \ \cdots \ \alpha_{\ell-1} = \varepsilon_{\ell-1} - \varepsilon_\ell, \ \alpha_\ell = 2\varepsilon_\ell.$$

Then $\theta = 2\varepsilon_1$. For each root α we choose a root vector X_α such that $[X_\alpha, X_{-\alpha}] = \alpha^\vee$. For the root vectors X_α we shall use the following notation:

$$\begin{array}{lll} X_{ij} & \text{or just } ij \text{ if } & \alpha = \varepsilon_i + \varepsilon_j \ , \ i \leq j \ , \\ X_{\underline{i}\underline{j}} & \text{or just } \underline{i}\underline{j} \text{ if } & \alpha = -\varepsilon_i - \varepsilon_j \ , \ i \geq j \ , \\ X_{i\underline{j}} & \text{or just } i\underline{j} \text{ if } & \alpha = \varepsilon_i - \varepsilon_j \ , \ i \neq j \ . \end{array}$$

With the previous notation $x_\theta = X_{11}$. We also write for $i = 1, \ldots, \ell$

$$X_{i\underline{i}} = \alpha_i^\vee \text{ or just } i\underline{i} \, .$$

These vectors X_{ab} form a basis B of $\mathfrak{g}$ which we shall write in a triangular scheme. For example, for $\ell = 3$ the basis B is

$$\begin{array}{llllll} 11 \\ 12 & 22 \\ 13 & 23 & 33 \\ 1\underline{3} & 2\underline{3} & 3\underline{3} & \underline{3}\underline{3} \\ 1\underline{2} & 2\underline{2} & 3\underline{2} & \underline{3}\underline{2} & \underline{2}\underline{2} \\ 1\underline{1} & 2\underline{1} & 3\underline{1} & \underline{3}\underline{1} & \underline{2}\underline{1} & \underline{1}\underline{1}. \end{array}$$

In general for the set of indices $\{1, 2, \cdots, \ell, \underline{\ell}, \cdots, \underline{2}, \underline{1}\}$ we use order

$$1 \succ 2 \succ \cdots \succ \ell - 1 \succ \ell \succ \underline{\ell} \succ \underline{\ell - 1} \succ \cdots \succ \underline{2} \succ \underline{1}$$

and a basis element X_{ab} we write in ath column and bth row,

$$B = \{X_{ab} \mid b \in \{1, 2, \cdots, \ell, \underline{\ell}, \cdots, \underline{2}, \underline{1}\}, \ a \in \{1, \cdots, b\}\}. \tag{20}$$

By using (20) we define on the basis B the corresponding lexicographical order

$$X_{ab} \succ X_{a'b'} \ if \ a \succ a' \ or \ a = a' \ and \ b \succ b' \ .$$

In other words, X_{ab} is larger than $X_{a'b'}$ if $X_{a'b'}$ lies in a column a' to the right of the column a, or X_{ab} and $X_{a'b'}$ are in the same column $a = a'$, but $X_{a'b'}$ is below X_{ab}. For $\ell = 1$ we get $C_1 \cong A_1$ with the basis B

$$\begin{array}{ll} 11 \\ 1\underline{1} & \underline{1}\underline{1}. \end{array}$$

For a simple Lie algebra $\mathfrak{g}$ of type C_ℓ, $\ell \geq 2$, we have the minuscule weight

$$\omega = \omega_\ell = \tfrac{1}{2}(\varepsilon_1 + \cdots + \varepsilon_\ell),$$

the corresponding $\Gamma = \{\varepsilon_i + \varepsilon_j \mid 1 \leq i \leq j \leq \ell\}$ and the basis

$$B_1 = \{X_\gamma \mid \gamma \in \Gamma\} = \{X_{ij} \mid 1 \leq i \leq j \leq \ell\} \subset B \tag{21}$$

of $\mathfrak{g}_1$. For C_3 the basis B_1 can be written as as the "upper" triangle in B:

11
12 22
13 23 33.

In the case C_1 the "upper" triangle in B is $B_1 = \{11\}$. In the case C_ℓ, $\ell \geq 2$, the subalgebra $\mathfrak{g}_0{}'$ is a simple Lie algebra of the type $A_{\ell-1}$ with a Cartan subalgebra $\mathfrak{h}_0' \subset \mathfrak{h}_0$ and the corresponding root system

$$\{\pm(\varepsilon_i - \varepsilon_j) \mid 1 \leq i < j \leq \ell\}$$

with the simple roots

$$\alpha_1 = \varepsilon_1 - \varepsilon_2,\ \alpha_2 = \varepsilon_2 - \varepsilon_3,\ \cdots\ \alpha_{\ell-1} = \varepsilon_{\ell-1} - \varepsilon_\ell.$$

4.3 *The Leading Terms of Relations for $W_{C_\ell^{(1)}}(k\Lambda_0)$*

For the finite dimensional $\mathfrak{g}_0{}'$-module of relations we have

$$U(\mathfrak{g}_0) \cdot x_\theta(z)^{k+1} \cong L_{A_{\ell-1}}((k+1)\theta|_{\mathfrak{h}_0'}) = L_{A_{\ell-1}}(2(k+1)\omega_1). \tag{22}$$

Note that $\mathfrak{g}_0{}'$-module $L_{A_{\ell-1}}(\omega_1)$ is the vector representation $\mathbb{C}^\ell$ of $\mathfrak{sl}_\ell$ with the canonical basis $e_1, \ldots, e_\ell$, the module $L_{A_{\ell-1}}(2\omega_1)$ is isomorphic to the symmetric power $S^2(\mathbb{C}^\ell)$ with a basis $e_i e_j \leftrightarrow X_{ij}$, and the module (22), as a vector space, is isomorphic to $S^{2(k+1)}(\mathbb{C}^\ell)$ with the canonical basis $e_1^{m_1}, \ldots, e_\ell^{m_\ell}$, $m_1 + \cdots + m_\ell = 2(k+1)$, which we view as multisets $\{1^{m_1}, \ldots, \ell^{m_\ell}\}$. By using the action of the root vectors $x_{-\alpha_1}, \ldots, x_{-\alpha_{\ell-1}}$ in $\mathfrak{g}_0$ on (11) for every multiset $\{1^{m_1}, \ldots, \ell^{m_\ell}\}$, $m_1 + \cdots + m_\ell = 2(k+1)$, we obtain a relation for Feigin–Stoyanovsky's type subspace $W_{C_\ell^{(1)}}(k\Lambda_0)$ (see [3], Proposition 1)

$$\sum_{\{i_1,\ldots,i_{k+1}\}\cup\{j_1,\ldots,j_{k+1}\}=\{1^{m_1},\ldots,\ell^{m_\ell}\}} C_{\mathbf{i},\mathbf{j}}\, X_{i_1,j_1}(z) \ldots X_{i_{k+1},j_{k+1}}(z) = 0,$$

with $C_{\mathbf{i},\mathbf{j}} \neq 0$, the sum runs over all such partitions of the multiset $\{1^{m_1}, \ldots, \ell^{m_\ell}\}$. The leading terms of these relations are most conveniently described as monomials (see [3], Proposition 2)

$$X_{i_t j_t}(-j-1)^{b_{i_t j_t}} \ldots X_{i_1 j_1}(-j-1)^{b_{i_1 j_1}} X_{i_s j_s}(-j)^{a_{i_s j_s}} \ldots X_{i_{t+1} j_{t+1}}(-j)^{a_{i_{t+1} j_{t+1}}}, \quad (23)$$

where $(i_p, j_p) \neq (i_{p+1}, j_{p+1})$ and

$$i_1 \leq \cdots \leq i_t \leq j_t \leq \cdots \leq j_1 \leq i_{t+1} \leq \cdots \leq i_s \leq j_s \leq \cdots \leq j_{t+1}, \quad (24)$$

$$b_{i_1 j_1} + \cdots + b_{i_t j_t} + a_{i_{t+1} j_{t+1}} + \cdots + a_{i_s j_s} = k+1.$$

The factors X_{pq} of leading terms (23) lie on diagonal paths in B_1 with $i = i_{t+1}$:

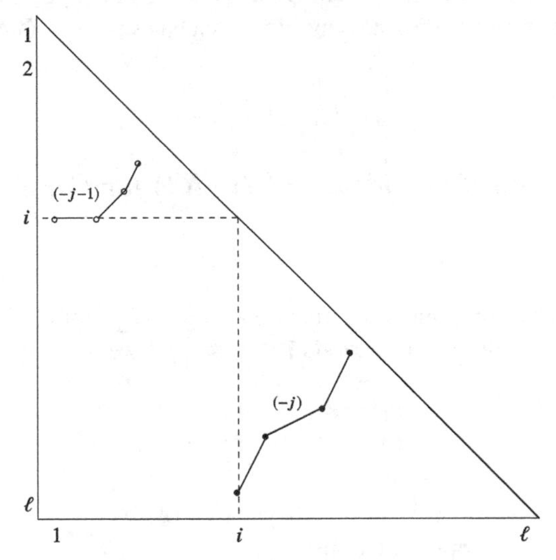

Note that there is ℓ choices for the upper triangle in B_1, from $i = 1$, when the upper triangle is $\{11\}$, all the way to $i = \ell$, when the upper triangle is the entire B_1. For example, for $\ell = 2$ the triangle B_1 is

11
12 22.

Then for given j and $i = 1$ we have two possible diagonal paths in the lower triangle, vertical and horizontal, with leading terms

$$\begin{aligned} &X_{11}(-j-1)^{b_{11}} X_{12}(-j)^{a_{12}} X_{11}(-j)^{a_{11}}, \qquad b_{11} + a_{12} + a_{11} = k+1, \\ &X_{11}(-j-1)^{b_{11}} X_{22}(-j)^{a_{22}} X_{12}(-j)^{a_{12}}, \qquad b_{11} + a_{22} + a_{12} = k+1, \end{aligned} \quad (25)$$

while for $i = 2$ we have two possible diagonal paths in the upper triangle, vertical and horizontal, with leading terms

$$\begin{aligned} &X_{12}(-j-1)^{b_{12}}X_{11}(-j-1)^{b_{11}}X_{22}(-j)^{a_{22}}, \quad b_{12}+b_{11}+a_{22}=k+1, \\ &X_{22}(-j-1)^{b_{22}}X_{12}(-j-1)^{b_{12}}X_{22}(-j)^{a_{22}}, \quad b_{22}+b_{12}+a_{22}=k+1. \end{aligned} \tag{26}$$

These leading terms are precisely the leading terms (16) which led to difference conditions (18) as a system of inequalities for all $j \in \mathbb{N}$. In a similar way for the monomial

$$\prod X_{pq}(-j)^{m_{pq;j}}$$

we can write difference conditions as a system of inequalities

$$m_{i_1j_1;j+1} + \cdots + m_{i_tj_t;j+1} + m_{i_{t+1}j_{t+1};j} + \cdots + m_{i_sj_s;j} \le k \tag{27}$$

for all $j \in \mathbb{N}$ and diagonal paths (24) in B_1. It is proved in [3] that the spanning set of vectors (14) satisfying difference conditions is a basis of $W_{C_\ell^{(1)}}(k\Lambda_0)$ for all $k \ge 1$ and $\ell \ge 2$.

4.4 Combinatorial Coincidence of $\ell t\left(\bar{R}_k^0\right)$ for $C_{2\ell}$ and $\ell t\left(\bar{R}_k\right)$ for C_ℓ

For $\ell = 1$ we have the combinatorial coincidence of $\ell t\left(\bar{R}_k^0\right)$ for $C_2 \cong B_2$ and $\ell t\left(\bar{R}_k\right)$ for $C_1 \cong A_1$ if we use the identification (19), that is, if we identify two bases

$$\begin{array}{l} 11 \\ 12\ 22 \end{array} \longleftrightarrow \begin{array}{l} 11 \\ 1\underline{1}\ \underline{11}. \end{array}$$

of A_1-modules. For $\ell \ge 2$ we can identify bases of two vector spaces for finite dimensional Lie algebra representations,

$$L_{A_{2\ell-1}}(2\omega_1) \qquad \text{and} \qquad L_{C_\ell}(\theta),$$

the basis (21) for $C_{2\ell}$ and the basis (20) for C_ℓ. For example, for $\ell = 3$ we identify two bases

$$\begin{array}{l} 11 \\ 12\ 22 \\ 13\ 23\ 33 \\ 14\ 24\ 34\ 44 \\ 15\ 25\ 35\ 45\ 55 \\ 16\ 26\ 36\ 46\ 56\ 66 \end{array} \longleftrightarrow \begin{array}{l} 11 \\ 12\ 22 \\ 13\ 23\ 33 \\ 1\underline{3}\ 2\underline{3}\ 3\underline{3}\ \underline{33} \\ 1\underline{2}\ 2\underline{2}\ 3\underline{2}\ \underline{32}\ \underline{22} \\ 1\underline{1}\ 2\underline{1}\ 3\underline{1}\ \underline{31}\ \underline{21}\ \underline{11}. \end{array} \tag{28}$$

We consider (cf. [4])

$$\mathfrak{sp}_{2\ell} \subset \mathfrak{sl}_{2\ell}.$$

Then the Weyl dimension formula gives the following:

Proposition 1 *For all $m \in \mathbb{N}$ the restriction of the representation $L_{A_{2\ell-1}}(2m\omega_1)$ for the Lie algebra $\mathfrak{sl}_{2\ell}$ to the subalgebra $\mathfrak{sp}_{2\ell}$ remains irreducible, i.e., as a vector space*

$$L_{C_\ell}(m\theta) = L_{A_{2\ell-1}}(2m\omega_1)|_{C_\ell} \cong S^{2m}(\mathbb{C}^{2\ell})|_{C_\ell}.$$

In particular, under the identification (28) of two bases and the respective orders, the Feigin–Stoyanovsky's type subspace $W_{C_{2\ell}^{(1)}}(k\Lambda_0)$ and the standard module $L_{C_\ell^{(1)}}(k\Lambda_0)$ have the same combinatorial description of the leading terms of relations for all $k \geq 1$ and $\ell \geq 1$.

Remark 1 The above argument gives an alternative construction of the set of leading terms of relations for standard modules $\ell t\left(\bar{R}_k\right)$ for C_ℓ, first obtained combinatorialy in [24].

Remark 2 So far it seems that it is easier to prove linear independence of spanning sets (14) for Feigin–Stoyanovsky's type subspaces than to prove linear independence of spanning sets (15) for standard modules. So the question is: can one use the results for $W_{C_{2\ell}^{(1)}}(k\Lambda_0)$ to study $L_{C_\ell^{(1)}}(k\Lambda_0)$ for $k \geq 2$?

Remark 3 D. Adamović pointed out that the isomorphism $L_{A_{2\ell-1}}(n\omega_1)|_{C_\ell} \cong L_{C_\ell}(n\omega_1)$ appears in [1] in the context of vertex algebras (cf. Proposition 7) and may be viewed as a consequence of the quantum Galois theory [7].

Acknowledgements This work is partially supported by the Croatian Science Foundation under the Project 2634 and by the QuantiXLie Centre of Excellence, a project cofinanced by the Croatian Government and European Union through the European Regional Development Fund—the Competitiveness and Cohesion Operational Programme (KK.01.1.1.01.0004).

References

1. Adamović, D., Perše, O.: The vertex algebra $M(1)^+$ and certain affine vertex algebras of level -1, SIGMA **8**, 040 (2012) 16 p
2. Baranović, I.: Combinatorial bases of Feigin-Stoyanovsky's type subspaces of level 2 standard modules for $D_4^{(1)}$. Commun. Algebra **39**, 1007–1051 (2011)
3. Baranović, I., Primc, M., Trupčević, G.: Bases of Feigin-Stoyanovsky's type subspaces for $C_\ell^{(1)}$. Ramanujan J. **45**, 265–289 (2018)
4. Bourbaki, N.: Algèbre Commutative. Hermann, Paris (1961)
5. Capparelli, S., Lepowsky, J., Milas, A.: The Rogers-Ramanujan recursion and intertwining operators. Commun. Contemp. Math. **5**, 947–966 (2003)
6. Capparelli, S., Lepowsky, J., Milas, A.: The Rogers-Selberg recursions, the Gordon-Andrews identities and intertwining operators. Ramanujan J. **12**, 379–397 (2006)
7. Dong, C., Mason, G.: On quantum Galois theory. Duke Math. J. **86**, 305–321 (1997)
8. Feigin, B., Jimbo, M., Loktev, S., Miwa, T., Mukhin, E.: Bosonic formulas for *(k, l)*-admissible partitions. Ramanujan J. **7**, 485–517; Addendum to 'Bosonic formulas for *(k, l)*-admissible partitions'. Ramanujan J. **7**(2003), 519–530 (2003)

9. Feigin, B., Kedem, R., Loktev, S., Miwa, T., Mukhin, E.: Combinatorics of the $\widehat{\mathfrak{sl}}_2$ spaces of coinvariants. Transform. Groups **6**, 25–52 (2001)
10. Stoyanovsky, A.V., Feigin, B.L.: Functional models of the representations of current algebras, and semi-infinite Schubert cells, (Russian) Funktsional. Anal. i Prilozhen. **28**(1), 68-90, 96 (1994); translation in Funct. Anal. Appl. **28**(1), 55-72 (1994); preprint Feigin, B., Stoyanovsky, A.: *Quasi-particles models for the representations of Lie algebras and geometry of flag manifold*, arXiv:hep-th/9308079, RIMS 942
11. Feigin, E.: The PBW filtration. Represent. Theory **13**, 165–181 (2009)
12. Kac, V.G.: Infinite-Dimensional Lie Algebras, 3rd edn. Cambridge University Press, Cambridge (1990)
13. Kang, S.-J., Kashiwara, M., Misra, K.C., Miwa, T., Nakashima, T., Nakayashiki, A.: Affine crystals and vertex models. Int. J. Modern Phys. A **7**, 449-484 (Suppl. 1A, Proceedings of the RIMS Research Project 1991, "Infinite Analysis", World Scientific, Singapore, (1992))
14. Lepowsky, J., Milne, S.: Lie algebraic approaches to classical partition identities. Adv. Math. **29**, 15–59 (1978)
15. Lepowsky, J., Primc, M.: Structure of the Standard Modules for the Affine Lie Algebra $A_1^{(1)}$, Contemporary Mathematics vol. 46. American Mathematical Society, Providence (1985)
16. Lepowsky, J., Wilson, R.L.: The structure of standard modules, I: Universal algebras and the Rogers-Ramanujan identities. Invent. Math. **77**, 199–290; II: The case $A_1^{(1)}$, principal gradation. Invent. Math. **79**(1985), 417–442 (1984)
17. Meurman, A., Primc, M.: Vertex operator algebras and representations of affine Lie algebras. Acta Appl. Math. **44**, 207–215 (1996)
18. Meurman, A., Primc, M.: Annihilating fields of standard modules of $sl(2, \mathbb{C})\tilde{}$ and combinatorial identities. Memoirs Am. Math. Soc. **137**(652) (1999)
19. Primc, M.: Vertex operator construction of standard modules for $A_n^{(1)}$. Pacific J. Math. **162**, 143–187 (1994)
20. Primc, M.: Basic representations for classical affine Lie algebras. J. Algebra **228**, 1–50 (2000)
21. Primc, M.: (k, r)-admissible configurations and intertwining operators Lie algebras, vertex operator algebras and their applications. pp. 425-434. Contemporary Mathematics, vol. 442. American Mathematical Society, Providence (2007)
22. Primc, M.: Combinatorial bases of modules for affine Lie algebra $B_2^{(1)}$. Cent. Eur. J. Math. **11**, 197–225 (2013)
23. Primc, M., Šikić, T.: Combinatorial bases of basic modules for affine Lie algebras $C_n^{(1)}$. J. Math. Phys. **57**, 1–19 (2016)
24. Primc, M., Šikić, T.: Leading terms of relations for standard modules of $C_n^{(1)}$. arXiv:math/1506.05026 (To appear in Ramanujan J)
25. Siladić, I.: Twisted $\mathfrak{sl}(3, \mathbb{C})\tilde{}$-modules and combinatorial identities, Glasnik Matematički **52**, 53-77 (2017). arXiv:math/0204042
26. Trupčević, G.: Combinatorial bases of Feigin-Stoyanovsky's type subspaces of level 1 standard $\tilde{\mathfrak{sl}}(\ell + 1, \mathbb{C})$-modules. Comm. Algebra **38**, 3913–3940 (2010)
27. Trupčević, G.: Combinatorial bases of Feigin-Stoyanovsky's type subspaces of higher-level standard $\tilde{\mathfrak{sl}}(\ell + 1, \mathbb{C})$-modules. J. Algebra **322**, 3744–3774 (2009)
28. Trupčević, G.: Bases of standard modules for affine Lie algebras of type $C_\ell^{(1)}$. Comm. Algebra **46**, 3663–3673 (2018)

CPSIA information can be obtained
at www.ICGtesting.com
Printed in the USA
LVHW050612151220
674155LV00009B/118